AF769559

Uncertainty Mathematics: Principles and Analysis

Uncertainty Mathematics: Principles and Analysis

Edited by **Matt Ferrier**

Willford Press

New York

Published by Willford Press,
118-35 Queens Blvd., Suite 400,
Forest Hills, NY 11375, USA
www.willfordpress.com

Uncertainty Mathematics: Principles and Analysis
Edited by Matt Ferrier

© 2016 Willford Press

International Standard Book Number: 978-1-68285-277-4 (Hardback)

This book contains information obtained from authentic and highly regarded sources. Copyright for all individual chapters remain with the respective authors as indicated. All chapters are published with permission under the Creative Commons Attribution License or equivalent. A wide variety of references are listed. Permission and sources are indicated; for detailed attributions, please refer to the permissions page and list of contributors. Reasonable efforts have been made to publish reliable data and information, but the authors, editors and publisher cannot assume any responsibility for the validity of all materials or the consequences of their use.

The publisher's policy is to use permanent paper from mills that operate a sustainable forestry policy. Furthermore, the publisher ensures that the text paper and cover boards used have met acceptable environmental accreditation standards.

Trademark Notice: Registered trademark of products or corporate names are used only for explanation and identification without intent to infringe.

Printed in China.

Contents

Preface

Every book is a source of knowledge and this one is no exception. The idea that led to the conceptualization of this book was the fact that the world is advancing rapidly; which makes it crucial to document the progress in every field. I am aware that a lot of data is already available, yet, there is a lot more to learn. Hence, I accepted the responsibility of editing this book and contributing my knowledge to the community.

Mathematical principles of uncertainty find applications across diverse fields, like engineering, information sciences, computer science and many other sectors as well. This book brings forth some of the most innovative concepts and elucidates the unexplored aspects of uncertainty in mathematics. Selected topics revolving around uncertainty have been presented in the text, such as optimization under uncertainty, applications of uncertainty analysis, etc. It is a vital tool for all researching or studying mathematics as it gives incredible insights into emerging trends. Mathematicians, engineers and students actively engaged in this field will find this book full of crucial and innovative concepts.

While editing this book, I had multiple visions for it. Then I finally narrowed down to make every chapter a sole standing text explaining a particular topic, so that they can be used independently. However, the umbrella subject sinews them into a common theme. This makes the book a unique platform of knowledge.

I would like to give the major credit of this book to the experts from every corner of the world, who took the time to share their expertise with us. Also, I owe the completion of this book to the never-ending support of my family, who supported me throughout the project.

Editor

1

Toward uncertain finance theory

Baoding Liu

Correspondence:
liu@tsinghua.edu.cn
Baoding Liu, Uncertainty Theory Laboratory, Department of Mathematical Sciences, Tsinghua University, Beijing 100084, China

Abstract

This paper first introduces a paradox of stochastic finance theory that shows the real stock price is impossible to follow any Ito's stochastic differential equation. After a survey on uncertainty theory, uncertain process, uncertain calculus, and uncertain differential equation, this paper discusses some possible applications of uncertain differential equations to financial markets. Finally, it is suggested that a new uncertain finance theory should be developed based on uncertainty theory and uncertain differential equation.

Keywords: Finance, Uncertainty theory, Uncertain process, Uncertain calculus, Uncertain differential equation

Review

When no samples are available to estimate a probability distribution, we have to invite some domain experts to evaluate their belief degree that each event will occur. Perhaps some people think that personal belief degree is subjective probability or fuzzy concept. However, Liu [1] declared that it is inappropriate because both probability theory and fuzzy set theory may lead to counterintuitive results in this case. In order to rationally deal with the belief degree, an uncertainty theory was founded by Liu [2] and subsequently studied by many scholars. Nowadays, uncertainty theory has become a branch of axiomatic mathematics for modeling human uncertainty.

Based on uncertainty theory, the concept of uncertain process was given by Liu [3] as a sequence of uncertain variables indexed by time. Besides, the concept of uncertain integral was also proposed by Liu [3] in order to integrate an uncertain process with respect to a canonical process. Furthermore, Liu [4] recast his work via the fundamental theorem of uncertain calculus and thus produced the techniques of chain rule, change of variables, and integration by parts. Since then, the theory of uncertain calculus was well developed.

After uncertain differential equation was proposed by Liu [3] as a differential equation involving uncertain process, an existence and uniqueness theorem of a solution of uncertain differential equation was proved by Chen and Liu [5] under linear growth condition and Lipschitz continuous condition. The theorem was verified again by Gao [6] under local linear growth condition and local Lipschitz continuous condition. In order to solve uncertain differential equations, Chen and Liu [5] obtained an analytic solution to linear uncertain differential equations. In addition, Liu [7] presented a spectrum of analytic methods to solve some special classes of nonlinear uncertain differential equations. More importantly, Yao and Chen [8] showed that the solution of an uncertain differential equation can be represented by a family of solutions of ordinary differential equations, thus relating uncertain differential equations and ordinary differential equations. On the basis of the Yao-Chen formula, a numerical method was also designed by Yao and Chen [8] for solving general uncertain differential equations. Furthermore, Yao [9] presented some formulas to calculate the extreme values, first hitting time and integral of solution of uncertain differential equation.

Uncertain differential equations were first introduced into finance by Liu [4] in which an uncertain stock model was proposed and European option price formulas were documented. Besides, Chen [10] derived American option price formulas for this type of uncertain stock model. In addition, Peng and Yao [11] presented a different uncertain stock model and obtained the corresponding option price formulas, and Yu [12] proposed an uncertain stock model with jumps. Uncertain differential equations were also employed to model uncertain currency markets by Liu and Chen [13] in which an uncertain currency model was proposed. Uncertain differential equations were used to simulate interest rate by Chen and Gao [14], and an uncertain interest rate model was presented. On the basis of this model, the price of zero-coupon bond was also produced. Uncertain differential equations were applied to optimal control by Zhu [15] in which Zhu's equation of optimality is proved to be a necessary condition for extremum of uncertain optimal control model.

This paper first introduces a paradox of stochastic finance theory. After a survey on uncertainty theory, uncertain process, uncertain calculus, and uncertain differential equation, this paper shows some possible applications of uncertain differential equations to financial markets. Finally, this paper suggests to develop an uncertain finance theory by using uncertainty theory and uncertain differential equation.

A paradox of stochastic finance theory

The origin of stochastic finance theory can be traced to Louis Bachelier's doctoral dissertation *Théorie de la Speculation* in 1900. However, Bachelier's work had little impact for more than a half century. After Kiyosi Ito invented stochastic calculus [16] and stochastic differential equation [17], stochastic finance theory was well developed among others by Samuelson [18], Black and Scholes [19], and Merton [20] during the 1960s and 1970s.

Traditionally, stochastic finance theory presumes that the stock price (including currency exchange rate and interest rate) follows an Ito's stochastic differential equation. Is it really reasonable? In fact, this widely accepted presumption was continuously challenged by many scholars. Let us assume that the stock price X_t follows the stochastic differential equation

$$dX_t = eX_t dt + \sigma X_t dW_t \tag{1}$$

where e is the log-drift, σ is the log-diffusion, and W_t is a Wiener process. Let us see what will happen with such an assumption. It follows from the stochastic differential equation (1) that X_t is a geometric Wiener process, i.e.,

$$X_t = X_0 \exp((e - \sigma^2/2)t + \sigma W_t) \tag{2}$$

from which we derive

$$W_t = \frac{\ln X_t - \ln X_0 - (e - \sigma^2/2)t}{\sigma} \tag{3}$$

whose increment is

$$\Delta W_t = \frac{\ln X_{t+\Delta t} - \ln X_t - (e - \sigma^2/2)\Delta t}{\sigma}. \tag{4}$$

Write

$$A = -\frac{(e - \sigma^2/2)\Delta t}{\sigma}. \tag{5}$$

Note that the real stock price X_t is actually a step function of time with a finite number of jumps although it looks like a curve. During a fixed period, without loss of generality, we assume that X_t is observed to have 100 jumps. Now we divide the period into 10,000 equal intervals. Then we may observe 10,000 samples of X_t. It follows from Equation 4 that ΔW_t has 10,000 samples that consist of 9,900 A's and 100 other numbers:

$$\underbrace{A, A, \cdots, A,}_{9,900}\ \underbrace{B, C, \cdots, Z.}_{100} \tag{6}$$

Nobody can believe that those 10,000 samples follow a normal probability distribution with expected value 0 and variance Δt. See Figure 1. This fact is in contradiction with the property of Wiener process that the increment ΔW_t is a normal random variable with expected value 0 and variance Δt. Therefore, the stock price X_t does not follow the stochastic differential equation.

Perhaps some people think that the stock price does behave like a geometric Wiener process (or Ornstein-Uhlenbeck process) in macroscopy although they recognize the paradox in microscopy. However, as the very core of stochastic finance theory, Ito's calculus is just built on the microscopic structure (i.e., the differential $\mathrm{d}W_t$) of Wiener process rather than macroscopic structure. More precisely, Ito's calculus is dependent on the presumption that $\mathrm{d}W_t$ is a normal random variable with expected value 0 and variance $\mathrm{d}t$. This unreasonable presumption is what causes the second order term in Ito's formula,

$$\mathrm{d}X_t = \frac{\partial h}{\partial t}(t, W_t)\mathrm{d}t + \frac{\partial h}{\partial w}(t, W_t)\mathrm{d}W_t + \frac{1}{2}\frac{\partial^2 h}{\partial w^2}(t, W_t)\mathrm{d}t. \tag{7}$$

In fact, the increment of stock price is impossible to follow any continuous probability distribution. On the basis of the above paradox, personally, I do not think Ito's calculus can play the essential tool of finance theory because Ito's stochastic differential equation is impossible to model real stock price.

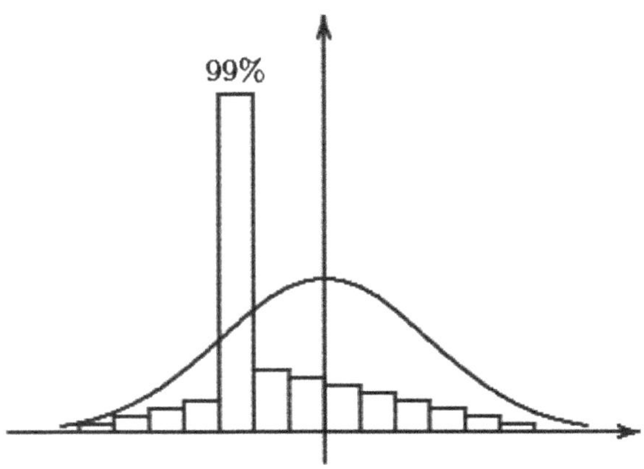

Figure 1 Normal distribution vs real frequency.

What is uncertainty theory?

Let Γ be a nonempty set, and $\mathcal{L}$ a σ-algebra over Γ. Each element Λ in $\mathcal{L}$ is called an event. A set function $\mathcal{M}$ from $\mathcal{L}$ to $[0,1]$ is called an *uncertain measure* if it satisfies the following axioms [2]:

Axiom 1. (Normality axiom) $\mathcal{M}\{\Gamma\} = 1$ for the universal set Γ;
Axiom 2. (Duality axiom) $\mathcal{M}\{\Lambda\} + \mathcal{M}\{\Lambda^c\} = 1$ for any event Λ;
Axiom 3. (Subadditivity axiom) For every countable sequence of events $\Lambda_1, \Lambda_2, \cdots$, we have

$$\mathcal{M}\left\{\bigcup_{i=1}^{\infty} \Lambda_i\right\} \leq \sum_{i=1}^{\infty} \mathcal{M}\{\Lambda_i\}. \tag{8}$$

The triplet $(\Gamma, \mathcal{L}, \mathcal{M})$ is called an *uncertainty space*. In order to obtain an uncertain measure of compound event, a product uncertain measure was defined by Liu [4], thus producing the fourth axiom of uncertainty theory:
Axiom 4. (Product axiom) Let $(\Gamma_k, \mathcal{L}_k, \mathcal{M}_k)$ be uncertainty spaces for $k = 1, 2, \cdots$ The product uncertain measure $\mathcal{M}$ is an uncertain measure satisfying

$$\mathcal{M}\left\{\prod_{k=1}^{\infty} \Lambda_k\right\} = \bigwedge_{k=1}^{\infty} \mathcal{M}_k\{\Lambda_k\} \tag{9}$$

where Λ_k are arbitrarily chosen events from $\mathcal{L}_k$ for $k = 1, 2, \cdots$, respectively.

An *uncertain variable* is defined by Liu [2] as a measurable function ξ from an uncertainty space $(\Gamma, \mathcal{L}, \mathcal{M})$ to the set of real numbers, i.e., for any Borel set B of real numbers, the set

$$\{\xi \in B\} = \{\gamma \in \Gamma \mid \xi(\gamma) \in B\} \tag{10}$$

is an event. In order to describe an uncertain variable in practice, the concept of *uncertainty distribution* is defined by Liu [2] as

$$\Phi(x) = \mathcal{M}\{\xi \leq x\}, \quad \forall x \in \Re. \tag{11}$$

Peng and Iwamura [21] proved that a function $\Phi : \Re \rightarrow [0,1]$ is an uncertainty distribution if and only if it is a monotone increasing function except $\Phi(x) \equiv 0$ and $\Phi(x) \equiv 1$.

An uncertainty distribution $\Phi(x)$ is said to be regular if it is a continuous and strictly increasing function with respect to x at which $0 < \Phi(x) < 1$, and

$$\lim_{x \to -\infty} \Phi(x) = 0, \qquad \lim_{x \to +\infty} \Phi(x) = 1. \tag{12}$$

Let ξ be an uncertain variable with regular uncertainty distribution $\Phi(x)$. Then the inverse function $\Phi^{-1}(\alpha)$ is called the inverse uncertainty distribution of ξ [22].

It is easy to verify that $\Phi^{-1}(\alpha)$ is a continuous and strictly increasing function with respect to $\alpha \in (0,1)$. Conversely, suppose $\Phi^{-1}(\alpha)$ is a continuous and strictly increasing function on $(0,1)$. Define

$$\Phi(x) = \begin{cases} 0, & \text{if } x \leq \lim_{\alpha \downarrow 0} \Phi^{-1}(\alpha) \\ \alpha, & \text{if } x = \Phi^{-1}(\alpha) \\ 1, & \text{if } x \geq \lim_{\alpha \uparrow 1} \Phi^{-1}(\alpha). \end{cases}$$

It follows that $\Phi(x)$ is an uncertainty distribution of some uncertain variable ξ. Then for each $\alpha \in (0,1)$, we have

$$\mathcal{M}\{\xi \le \Phi^{-1}(\alpha)\} = \Phi(\Phi^{-1}(\alpha)) = \alpha.$$

Thus, $\Phi^{-1}(\alpha)$ is just the inverse uncertainty distribution of the uncertain variable ξ. Hence, we have a sufficient and necessary condition of inverse uncertainty distribution: A function $\Phi^{-1}(\alpha) : (0,1) \to \Re$ is an inverse uncertainty distribution if and only if it is a continuous and strictly increasing function with respect to α.

The *expected value* of an uncertain variable ξ is defined by Liu [2] as an average value of the uncertain variable in the sense of uncertain measure, i.e.,

$$E[\xi] = \int_0^{+\infty} \mathcal{M}\{\xi \ge r\}\mathrm{d}r - \int_{-\infty}^0 \mathcal{M}\{\xi \le r\}\mathrm{d}r \tag{13}$$

provided that at least one of the two integrals is finite. If ξ has an uncertainty distribution Φ, then the expected value may be calculated by

$$E[\xi] = \int_0^{+\infty} (1 - \Phi(x))\mathrm{d}x - \int_{-\infty}^0 \Phi(x)\mathrm{d}x. \tag{14}$$

Independence is an extremely important concept in uncertainty theory. The uncertain variables $\xi_1, \xi_2, \cdots, \xi_n$ are said to be independent [4] if

$$\mathcal{M}\left\{\bigcap_{i=1}^{n}(\xi_i \in B_i)\right\} = \bigwedge_{i=1}^{n} \mathcal{M}\{\xi_i \in B_i\} \tag{15}$$

for any Borel sets $B_1, B_2, \cdots, B_n$ of real numbers. Equivalently, those uncertain variables are independent if and only if

$$\mathcal{M}\left\{\bigcup_{i=1}^{n}(\xi_i \in B_i)\right\} = \bigvee_{i=1}^{n} \mathcal{M}\{\xi_i \in B_i\}. \tag{16}$$

Let $\xi_1, \xi_2, \cdots, \xi_n$ be independent uncertain variables with uncertainty distributions $\Phi_1, \Phi_2, \cdots, \Phi_n$, respectively. If the function $f(x_1, x_2, \cdots, x_n)$ is strictly increasing with respect to $x_1, x_2, \cdots, x_m$ and strictly decreasing with respect to $x_{m+1}, x_{m+2}, \cdots, x_n$, then $\xi = f(\xi_1, \xi_2, \cdots, \xi_n)$ is an uncertain variable with inverse uncertainty distribution

$$\Psi^{-1}(\alpha) = f(\Phi_1^{-1}(\alpha), \cdots, \Phi_m^{-1}(\alpha), \Phi_{m+1}^{-1}(1-\alpha), \cdots, \Phi_n^{-1}(1-\alpha)). \tag{17}$$

Then Liu and Ha [23] proved that the uncertain variable $\xi = f(\xi_1, \xi_2, \cdots, \xi_n)$ has an expected value

$$E[\xi] = \int_0^1 f(\Phi_1^{-1}(\alpha), \cdots, \Phi_m^{-1}(\alpha), \Phi_{m+1}^{-1}(1-\alpha), \cdots, \Phi_n^{-1}(1-\alpha))\mathrm{d}\alpha. \tag{18}$$

For exploring the details of uncertainty theory, the readers may consult Liu [24].

Uncertain process

Let T be a totally ordered set (that is usually "time"), and let $(\Gamma, \mathcal{L}, \mathcal{M})$ be an uncertainty space. An *uncertain process* is defined by Liu [3] as a measurable function from $T \times (\Gamma, \mathcal{L}, \mathcal{M})$ to the set of real numbers, i.e., for each $t \in T$ and any Borel set B of real numbers, the set

$$\{X_t \in B\} = \{\gamma \in \Gamma \mid X_t(\gamma) \in B\} \tag{19}$$

is an event. In other words, an uncertain process is a sequence of uncertain variables indexed by time.

Note that if the index set T becomes a partially ordered set (e.g., time × space, or a surface), then X_t is called an *uncertain field* provided that X_t is an uncertain variable at each point t. That is, an uncertain field is a generalization of an uncertain process.

An uncertain process X_t is said to have an uncertainty distribution $\Phi_t(x)$ if at each time t, the uncertain variable X_t has the uncertainty distribution $\Phi_t(x)$. It is easy to prove that $\Phi_t(x)$ is a monotone increasing function with respect to x and $\Phi_t(x) \not\equiv 0$, $\Phi_t(x) \not\equiv 1$. Conversely, if at each time t, $\Phi_t(x)$ is a monotone increasing function except $\Phi_t(x) \equiv 0$ and $\Phi_t(x) \equiv 1$, it follows that there exists an uncertain variable ξ_t whose uncertainty distribution is just $\Phi_t(x)$. Define

$$X_t = \xi_t, \quad \forall t \in T.$$

Then X_t is an uncertain process and has the uncertainty distribution $\Phi_t(x)$. Thus, a function $\Phi_t(x) : T \times \Re \to [0,1]$ is an uncertainty distribution of uncertain process if and only if at each time t, it is a monotone increasing function except $\Phi_t(x) \equiv 0$ and $\Phi_t(x) \equiv 1$.

An uncertainty distribution $\Phi_t(x)$ is said to be regular if at each time t, it is a continuous and strictly increasing function with respect to x at which $0 < \Phi_t(x) < 1$, and

$$\lim_{x\to-\infty} \Phi_t(x) = 0, \qquad \lim_{x\to+\infty} \Phi_t(x) = 1. \tag{20}$$

Let X_t be an uncertain process with regular uncertainty distribution $\Phi_t(x)$. Then the inverse function $\Phi_t^{-1}(\alpha)$ is called the inverse uncertainty distribution of X_t. It is easy to prove that $\Phi_t^{-1}(\alpha)$ is a continuous and strictly increasing function with respect to $\alpha \in (0,1)$. Conversely, if $\Phi_t^{-1}(\alpha)$ is a continuous and strictly increasing function with respect to $\alpha \in (0,1)$, it follows that there exists an uncertain variable ξ_t whose inverse uncertainty distribution is just $\Phi_t^{-1}(\alpha)$. Define

$$X_t = \xi_t, \quad \forall t \in T.$$

Then X_t is an uncertain process and has the inverse uncertainty distribution $\Phi_t^{-1}(\alpha)$. Hence, a function $\Phi_t^{-1}(\alpha) : T \times (0,1) \to \Re$ is an inverse uncertainty distribution of uncertain process if and only if at each time t, it is a continuous and strictly increasing function with respect to α.

An uncertain process X_t is said to have *independent increments* if

$$X_{t_0}, X_{t_1} - X_{t_0}, X_{t_2} - X_{t_1}, \cdots, X_{t_k} - X_{t_{k-1}} \tag{21}$$

are independent uncertain variables where t_0 is the initial time and $t_1, t_2, \cdots, t_k$ are any times with $t_0 < t_1 < \cdots < t_k$. That is, an independent increment process means that its increments are independent uncertain variables whenever the time intervals do not overlap. Let X_t be a sample-continuous independent increment process with an uncertainty distribution $\Phi_t(x)$ at each time t. When f is a strictly increasing function, Liu [25] proved that the supremum

$$\sup_{0\le t\le s} f(X_t) \tag{22}$$

has an uncertainty distribution

$$\Psi(x) = \inf_{0\le t\le s} \Phi_t(f^{-1}(x)). \tag{23}$$

This result is called the *extreme value theorem* of uncertain process.

An uncertain process X_t is said to have *stationary increments* if its increments are identically distributed uncertain variables whenever the time intervals have the same length, i.e., for any given $t > 0$, the increments $X_{s+t} - X_s$ are identically distributed uncertain variables for all $s > 0$.

Let X_t be a stationary independent increment process with a crisp initial value X_0. Liu [22] showed that there exist two real numbers a and b such that the expected value

$$E[X_t] = a + bt \tag{24}$$

for any time $t \geq 0$. Furthermore, Chen [26] verified that there exists a real number c such that the variance

$$V[X_t] = ct^2 \tag{25}$$

for any time $t \geq 0$.

As an important type of uncertain process, a canonical process is a stationary independent increment process whose increments are normal uncertain variables. More precisely, an uncertain process C_t is called a *canonical process* by Liu [4] if (1) $C_0 = 0$ and almost all sample paths are Lipschitz continuous, (2) C_t has stationary and independent increments, and (3) every increment $C_{s+t} - C_s$ is a normal uncertain variable with expected value 0 and variance t^2.

It is easy to verify that the canonical process C_t is a normal uncertain variable with expected value 0 and variance t^2 and has an uncertainty distribution

$$\Phi(x) = \left(1 + \exp\left(-\frac{\pi x}{\sqrt{3}t}\right)\right)^{-1} \tag{26}$$

at each time $t > 0$. In addition, for each time $t > 0$, the ratio C_t/t is a normal uncertain variable with expected value 0 and variance 1. That is,

$$\frac{C_t}{t} \sim \mathcal{N}(0, 1) \tag{27}$$

for any $t > 0$.

What is the difference between canonical process and the Wiener process? First, canonical process is an uncertain process while the Wiener process is a stochastic process. Second, almost all sample paths of canonical process are Lipschitz continuous functions while almost all sample paths of the Wiener process are continuous but non-Lipschitz functions. Third, canonical process has a variance t^2 while the Wiener process has a variance t at each time t.

Uncertain calculus

Uncertain calculus is a branch of mathematics that deals with differentiation and integration of uncertain processes. The key concept in uncertain calculus is the uncertain integral that allows us to integrate an uncertain process (the integrand) with respect to the canonical process (the integrator). The result of the uncertain integral is another uncertain process.

Let X_t be an uncertain process and let C_t be a canonical process. For any partition of closed interval $[a, b]$ with $a = t_1 < t_2 < \cdots < t_{k+1} = b$, the mesh is written as

$$\Delta = \max_{1 \leq i \leq k} |t_{i+1} - t_i|. \tag{28}$$

Then the *uncertain integral* of X_t with respect to C_t is defined by Liu [4] as

$$\int_a^b X_t \mathrm{d}C_t = \lim_{\Delta \to 0} \sum_{i=1}^{k} X_{t_i} \cdot (C_{t_{i+1}} - C_{t_i}) \tag{29}$$

provided that the limit exists almost surely and is finite. Since X_t and C_t are uncertain variables at each time t, the limit in Equation 29 is also an uncertain variable.

Let Z_t be an uncertain process. If there exist two uncertain processes μ_t and σ_t such that

$$Z_t = Z_0 + \int_0^t \mu_s \mathrm{d}s + \int_0^t \sigma_s \mathrm{d}C_s \tag{30}$$

for any $t \geq 0$, then we say Z_t has an *uncertain differential*

$$\mathrm{d}Z_t = \mu_t \mathrm{d}t + \sigma_t \mathrm{d}C_t. \tag{31}$$

In this case, Z_t is called an uncertain process with drift μ_t and diffusion σ_t. It is clear that uncertain integral and differential are mutually inverse operations. Please also note that an uncertain differential of an uncertain process has two parts, the "$\mathrm{d}t$" part and the "$\mathrm{d}C_t$" part.

Let $h(t, c)$ be a continuously differentiable function. Liu [4] showed that the uncertain process $Z_t = h(t, C_t)$ has an uncertain differential

$$\mathrm{d}Z_t = \frac{\partial h}{\partial t}(t, C_t)\mathrm{d}t + \frac{\partial h}{\partial c}(t, C_t)\mathrm{d}C_t. \tag{32}$$

This result is called the *fundamental theorem of uncertain calculus.*

Example 1. Let us calculate the uncertain differential of tC_t. In this case, we have $h(t, c) = tc$ whose partial derivatives are

$$\frac{\partial h}{\partial t}(t, c) = c, \quad \frac{\partial h}{\partial c}(t, c) = t.$$

It follows from the fundamental theorem of uncertain calculus that

$$\mathrm{d}(tC_t) = C_t \mathrm{d}t + t\mathrm{d}C_t. \tag{33}$$

Example 2. Let us calculate the uncertain differential of C_t^2. In this case, we have $h(t, c) = c^2$ whose partial derivatives are

$$\frac{\partial h}{\partial t}(t, c) = 0, \quad \frac{\partial h}{\partial c}(t, c) = 2c.$$

It follows from the fundamental theorem of uncertain calculus that

$$\mathrm{d}C_t^2 = 2C_t \mathrm{d}C_t. \tag{34}$$

Example 3. Let $f(c)$ be a continuously differentiable function. Then we have

$$\frac{\partial f}{\partial t}(c) = 0, \quad \frac{\partial f}{\partial c}(c) = f'(c).$$

It follows from the fundamental theorem of uncertain calculus that the uncertain process $f(C_t)$ has an uncertain differential

$$\mathrm{d}f(C_t) = f'(C_t)\mathrm{d}C_t. \tag{35}$$

This formula is also called the *chain rule* of uncertain calculus.

As supplements to uncertain integral, Liu and Yao [27] suggested an uncertain integral with respect to multiple canonical processes. More generally, Chen and Ralescu [28] presented an uncertain integral with respect to the general Liu process.

Uncertain differential equation

The study of uncertain differential equation was pioneered by Liu [3]. Nowadays, uncertain differential equation has achieved fruitful results in both theory and practice. Let f and g be two functions. Then

$$dX_t = f(t, X_t)dt + g(t, X_t)dC_t \tag{36}$$

is called an *uncertain differential equation*. A solution is an uncertain process X_t that satisfies Equation 36 identically in t.

Some analytic methods have been proposed for solving uncertain differential equations. For example, Chen and Liu [5] showed that the linear uncertain differential equation

$$dX_t = (u_{1t}X_t + u_{2t})dt + (v_{1t}X_t + v_{2t})dC_t \tag{37}$$

has a solution

$$X_t = U_t\left(X_0 + \int_0^t \frac{u_{2s}}{U_s}ds + \int_0^t \frac{v_{2s}}{U_s}dC_s\right) \tag{38}$$

where

$$U_t = \exp\left(\int_0^t u_{1s}ds + \int_0^t v_{1s}dC_s\right). \tag{39}$$

In addition, Liu [7] verified that the nonlinear uncertain differential equation like

$$dX_t = f(t, X_t)dt + \sigma_t X_t dC_t \tag{40}$$

has a solution

$$X_t = Y_t^{-1}Z_t \tag{41}$$

where

$$Y_t = \exp\left(-\int_0^t \sigma_s dC_s\right) \tag{42}$$

and Z_t is the solution of uncertain differential equation

$$dZ_t = Y_t f(t, Y_t^{-1}Z_t)dt \tag{43}$$

with initial value $Z_0 = X_0$.

An important contribution to uncertain differential equation is the existence and uniqueness theorem by Chen and Liu [5]. An uncertain differential equation has a unique solution if the coefficients $f(t, x)$ and $g(t, x)$ satisfy linear growth condition

$$|f(t, x)| + |g(t, x)| \le L(1 + |x|), \quad \forall x \in \Re, t \ge 0 \tag{44}$$

and Lipschitz condition

$$|f(t, x) - f(t, y)| + |g(t, x) - g(t, y)| \le L|x - y|, \quad \forall x, y \in \Re, t \ge 0 \tag{45}$$

for some constant L. Moreover, the solution is sample-continuous.

The concept of stability was given by Liu [4]. An uncertain differential equation is said to be *stable* if for any two solutions X_t and Y_t, we have

$$\lim_{|X_0 - Y_0| \to 0} \mathcal{M}\{|X_t - Y_t| > \varepsilon\} = 0, \quad \forall t > 0 \tag{46}$$

for any given number $\varepsilon > 0$. Yao et al. [29] proved that the uncertain differential equation is stable if the coefficients $f(t,x)$ and $g(t,x)$ satisfy linear growth condition

$$|f(t,x)| + |g(t,x)| \leq K(1+|x|), \quad \forall x \in \Re, t \geq 0 \tag{47}$$

for some constant K and strong Lipschitz condition

$$|f(t,x) - f(t,y)| + |g(t,x) - g(t,y)| \leq L(t)|x-y|, \quad \forall x, y \in \Re, t \geq 0 \tag{48}$$

for some bounded and integrable function $L(t)$ on $[0, +\infty)$.

Uncertain differential equation has been extended by many scholars. For example, uncertain delay differential equation was studied among others by Barbacioru [30], Ge and Zhu [31], and Liu and Fei [32]. In addition, uncertain differential equation with jumps was suggested by Yao [33], and backward uncertain differential equation was discussed by Ge and Zhu [34].

Numerical method

Let α be a number with $0 < \alpha < 1$. An uncertain differential equation

$$\mathrm{d}X_t = f(t, X_t)\mathrm{d}t + g(t, X_t)\mathrm{d}C_t \tag{49}$$

is said to have an α-path X_t^α if it solves the corresponding ordinary differential equation

$$\mathrm{d}X_t^\alpha = f(t, X_t^\alpha)\mathrm{d}t + |g(t, X_t^\alpha)|\Phi^{-1}(\alpha)\mathrm{d}t \tag{50}$$

where $\Phi^{-1}(\alpha)$ is the inverse uncertainty distribution of standard normal uncertain variable, i.e.,

$$\Phi^{-1}(\alpha) = \frac{\sqrt{3}}{\pi} \ln \frac{\alpha}{1-\alpha}. \tag{51}$$

Then

$$\mathcal{M}\{X_t \leq X_t^\alpha, \forall t\} = \alpha, \tag{52}$$

$$\mathcal{M}\{X_t > X_t^\alpha, \forall t\} = 1 - \alpha. \tag{53}$$

This result is called the *Yao-Chen formula* [8]. In addition, at each time t, the solution X_t has an inverse uncertainty distribution

$$\Psi_t^{-1}(\alpha) = X_t^\alpha. \tag{54}$$

Furthermore, for any monotone (increasing or decreasing) function J, we have

$$E[J(X_t)] = \int_0^1 J(X_t^\alpha)\mathrm{d}\alpha. \tag{55}$$

The Yao-Chen formula relates uncertain differential equations and ordinary differential equations, just like that Feynman-Kac formula relates stochastic differential equations and partial differential equations.

It is almost impossible to find analytic solutions for general uncertain differential equations. This fact provides a motivation to design a numerical method to solve general uncertain differential equation

$$\mathrm{d}X_t = f(t, X_t)\mathrm{d}t + g(t, X_t)\mathrm{d}C_t. \tag{56}$$

In order to do so, a key point is to obtain an inverse uncertainty distribution $\Psi_t^{-1}(\alpha)$ of its solution X_t at any given time t. For this purpose, Yao and Chen [8] designed the following algorithm:

Step 1. Fix α on $(0, 1)$.
Step 2. Solve $\mathrm{d}X_t^\alpha = f(t, X_t^\alpha)\mathrm{d}t + |g(t, X_t^\alpha)|\Phi^{-1}(\alpha)\mathrm{d}t$ by any method of ordinary differential equation and obtain the α-path X_t^α, for example, by using the recursion formula

$$X_{i+1}^\alpha = X_i^\alpha + f(t_i, X_i^\alpha)h + |g(t_i, X_i^\alpha)|\Phi^{-1}(\alpha)h \tag{57}$$

where Φ is the standard normal uncertainty distribution and h is the step length.
Step 3. The inverse uncertainty distribution of the solution X_t is determined by

$$\Psi_t^{-1}(\alpha) = X_t^\alpha. \tag{58}$$

Uncertain stock model

Uncertain differential equations were first introduced into finance by Liu [4] in which an *uncertain stock model* was proposed,

$$\begin{cases} \mathrm{d}X_t = rX_t\mathrm{d}t \\ \mathrm{d}Y_t = eY_t\mathrm{d}t + \sigma Y_t\mathrm{d}C_t \end{cases} \tag{59}$$

where X_t is the bond price, Y_t is the stock price, r is the riskless interest rate, e is the log-drift, σ is the log-diffusion, and C_t is a canonical process.

A *European call option* is a contract that gives the holder the right to buy a stock at an expiration time s for a strike price K. The payoff from a European call option is $(Y_s - K)^+$ since the option is rationally exercised if and only if $Y_s > K$. Considering the time value of money resulted from the bond, the present value of this payoff is $\exp(-rs)(Y_s - K)^+$. Hence, the European call option price should be the expected present value of the payoff, i.e.,

$$f_c = \exp(-rs)E[(Y_s - K)^+]. \tag{60}$$

Liu [4] proved that

$$f_c = \exp(-rs)Y_0 \int_{K/Y_0}^{+\infty} \left(1 + \exp\left(\frac{\pi(\ln y - es)}{\sqrt{3}\sigma s}\right)\right)^{-1} \mathrm{d}y. \tag{61}$$

A *European put option* is a contract that gives the holder the right to sell a stock at an expiration time s for a strike price K. The payoff from a European put option is $(K - Y_s)^+$ since the option is rationally exercised if and only if $Y_s < K$. Considering the time value of money resulted from the bond, the present value of this payoff is $\exp(-rs)(K - Y_s)^+$. Hence, the European put option price should be the expected present value of the payoff, i.e.,

$$f_p = \exp(-rs)E[(K - Y_s)^+]. \tag{62}$$

Liu [4] proved that

$$f_p = \exp(-rs)Y_0 \int_0^{K/Y_0} \left(1 + \exp\left(\frac{\pi(es - \ln y)}{\sqrt{3}\sigma s}\right)\right)^{-1} \mathrm{d}y. \tag{63}$$

An *American call option* is a contract that gives the holder the right to buy a stock at any time prior to an expiration time s for a strike price K. It is clear that the payoff from an American call option is the supremum of $(Y_t - K)^+$ over the time interval $[0, s]$.

Considering the time value of money resulted from the bond, the present value of this payoff is

$$\sup_{0\le t\le s} \exp(-rt)(Y_t - K)^+. \tag{64}$$

Hence, the American call option price should be the expected present value of the payoff, i.e.,

$$f_c = E\left[\sup_{0\le t\le s} \exp(-rt)(Y_t - K)^+\right]. \tag{65}$$

Chen [10] proved that

$$f_c = \exp(-rs)Y_0 \int_{K/Y_0}^{+\infty} \left(1 + \exp\left(\frac{\pi(\ln y - es)}{\sqrt{3}\sigma s}\right)\right)^{-1} \mathrm{d}y. \tag{66}$$

An *American put option* is a contract that gives the holder the right to sell a stock at any time prior to an expiration time s for a strike price K. It is clear that the payoff from an American put option is the supremum of $(K - Y_t)^+$ over the time interval $[0, s]$. Considering the time value of money resulted from the bond, the present value of this payoff is

$$\sup_{0\le t\le s} \exp(-rt)(K - Y_t)^+. \tag{67}$$

Hence, the American put option price should be the expected present value of the payoff, i.e.,

$$f_p = E\left[\sup_{0\le t\le s} \exp(-rt)(K - Y_t)^+\right]. \tag{68}$$

Chen [10] proved that

$$f_p = \int_0^{K\exp(-rs)} \sup_{0\le t\le s} \left(1 + \exp\left(\frac{e}{\sqrt{3}\sigma} + \frac{\pi}{\sqrt{3}\sigma t}\ln\frac{Y_0}{K - y\exp(rt)}\right)\right)^{-1} \mathrm{d}y.$$

It is emphasized that other stock models were also actively investigated by Peng and Yao [11], Yu [12], and Chen et al. [35], among others.

Uncertain currency model

Liu and Chen [13] assumed that the exchange rate follows an uncertain differential equation and then proposed an *uncertain currency model*,

$$\begin{cases} \mathrm{d}X_t = uX_t\mathrm{d}t & \text{(Domestic currency)} \\ \mathrm{d}Y_t = vY_t\mathrm{d}t & \text{(Foreign currency)} \\ \mathrm{d}Z_t = eZ_t\mathrm{d}t + \sigma Z_t\mathrm{d}C_t & \text{(Exchange rate)} \end{cases} \tag{69}$$

where X_t represents the domestic currency with domestic interest rate u, Y_t represents the foreign currency with foreign interest rate v, and Z_t represents the exchange rate, that is, the domestic currency price of one unit of foreign currency at time t.

A *currency option* is a contract that gives the holder the right to exchange one unit of foreign currency at an expiration time s for K units of domestic currency. Suppose that the price of this contract is f in domestic currency. Then the investor pays f for buying the contract at time 0 and receives $(Z_s - K)^+$ in domestic currency at the expiration time s. Thus, the expected return of the investor is

$$-f + \exp(-us)E[(Z_s - K)^+]. \tag{70}$$

On the other hand, the bank receives f for selling the contract at time 0 and pays $(1 - K/Z_s)^+$ in foreign currency at the expiration time s. Thus, the expected return of the bank is

$$f - Z_0 \exp(-vs)E[(1 - K/Z_s)^+]. \tag{71}$$

The fair price of this contract should make the investor and the bank have an identical expected return, i.e.,

$$-f + \exp(-us)E[(Z_s - K)^+] = f - Z_0 \exp(-vs)E[(1 - K/Z_s)^+]. \tag{72}$$

Thus, the currency option price is

$$f = \frac{1}{2}\exp(-us)E[(Z_s - K)^+] + \frac{1}{2}\exp(-vs)Z_0E[(1 - K/Z_s)^+]. \tag{73}$$

Liu and Chen [13] proved that

$$f = \frac{1}{2}\exp(-us)Z_0 \int_{K/Z_0}^{+\infty} \left(1 + \exp\left(\frac{\pi(\ln y - es)}{\sqrt{3}\sigma s}\right)\right)^{-1} \mathrm{d}y$$

$$+ \frac{1}{2}\exp(-vs)Z_0 \int_0^1 \left(1 + \exp\left(\frac{\pi(\ln(K/Z_0) - \ln y - es)}{\sqrt{3}\sigma s}\right)\right)^{-1} \mathrm{d}y.$$

Uncertain interest rate model

Real interest rates do not remain unchanged. Chen and Gao [14] assumed that the interest rate X_t follows an uncertain differential equation,

$$\mathrm{d}X_t = (m - aX_t)\mathrm{d}t + \sigma \mathrm{d}C_t \tag{74}$$

where m, a, and σ are positive numbers, and C_t is a canonical process.

A *zero-coupon bond* is a bond bought at a price lower than its *face value*, that is, the amount it promises to pay at the maturity date. For simplicity, we assume that the face value is always US$1. Then the price of a zero-coupon bond with a maturity date s is

$$f = E\left[\exp\left(-\int_0^s X_t \mathrm{d}t\right)\right]. \tag{75}$$

Chen and Gao [14] proved that

$$f = \frac{\sqrt{3}\sigma}{a}(s - g)\exp\left(-\frac{ms}{2a} - \left(r_0 - \frac{m}{a}\right)g\right)\csc\left(\frac{\sqrt{3}\sigma}{a}(s - g)\right) \tag{76}$$

where

$$g = \frac{1}{a}(1 - \exp(-as)). \tag{77}$$

Uncertain finance theory

At the beginning of this paper, a paradox was proposed to show that the real stock price is impossible to follow an Ito's stochastic differential equation. It follows from Figure 1 that the increments behave like an uncertain variable rather than a random variable. This fact motives us to model stock prices by uncertain differential equations. Personally, I think uncertain calculus may play a potential mathematical foundation of finance theory.

If we say that the classical finance theory is a methodology dealing with financial markets by using probability theory, then uncertain finance theory is a methodology dealing with financial markets by using uncertainty theory. In addition to the uncertain stock models shown above, we may also accept other uncertain differential equations, for example,

$$dX_t = (m - aX_t)dt + \sigma X_t dC_t, \tag{78}$$

$$dX_t = (m - aX_t)dt + \sigma \sqrt{X_t} dC_t, \tag{79}$$

$$dX_t = (m - aX_t)dt + \sigma \sqrt{b + X_t} dC_t. \tag{80}$$

Conclusions

At first, a paradox of stochastic finance theory was introduced in this paper. After a survey on uncertainty theory, uncertain process, uncertain calculus, and uncertain differential equation, this paper summarized uncertain stock model, uncertain currency model, and uncertain interest model by using the tool of uncertain differential equation. Finally, it was suggested that an uncertain finance theory should be developed based on uncertainty theory.

Competing interests

The author declares that he has no competing interests.

Acknowledgements

This work was supported by the National Natural Science Foundation of China, grant no. 61273044.

References

1. Liu, B: Why is there a need for uncertainty theory? J. Uncertain Syst. **6**(1), 3–10 (2012)
2. Liu, B: Uncertainty Theory. 2nd edn. Springer, Berlin (2007)
3. Liu, B: Fuzzy process, hybrid process and uncertain process. J. Uncertain Syst. **2**(1), 3–16 (2008)
4. Liu, B: Some research problems in uncertainty theory. J. Uncertain Syst. **3**(1), 3–10 (2009)
5. Chen, XW, Liu, B: Existence and uniqueness theorem for uncertain differential equations. Fuzzy Optimization Decis. Mak. **9**(1), 69–81 (2010)
6. Gao, Y: Existence and uniqueness theorem on uncertain differential equations with local Lipschitz condition. J. Uncertain Syst. **6**(3), 223–232 (2012)
7. Liu, YH: An analytic method for solving uncertain differential equations. J. Uncertain Syst. **6**(4), 244–249 (2012)
8. Yao, K, Chen, XW: A numerical method for solving uncertain differential equations. J. Intell. Fuzzy Syst. (2013, in press)
9. Yao, K: Extreme values and integral of solution of uncertain differential equation. J. Uncertainty Anal. Appl. **1**, Article 3 (2013)
10. Chen, XW: American option pricing formula for uncertain financial market. Int. J. Oper. Res. **8**(2), 32–37 (2011)
11. Peng, J, Yao, K: A new option pricing model for stocks in uncertainty markets. Int J. Oper. Res. **8**(2), 18–26 (2011)
12. Yu, XC: A stock model with jumps for uncertain markets. Int. J. Uncertainty Fuzziness Knowledge-Based Syst. **20**(3), 421–432 (2012)
13. Liu, YH, Chen, XW: Uncertain currency model and currency option pricing. http://orsc.edu.cn/online/091010.pdf (2013)
14. Chen, XW, Gao, J: Uncertain term structure model of interest rate. Soft Comput. **17**(4), 597–604 (2013)
15. Zhu, Y: Uncertain optimal control with application to a portfolio selection model. Cybern. Syst. **41**(7), 535–547 (2010)
16. Ito, K: Stochastic integral. Proc. Jpn. Acad. Ser. A. **20**(8), 519–524 (1944)
17. Ito, K: On stochastic differential equations. Mem. Am. Math. Soc. **4**, 1–51 (1951)
18. Samuelson, PA: Rational theory of warrant pricing. Ind. Manage. Rev. **6**, 13–31 (1965)
19. Black, F, Scholes, M: The pricing of option and corporate liabilities. J. Pol. Econ. **81**, 637–654 (1973)
20. Merton, RC: Theory of rational option pricing. Bell J. Econ. Manage. Sci. **4**, 141–183 (1973)
21. Peng, ZX, Iwamura, K: A sufficient and necessary condition of uncertainty distribution. J. Interdiscip. Math. **13**(3), 277–285 (2010)
22. Liu, B: Uncertainty Theory: A Branch of Mathematics for Modeling Human Uncertainty. Springer, Berlin (2010)
23. Liu, YH, Ha, MH: Expected value of function of uncertain variables. J. Uncertain Syst. **4**(3), 181–186 (2010)
24. Liu, B: Uncertainty Theory. 4th edn. http://orsc.edu.cn/liu/ut.pdf (2013)

25. Liu, B: Extreme value theorems of uncertain process with application to insurance risk model. Soft Comput. **17**(4), 549–556 (2013)
26. Chen, XW: Variation analysis of uncertain stationary independent increment process. Eur. J. Oper. Res. **222**(2), 312–316 (2012)
27. Liu, B, Yao, K: Uncertain integral with respect to multiple canonical processes. J. Uncertain Syst. **6**(4), 250–255 (2012)
28. Chen, XW, Ralescu, DA: Liu process and uncertain calculus. J. Uncertainty Anal. Appl. **1**, Article 2 (2013)
29. Yao, K, Gao, J, Gao, Y: Some stability theorems of uncertain differential equation. Fuzzy Optimization Decis. Mak. **12**(1), 3–13 (2013)
30. Barbacioru, IC: Uncertainty functional differential equations for finance. Surv Math. Appl. **5**, 275–284 (2010)
31. Ge, XT, Zhu, Y: Existence and uniqueness theorem for uncertain delay differential equations. J. Comput. Inf. Syst. **8**(20), 8341–8347 (2012)
32. Liu, HJ, Fei, WY: Neutral uncertain delay differential equations. Inf. Int. Interdiscip. J. **16**(2), 1225–1232 (2013)
33. Yao, K: Uncertain calculus with renewal process. Fuzzy Optimization Decis. Mak. **11**(3), 285–297 (2012)
34. Ge, XT, Zhu, Y: A necessary condition of optimality for uncertain optimal control problem. Fuzzy Optimization Decis. Mak. **12** (2013)
35. Chen, XW, Liu, YH, Ralescu, DA: Uncertain stock model with periodic dividends. Fuzzy Optimization Decis. Mak. **12**(1), 111–123 (2013)

2

Distance-based test for uncertainty hypothesis testing

Sundaram Sampath* and Balu Ramya

* Correspondence:
sampath1959@yahoo.com
Department of Statistics, University of Madras, Chennai 600005, India

Abstract

Background: If an appropriate probability distribution cannot be identified for a given situation, it becomes extremely difficult to draw reliable inferences about the given domain of study under investigation. This is due to the fact that statistical theory of testing of hypothesis cannot be meaningfully employed in those cases. To deal with such situations, Uncertainty theory is recommended as an alternative by Liu (2007) and testing the validity of the hypotheses about uncertainty distributions is currently receiving the attention of researchers.

Methods: In this paper, for testing uncertain hypotheses about the true uncertainty distribution function, a new test procedure based on the inputs given by one or more domain experts is suggested. The proposed method can also be used for testing uncertain hypotheses about the equality of two uncertainty distribution functions.

Results: Illustrative examples are also provided in support of the test procedure suggested in this paper to demonstrate the utility of the same.

Conclusions: The same methodology can be used for testing the equality of two uncertainty distributions by making use of the ratio used in the construction of the test.

Keywords: Uncertain measure; Uncertainty distribution; Empirical uncertainty distribution; Testing uncertain hypothesis

Background

Testing of statistical hypotheses is a major branch of study in classical statistical inference. It deals with the process of developing appropriate test procedures for testing the validity of statistical hypotheses. Statistical hypotheses are statements about characteristics of real-life situations modeled in terms of probability distributions and a statistical test helps the decision maker whether to accept or reject the given hypothesis based on sampled observations. The theory of testing of statistical hypotheses revolves around the probability theory.

There are several real-life situations where it would be very difficult to identify appropriate models (probability distributions) describing the probabilistic properties of the given phenomena. Further, collection of adequate information in the form of sampled data to explain fully the probability distribution is not always viable. To deal with these situations, [1] introduced a new theory called the *Theory of Uncertainty*. Further refinements on the *Theory of Uncertainty* have been carried out by Liu

[2]. For more details about the *Theory of Uncertainty* and its applications in various fields of research, one can refer to [2]. The online resource of [3] is an excellent source of information on the latest status of various aspects related to *Uncertainty Theory.*

It is well known that probability distributions are the backbone of the theory of statistical inference that helps practitioners to study about the inherent characteristics of the given situation. Inferences related to the given system require the knowledge of parameters involved in the underlying probability distributions for which several solutions are available in the literature. Similar to probability distributions playing a crucial role in the stochastic situations, uncertainty distributions play a significant role in the *Theory of Uncertainty.* Uncertainty distributions model the nature of uncertainty present in the given system. Several uncertainty distributions and their properties are available in [3]. These distributions have certain unknown constants, and practitioners require the knowledge of these quantities to study the nature of uncertainty. Liu [2] suggested an estimation procedure for the estimation of parameters in an uncertainty distribution. It was followed by the works of Wang and Peng [4] and Wang et al. [5]. Recently, Wang et al. [6] introduced an uncertain hypothesis testing procedure to test the equality of two uncertainty distributions.

In this paper, a new test procedure is introduced for testing whether a specified uncertainty distribution function can be the true uncertainty distribution function of the given system. The proposed test procedure makes use of a distance based on empirical comprehensive uncertainty distribution defined by Liu [2]. The suggested procedure can be modified suitably for handling the situation wherein one will be interested in testing the equality of two uncertainty distributions. The paper is organized as follows. The second section of the paper introduces the uncertainty theory and uncertainty distributions briefly. The third section of the paper explains Wang et al. [6] test procedure and introduces the new test for testing hypotheses about uncertainty distributions. Illustrations are given in the fourth section, and conclusions are provided in fifth section.

Methods

Uncertainty distributions

Let Γ be a nonempty set and L be a σ- algebra over Γ. Elements of L are known as events. Uncertainty measure M is a function from L to $[0,1]$ which measures the degree of belief associated with an event. Initially, it was introduced as a function from L to $[0,1]$, satisfying the axioms such as normality, monotonicity, self-duality, and countable subadditivity [1]. Later on, Liu [3] refined the definition of uncertainty measure and defined it as a measure satisfying normality, duality, and subadditivity axioms. A measureable function ξ from the uncertainty space (Γ,L,M) to the set of real numbers is defined as uncertain variable. The uncertainty distribution $\Phi{:}R \to [0,1]$ of an uncertain variable ξ is defined by $\Phi(x) = M\{\xi \le x\}$, for any $x \in R$. According to Peng and Iwamura [7], a sufficient and necessary condition for a function $\Phi{:}R \to [0,1]$ to be an uncertainty distribution function is that the function is an increasing function except for the choices $\Phi(x) \equiv 0$ and $\Phi(x) \equiv 1$.

Example of some uncertainty distribution functions [3] are

1. The uncertainty normal distribution denoted by $N(c,\sigma), c \in R, \sigma > 0$ is defined as

$$\Phi(x) = \left[1 + e^{\frac{\pi(c-x)}{\sqrt{3}\sigma}}\right]^{-1}, \quad x \in R. \tag{1}$$

2. The uncertainty lognormal distribution denoted by $LOGN(c,\sigma)$, $\in c \ R, \sigma > 0$ is defined as

$$\Phi(x) = \left[1 + e^{\frac{\pi(c-\log x)}{\sqrt{3}\sigma}}\right]^{-1}, \quad x \geq 0. \tag{2}$$

Liu [2] has given a method of computing empirical uncertainty distribution function using the data collected from an expert. Assume that the set of expert's experimental data $(x_1,\alpha_1),(x_2,\alpha_2),\ldots,(x_n,\alpha_n)$meets the consistent condition $x_1 < x_2 < \ldots < x_n$, $0 \leq \alpha_1 \leq \alpha_2 \leq \ldots \leq \alpha_n \leq 1$.

Then, the empirical uncertainty distribution is computed by

$$\hat{\Phi}(x) = \begin{cases} 0 & \text{if } x < x_1 \\ \alpha_i + \dfrac{(\alpha_{i+1}-\alpha_i)(x-x_i)}{x_{i+1}-x_i} & \text{if } x_i \leq x \leq x_{i+1}, \ 1 \leq i < n. \\ 1 & \text{if } x > x_n \end{cases} \tag{3}$$

To distinguish the empirical uncertainty distribution function from the true uncertainty distribution Φ, we use the symbol ^ on top of Φ for empirical uncertainty distribution function. When the experimental data is collected from m experts, empirical comprehensive uncertainty distribution function is obtained using the convex combination of the empirical uncertainty distribution computed for each expert. That is, we compute the empirical uncertainty distribution functions $\hat{\Phi}_1$, $\hat{\Phi}_2$,..., $\hat{\Phi}_m$ using the above definition of empirical uncertainty distribution function for the data collected from the m experts. Then, these m empirical uncertainty distribution functions are combined in the following manner to get the empirical comprehensive uncertainty distribution function

$$\hat{\Phi}(x) = w_1\hat{\Phi}_1(x) + w_2\hat{\Phi}_2(x) + \ldots + w_m\hat{\Phi}_m(x), \tag{4}$$

where $\sum_{i=1}^{m} w_i = 1, w_i \geq 0, i = 1, 2, \ldots, m.$

It is pertinent to note that this convex combination is also an empirical uncertainty distribution as proved by Peng and Iwamura [7].

Test for uncertainty distribution hypotheses

In this section, we develop a test procedure for testing hypotheses about uncertainty distributions. An uncertain hypothesis is a hypothesis about uncertainty distributions that characterize uncertain situations.

Wang et al. [4] suggested a method of uncertain hypotheses testing based on uncertainty theory to test whether two uncertainty distributions are equal or not. They considered the two sets of expert's data of the form $(x_1^1, \alpha_1^1), (x_2^1, \alpha_2^1), \ldots, (x_m^1, \alpha_m^1)$ and $(x_1^2, \alpha_1^2), (x_2^2, \alpha_2^2), \ldots, (x_n^2, \alpha_n^2)$ that meet the consistent conditions

$$x_1^1 < x_2^1 < \ldots < x_m^1, \quad 0 \le \alpha_1^1 \le \alpha_2^1 \le \ldots \le \alpha_m^1 \le 1$$
$$x_1^2 < x_2^2 < \ldots < x_n^2, \quad 0 \le \alpha_1^2 \le \alpha_2^2 \le \ldots \le \alpha_n^2 \le 1.$$

It is presumed that the two theoretical uncertainty distributions with respect to the expert's data are $F_1(x)$ and $F_2(x)$. To test the null uncertainty hypothesis $H_0{:}F_1(x) = F_2(x)$ for any $x \in R$ against the alternative uncertainty hypothesis $H_1{:}F_1(x) \neq F_2(x)$ for some $x \in R$, Wang et al. [4] constructed a test procedure based on randomly generated points from two empirical uncertainty distribution functions corresponding to the two experts' data In this paper, we construct a test for testing the uncertain hypothesis that a given function F_1 can be the true uncertainty distribution function for the given situation of interest against the alternative hypothesis that F_2 is the true uncertainty distribution function. Consider the problem of testing the hypothesis $H_0{:}F(x) = F_1(x)\ \forall x$ against the alternative hypothesis $H_1{:}F(x) = F_2(x)\ \forall x$, where F_1 and F_2 are known theoretical uncertainty distribution functions.

The test procedure assumes that the data related to the given testing problem are collected from m experts. Each expert gives his opinion in terms of a sequence of numbers along with the corresponding belief levels, where the lengths of the sequences are not necessarily the same for all experts but the numbers in every sequence are expected to cover possible values of the subject of interest. The data collected from the m experts can be mathematically described as given in Table 1. It is pertinent to note as in the case of formulating hypotheses in the classical theory of testing statistical hypotheses, the functions specified under null and alternative hypothesis are determined judiciously in an appropriate manner by taking into account the chance environment and experts' opinion. Issues related to this aspect have been discussed in the illustrative examples.

Corresponding to the information given in the m rows of Table 1, we compute the empirical uncertainty distribution functions $\hat{\Phi}_1$, $\hat{\Phi}_2$ ... $\hat{\Phi}_m$ using the definition given in the previous section and combine these m empirical functions, using a convex combination, to get the empirical comprehensive uncertainty distribution function defined in Eq. (4).

Let A_r be the sequence of numbers reported by the expert $r(r = 1,2,\ldots,m)$ and $S = \cup_{r=1}^{m} A_r$. For every $x \in A_r$, we compute the distance between the empirical comprehensive uncertainty distribution and the empirical uncertainty distribution function given under the null hypothesis, namely,

$$d(\hat{\Phi}, F_1) = \sum_{x \in S} d_x(\hat{\Phi}(x), F_1(x)),$$

where $d_x(\hat{\Phi}(x), F_1(x)) = (\hat{\Phi}(x) - F_1(x))^2\ \forall x \in S$.

Table 1 Experimental data for *m* experts

Expert	Values and degree of belief
1	$(x_1^1, a_1^1), (x_2^1, a_2^1), \ldots, (x_{n_1}^1, a_{n_1}^1)$
2	$(x_1^2, a_1^2), (x_2^2, a_2^2), \ldots, (x_{n_2}^2, a_{n2}^2)$
M	$(x_1^m, a_1^m), (x_2^m, a_2^m), \ldots, (x_{n_m}^m, a_{n_m}^m)$

Similarly, we find the distance between the empirical comprehensive uncertainty distribution and the uncertainty distribution function given under the alternative hypothesis using

$$d(\hat{\Phi}, F_2) = \sum_{x \in S} d_x(\hat{\Phi}(x), F_2(x)),$$

where $d_x(\hat{\Phi}(x), F_2(x)) = (\hat{\Phi}(x) - F_2(x))^2 \ \forall x \in S$.

If we treat the empirical comprehensive uncertainty distribution function based on the opinion of several domain experts as an appropriate estimate of the true distribution, then it is reasonable to develop a test based on the same. If the distribution mentioned under the null hypothesis is to be supported, then we expect the distance between the empirical comprehensive uncertainty distribution function $\hat{\Phi}$ and F_1 to be small. Hence, it is reasonable to define the rejection rule for rejecting the null hypothesis as

$$\frac{d(\hat{\Phi}, F_1)}{d(\hat{\Phi}, F_2)} > k, \tag{5}$$

where k is the largest real number satisfying the inequality

$\sigma(R_k) \geq \alpha$.

Here, $R_k = \left\{ x \middle| \frac{d_x(\hat{\Phi}(x) - F_1(x))}{d_x(\hat{\Phi}(x) - F_2(x))} > k \right\}$ and $\sigma(R_k) = \frac{N(R_k)}{N(S)}$, N being the number of elements in the given set. The constant α is predetermined by the user. It is nothing but the proportion of items in $S = \bigcup_{r=1}^{m} A_r$ for which the ratio of the distance between the empirical comprehensive uncertainty distribution function and the distribution specified under null hypothesis to the corresponding distance based on the distribution mentioned under the alternative hypothesis, exceeding the threshold value k. It may be noted that the value of α closer to 1 will increase the number of cases where the condition mentioned in R_k will be satisfied leading higher chance of rejection. Similarly, a value of α closer to 0 will decrease the number of cases where the condition mentioned in R_k will be satisfied leading to lower chance of rejection. The practitioner has to decide the choice of α in a judicious manner striking a balance between the rejection rate and the chance of taking a correct decision.

Results and discussion

To illustrate the process of developing a test procedure using the above method, two examples are considered.

Example 1

The data given in [6] are based on knowledge and experience of three teachers who performed an analysis about the degree of difficulty of a higher mathematics examination. The experimental data describing their estimated average scores and belief degrees are given below.

- Teacher 1: (60, 0.05), (70, 0.15), (80, 0.55), (85, 0.85), (90, 0.95)
- Teacher 2: (60, 0.08), (70, 0.17), (75, 0.36), (80, 0.58), (85, 0.85), (90, 0.95)
- Teacher 3: (50, 0.2), (60, 0.3), (70, 0.4), (80, 0.8), (85, 1)

As mentioned earlier, the most important task in testing of uncertain hypotheses is the formulation of null and alternative hypotheses in a meaningful manner. It has two stages, namely, identifying a suitable uncertainty distribution (e.g., zigzag, normal, and lognormal distributions) for the given situation and the parametric values to be used under the null and alternative hypotheses. This can be accomplished using the works of Liu [3] and Wang and Peng [4] related to estimation of uncertainty distributions.

Using the Matlab Uncertainty Toolbox (http://orsc.edu.cn/liu/resources.htm), normal and lognormal uncertainty distributions were fitted for three data sets using the least squares method of Liu [3]. Since the errors corresponding to lognormal are small (for all the three data sets) when compared to normal uncertainty distributions, we have decided to use the lognormal distribution under both null and alternative hypotheses. Figures 1, 2, and 3 give shapes of the fitted lognormal uncertainty distribution functions as well as the pairs of experimental data corresponding to the true teachers. For the three data sets, the least squares estimated values corresponding to lognormal fit for teacher 1, teacher 2, and teacher 3 are ($C = 4.3605$, $\sigma = 0.1010$), ($C = 4.3520$, $\sigma = 0.1095$), and ($C = 4.2227$, $\sigma = 0.2271$), respectively. As in the case of classical statistical theory of testing of hypotheses, it is left to the discretion of the practitioner to decide the choices to be used under null and alternative hypotheses based on his belief and observation of the given system. The distributions to be used under the null and alternative uncertainty hypotheses have been chosen by taking into account the distances between the fitted distributions and the empirical comprehensive uncertainty distribution as defined in the third section . One of the distributions having the two smallest distances has been used in the null hypothesis and the other in the alternative hypothesis. Thus, we consider testing the null uncertainty hypothesis $H_0 : \Phi(x) = \left[1 + e^{\frac{\pi(4.3605 - \ln x)}{\sqrt{3}(0.1010)}}\right]^{-1}$ against the alternative uncertainty hypothesis

$$H_1 : \Phi(x) = \left[1 + e^{\frac{\pi(4.3520 - \ln x)}{\sqrt{3}(0.1095)}}\right]^{-1}$$

We start the process of formulating the test procedure by computing the empirical comprehensive uncertainty distribution using the experimental data obtained from the three different teachers, where empirical comprehensive uncertainty distribution based on three experts' opinion are given by

$$\hat{\Phi}(x) = w_1\hat{\Phi}_1(x) + w_2\hat{\Phi}_2(x) + w_3\hat{\Phi}_3(x)$$

Here, $\hat{\Phi}_1$, $\hat{\Phi}_2$, and $\hat{\Phi}_3$ are empirical distributions based on first, second, and third teachers. It may be noted that the weightsw_1, w_2, and w_3 are non-negative quantities satisfying, $w_1 + w_2 + w_3 = 1$. For the sake of simplicity, we assume w_1, w_2, and w_3 are $\frac{1}{3}$. For the given data, we have $A_1 = \{60,70,80,85,90\}$,$A_2 = \{60,70,75,80,85,90\}$, and $A_3 = \{50,60,70,80,85\}$ $S = \{50,60,70,75,80,85,90\}$.

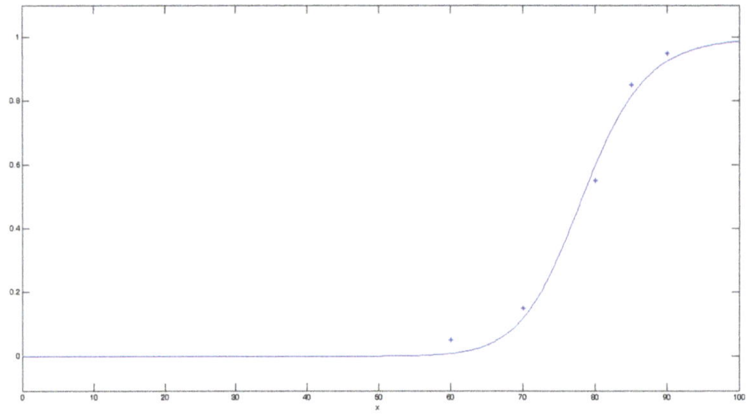

Figure 1 Fitted lognormal for Wang et al. [6] data of teacher 1 with $C = 4.3605$ and $\sigma = 0.1010$.

Therefore, we get

$$d(\hat{\Phi}, F_1) = \sum_{x \in S} d_x(\hat{\Phi}(x), F_1(x)) = 0.06359114$$

$$d(\hat{\Phi}, F_2) = \sum_{x \in S} d_x(\hat{\Phi}(x), F_2(x)) = 0.04387163.$$

Hence, $\frac{d(\hat{\Phi}, F_1)}{d(\hat{\Phi}, F_2)} = 1.449482$.

Table 2 gives $\sigma(R_k)$ for different values of k. If the user decides to choose $\alpha = 0.4$, then k is taken as 2.5. This choice leads to the acceptance of the hypothesis $H_0 : \Phi(x) = \left[1 + e^{\frac{\pi(4.3605 - \ln x)}{\sqrt{3}(0.1010)}}\right]^{-1}$.

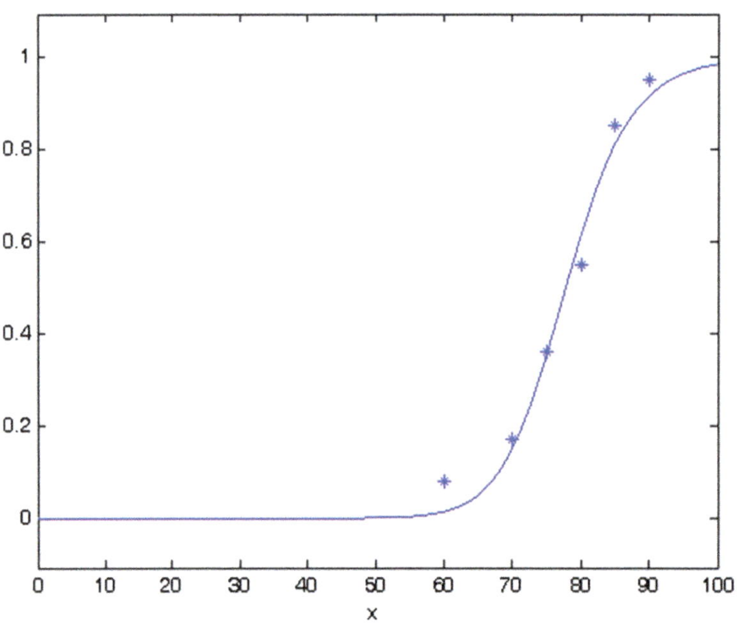

Figure 2 Fitted lognormal for Wang et al. [6] data of teacher 2 with $C = 4.3546$ and $\sigma = 0.1109$.

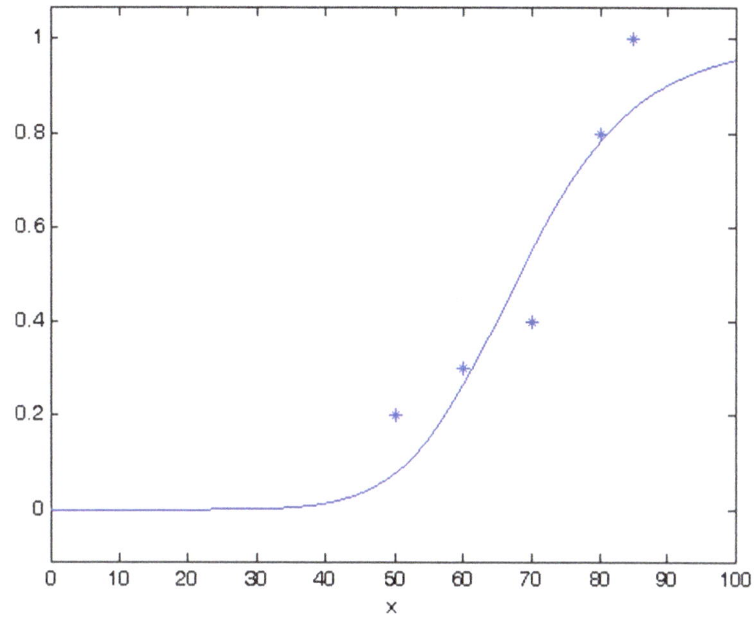

Figure 3 **Fitted lognormal for Wang et al. [6] data of teacher 3 with $C = 4.2227$ and $\sigma = 0.2271$.**

Example 2

In Example 1, since lognormal uncertainty distribution fitted well for all the three data sets, we have used the same distribution under the null and alternative hypotheses. Now, we consider a testing problem where different distributions are used under the null and alternative hypotheses.

This example is based on the data provided in Chapter 4 of [3]. The expert's experimental data is given below:

(0.6, 0.1), (1.0, 0.3), (1.5, 0.4), (2.0, 0.6), (2.8, 0.8), (3.6, 0.9).

Using the Matlab Uncertainty Toolbox (http://orsc.edu.cn/liu/resources.htm), lognormal uncertainty distribution and normal uncertainty distribution were fitted for the above data set using the least squares method. Figures 4 and 5 explain the observed data points as well as the fitted distributions for the expert's experimental data set.

The least squares estimated values corresponding to lognormal and normal fit for data set are ($C = 0.4825$, $\sigma = 0.7852$) and ($C = 1.7690$, $\sigma = 1.2953$), respectively. The errors corresponding to the lognormal and normal fit are 0.0081 and 0.0074. It is decided to test the null uncertain hypothesis $H_0 : \Phi(x) = \left[1 + e^{\frac{\pi(0.4825 - \ln x)}{\sqrt{3}(0.7852)}}\right]^{-1}$ against alternative uncertain hypothesis $H_0 : \Phi(x) = \left[1 + e^{\frac{\pi(0.4825 - \ln x)}{\sqrt{3}(0.7852)}}\right]^{-1}$.

It is to be noted that a testing problem of this kind becomes meaningful in these types of situations since the errors do not show a huge difference. If the difference between the

Table 2 Range of k and σ (R_k) for Wang et al. [6] data

k	>5	[2.6,4.9]	[2.0,2.5]	[1.1,1.9]	1.09	[1.02,1.08]	[1.0,1.01]	<1
$\sigma(R_k)$	0	$\frac{1}{7}$	$\frac{2}{7}$	$\frac{3}{7}$	$\frac{4}{7}$	$\frac{5}{7}$	$\frac{3}{7}$	$\frac{3}{7}$

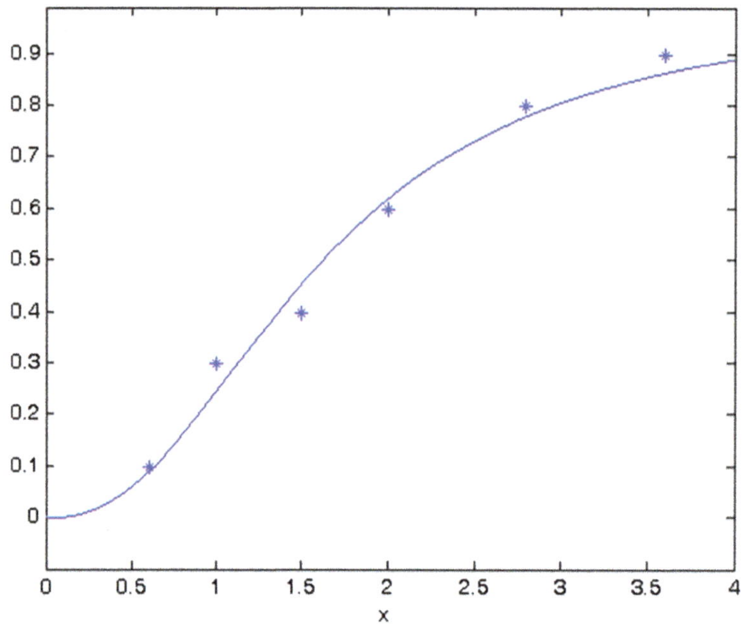

Figure 4 Fitted lognormal for Liu [3] data with $C = 0.4825$ and $\sigma = 0.7852$.

errors is considerably large, then one can use the distribution function corresponding to the smaller error as the one suitable for the given uncertain situation without depending on any test procedure.

Since the data is based on only one expert, the test makes use of the empirical uncertainty distribution using the experimental data obtained from the expert.

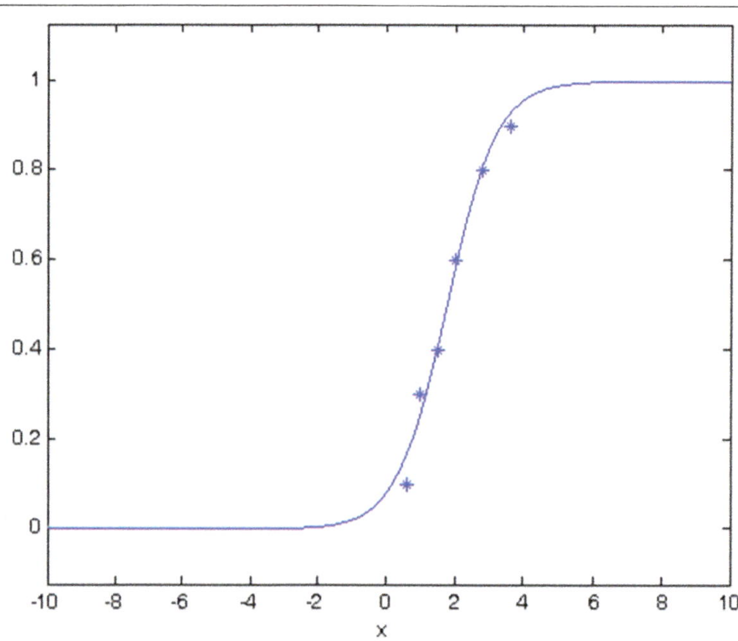

Figure 5 Fitted lognormal for Liu [3] data with $C = 1.7690$ and $\sigma = 1.2953$.

Table 3 Range of k and $\sigma(R_k)$ for Liu [3] data

K	>55.6	[1.1, 55]	[0.8,1]	0.7	[0.2, 0.6]	0.1	0
$\sigma(R_k)$	0	$\frac{1}{6}$	$\frac{2}{6}$	$\frac{3}{6}$	$\frac{4}{6}$	$\frac{5}{6}$	$\frac{6}{6}$

Here, $A = S = \{0.6,1,1.5,2,2.8\}$,

$$d(\hat{\Phi}, F_1) = \sum_{x \in S} d_x(\hat{\Phi}(x), F_1(x)) = 0.0073954$$

$$d(\hat{\Phi}, F_2) = \sum_{x \in S} d_x(\hat{\Phi}(x), F_2(x)) = 0.0080926.$$

Therefore, $\frac{d(\hat{\Phi},F_1)}{d(\hat{\Phi},F_2)} = 0.913856.$

Table 3 gives $\sigma(R_k)$ for different values of k. If the user decides to choose $\alpha = 0.4$, then k must be taken as 0.7. This choice leads to the rejection of the *hypothesis* $H_0 : \Phi(x) = \left[1 + e^{\frac{\pi(0.4825 - \ln x)}{\sqrt{3}(0.7852)}}\right]^{-1}$.

Conclusions

In this paper, a new test procedure that makes use of the data gathered from one or more domain experts has been developed for testing whether a specified uncertainty distribution can be the true uncertainty distribution function of the given situation. Two illustrative examples are also provided by making use of the data sets available in [4] and [3]. The first example deals with the case where both the null and alternative hypotheses use the lognormal uncertainty distribution, whereas the second example considers the testing problem where lognormal uncertainty and normal uncertainty distributions are used under null and alternative hypotheses, respectively.

It is pertinent to note that the same methodology can be used for testing the equality of two uncertainty distributions by making use of the ratio used in the construction of the test explained in the third section. Decision regarding the acceptance or rejection of the null hypotheses can be made by making use of the same ratio, namely, $\frac{d(\hat{\Phi},F_1)}{d(\hat{\Phi},F_2)}$. However, the null hypothesis will be rejected if the ratio $\frac{d(\hat{\Phi},F_1)}{d(\hat{\Phi},F_2)}$ is either very small or very large.

Competing interests

The authors declare that they have no competing interests.

Authors' contribution

SS has initiated the research work, devised the testing criteria introduced in this paper, and guided RB in working out the illustrative examples. RB has prepared the initial draft of the manuscript and contributed in numerical work. SS has suggested the sequence alignment and helped in preparing the final version of the manuscript. Both authors read and approved the final manuscript.

Authors' information

First author SS is currently holding the position of Professor of Statistics in the University of Madras, Chennai, India. He has more than 30 years experience in teaching and research. His main areas of research are Inference for Finite Populations, Classical Statistical Inference, and Data Mining. He has to his credit more than 30 research articles in highly rated research journals. Second author BR is a full-time research scholar working for her Ph.D. degree in the Department of Statistics, University of Madras, Chennai, India. Her research interest includes Fuzzy and Rough Set theories.

Acknowledgments
The authors wish to thank the referees for their comments and suggestions that lead to considerable amount of improvement in the contents as well as the overall organization of the manuscript.

References

1. Liu, B: Uncertainty theory, 2nd edn. Springer-Verlag, Berlin (2007)
2. Liu, B: A branch of mathematics for modeling human uncertainty. Springer-Verlag, Berlin (2011)
3. Liu, B: Uncertainty theory, 4th edn. (2013). http://orsc.edu.cn/liu/ut.pdf. Accessed January 2013
4. Wang, X, Peng, Z: Method of moments for estimation uncertainty distribution. (2012). http://orsc.edu.cn/online/100408.pdf. Accessed January 2013
5. Wang, X, Gao, Z, Guo, H: Delphi method for estimating uncertainty distributions. Information: An International Interdisciplinary Journal **12**(2), 449–460 (2012)
6. Wang, X, Gao, Z, Guo, H: Uncertain hypothesis testing for expert's empirical data. Math. Comput. Model. **55**, 1478–1482 (2012)
7. Peng, Z, Iwamura, K: A sufficient and necessary condition of uncertainty distribution. J. Interdiscipl. Math. **13**, 277–285 (2010)

3

Robust design under mixed aleatory/epistemic uncertainties using gradients and surrogates

Markus P Rumpfkeil

Correspondence:
Markus.Rumpfkeil@udayton.edu
Department of Mechanical and Aerospace Engineering, University of Dayton, Dayton, Ohio, USA

Abstract

In this paper, mixed aleatory/epistemic uncertainties in a robust design problem are propagated *via* the use of box-constrained optimizations and surrogate models. The assumption is that the uncertain input parameters can be divided into a set only containing aleatory uncertainties and a set with only epistemic uncertainties. Uncertainties due to the epistemic inputs can then be propagated *via* a box-constrained optimization approach, while the uncertainties due to aleatory inputs can be propagated *via* sampling. A statistics-of-intervals approach is used in which the box-constrained optimization results are treated as a random variable and multiple optimizations need to be performed to quantify the aleatory uncertainties *via* sampling. A Kriging surrogate is employed to model the variation of the optimization results with respect to the aleatory variables enabling exhaustive Monte-Carlo sampling to determine the desired statistics for each robust design iteration. This approach is applied to the robust design of a transonic NACA 0012 airfoil where shape design variables are assumed to have epistemic uncertainties and the angle of attack and Mach number are considered to have aleatory uncertainties. The very good scalability of the framework in the number of epistemic variables is demonstrated as well.

Introduction and motivation

Computational methods have been playing an increasingly important role in science and engineering analysis and design over the last several decades, due to the rapidly advancing capabilities of computer hardware, as well as increasingly sophisticated and capable numerical algorithms. However, in spite of the rapid advances and acceptance of numerical simulations, serious deficiencies remain in terms of accuracy, uncertainty, and validation for many applications. Many real-world problems involve input data that is noisy or uncertain, due to measurement or modeling errors, approximate modeling parameters [1], manufacturing tolerances [2], in-service wear-and-tear, or simply the unavailability of the information at the time of the decision [3]. These imprecise or unknown inputs are important in the design process and need to be quantified in some fashion. To this end, uncertainty quantification (UQ) has emerged as an important area in modern computational engineering. Today, it is no longer sufficient to predict specific objectives using a particular physical model with deterministic inputs. Rather, a probability distribution function (PDF) or interval bound of the simulation objectives is required depending on whether aleatory or epistemic uncertainties are involved [4]. Epistemic uncertainty (or type B, or reducible uncertainty) represents a lack of knowledge about the appropriate value to use for a quantity, i.e. there is a single correct (but unknown value) [5]. This may be, for example, because a quantity has not been measured sufficiently accurately or because the model neglects certain effects. In contrast, uncertainty characterized by inherent natural randomness is called aleatory uncertainty (or type A, or irreducible uncertainty). For discrete variables, this randomness is parameterized by the probability of each possible value. For continuous variables, the randomness is parameterized by a PDF. Regulatory agencies and design teams are increasingly being asked to specifically characterize and quantify epistemic uncertainty and separate its effect from that of aleatory uncertainty [6].

Probabilistic assessment of uncertainty in computational models consists of three major phases [7,8]:

1. Data assimilation in which the input parameters are characterized as aleatory or epistemic (*via* appropriate PDFs or interval bounds) from observations and physical evidence
2. Uncertainty propagation in which the input variabilities are propagated through the mathematical model
3. Characterization of the outputs of the numerical simulation in terms of their statistical properties

Arguably, the computationally most expensive part of UQ is the second phase. A mixed aleatory/epistemic UQ typically relies on a nested sampling strategy. Although the required number of samples grows extremely fast, these strategies are conceptually easy to understand and are capable of separating the effects of each type of uncertainty [8,9]. For nested strategies, samples are typically first drawn from the epistemic variables and for each set of epistemic variables, the distribution of the output due to the aleatory variables is determined using sampling of the aleatory variables. The simplest approach for sampling is the Monte-Carlo (MC) method [10] which for expensive mathematical models (e.g., high-fidelity physics-based simulations) becomes very quickly prohibitively expensive due to the large number of model evaluations. For example, the number of samples required for the epistemic uncertainty grows exponentially fast with the number of epistemic variables [8]. To alleviate some of the cost, surrogates can be created as a function of all variables and samples extracted according to a nested strategy. For relatively low dimensions, this strategy can be effective and, when combined with gradient-enhancement, could be applied to problems of moderate dimension [11]. However, once the number of epistemic variables increases sufficiently, surrogate-based approaches will again become prohibitively expensive as the required number of training points increases exponentially fast for an accurate surrogate model known as "curse of dimensionality". In order to address this concern, combinations of sampling and optimization approaches have been explored [9,12]. The idea is that for mixed aleatory/epistemic problems, the goal of the uncertainty quantification is to produce a region in which the function is contained with a specific level of confidence, known as a P-Box [8] or horsetail as shown in Figure 1. The bounds of the confidence interval of the output distribution must itself be an interval in order to account for the epistemic uncertainties. Because only the bounds of this box are required, the sampling with respect to the epistemic variables can be replaced by one maximization and one minimization problem.

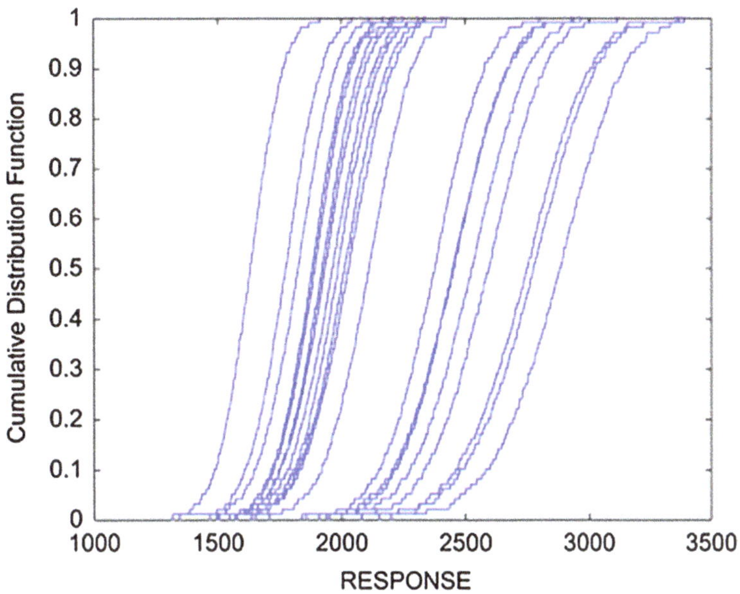

Figure 1 A P-Box where every line corresponds to a different set of values of the epistemic variables.

In principle, these mixed sampling/optimization approaches may be posed in two ways: determining intervals of statistics and determining statistics of intervals:

- *Intervals-of-statistics* can be viewed as an optimization under uncertainty problem with the metric of the optimization defined as a relevant statistic of the aleatory distribution, such as the mean and variance, bounds on a confidence interval, or a reliability index [9,13]. For each step in the optimization, the aleatory uncertainty is quantified, and the relevant statistics of the distribution are calculated and used as the objective function for the optimization.
- *Statistics-of-interval* poses an optimization problem for each set of aleatory variables, and repeated optimization evaluations over the epistemic design space can be used to determine the relevant statistics of the interval [12].

In the statistics-of-interval approach, gradient-based optimization methods can be employed, assuming that the global extrema in the epistemic design space can be found this way, reducing the cost of each optimization and ensuring very good scaling as the number of epistemic variables increases if adjoint capabilities [14,15] are used. To reduce the number of required optimizations for low statistical errors, a surrogate model of the optimization results can be constructed with respect to the aleatory variables which can then be sampled exhaustively, ensuring that fewer optimizations are required to characterize the statistics of the interval accurately.

A last important observation for the work in this paper is that deterministic optimization tools are widely used in engineering practice; however, engineering designs do not operate exactly at their design point due to physical variability in the environment. These small variations can deteriorate the performance of deterministically optimized designs. It is, therefore, necessary to account for these uncertainties in the optimization process using optimization under uncertainty (OUU) techniques, which implies that UQ is used in the optimization loop instead of a deterministic simulation. Beginning with the seminal works of Beale [16], Dantzig [17], and Tintner [18], OUU has experienced rapid development in both theory and algorithms. Dantzig considers planning under uncertainty as one of the most important open problems in optimization [19,20]. Good overviews of the state of the art in the field of OUU are provided by Beyera et al. [21], Sahinidis [19], Giunta et al. [22] and Li [23]. An important subfield in OUU is robust optimization (RO) [24,25], which can be subdivided into robust-design-based methods and reliability-based methods [26]. Robust design improves the quality of a product by minimizing the effect of the causes of variation without eliminating these causes. The objective here is to optimize the mean performance and minimize its variation, while maintaining feasibility with probabilistic constraints, hence the robust design concentrates on the probability distribution near the mean values. The ability to identify and catalog overly conservative design margins resulting from applying safety factors on top of other safety factors, for example, is an important application for the robust design, which is being increasingly viewed as an enabling technology for design of aerospace, civil, and automotive structures subject to uncertainty [27-30]. The reliability-based methods, on the other hand, are predominantly used for risk analysis by computing the probability of failure of a system. Thus, reliability approaches concentrate on the rare events at the tails of the probability distribution.

The outline of the remainder of this paper is as follows: section 'Optimization with mixed aleatory/epistemic uncertainty' describes the employed OUU approach for mixed aleatory/epistemic uncertainties in detail. Application results of the presented approach are given in section 'Robust design of a transonic airfoil' and section 'Conclusions' concludes this paper.

Optimization with mixed aleatory/epistemic uncertainty

A conventional constrained optimization problem for an objective function, J, that is a function of input variables, D, state variables, $q(D)$, and simulation outputs, $f(D) = F(q(D), D)$, can be written as

$$\begin{aligned} \min_{D} \qquad & J = J(f, q, D) \\ s.t. \qquad & 0 = R(q, D) \\ & 0 \leq g(f, q, D). \end{aligned} \tag{1}$$

Here, the state equation residuals, R, are expressed as an equality constraint, and other system constraints, g, are represented as general inequality constraints. Note, that R (and g) could represent any class of models, however, if gradient information is to be used the models must be differentiable and if surrogate models are to be employed successfully the models must also be relatively smooth. In the case where the input variables are precisely known, all functions dependent on D are deterministic. However, given uncertain inputs all functions in Equation (1) can no longer be treated deterministically.

Objective function evaluation

In this work, the design variables are assumed to have only aleatory or only epistemic uncertainty. Let α represent the variables associated with aleatory uncertainties and β represent variables with epistemic uncertainties, for example, geometric shape variables subject to manufacturing tolerances, or flow boundary conditions subject to random fluctuations. The design variables $D = (D_\alpha, D_\beta)$ are considered to be either the mean values of aleatory uncertainties which are assumed to be statistically independent and normally distributed with $\alpha \sim \mathcal{N}(D_\alpha, \sigma_D^2)$, or the midpoint of bounds on epistemic uncertainties with $\beta \in I(D)$, where $I(D) = [D_\beta - s_D, D_\beta + s_D]$. Note that σ_D and s_D are treated as fixed but this could be easily changed. One could also derive equations for correlated and/or non-normally distributed aleatory variables; however, the analysis and resulting equations become more complex [31] and are beyond the scope of this paper.

In order to account for both types of uncertainty, sampling is performed for the aleatory variables while optimization is performed over the epistemic variables as described in the introduction. Let $f(D) = f(\alpha, \beta)$ represent the output of interest of a simulation with uncertain inputs then the optimization can be represented mathematically as follows:

$$f_{\max}(\alpha) = \max_{\beta \in I(D)} f(\alpha, \beta) \tag{2}$$

$$f_{\min}(\alpha) = \min_{\beta \in I(D)} f(\alpha, \beta). \tag{3}$$

The functional outputs $f_{\max}$ and $f_{\min}$ can now be treated as random variables, since their only inputs are random variables with associated probability distributions. In the remainder of this paper, the subscript ext (for extrema) will be used as a placeholder for either max or min. To characterize the probability distribution of f_{ext}, one must extract repeated samples of f_{ext} according to the underlying PDF of α. Each sampling entails solving the appropriate optimization problem, Equation (2) or (3), for the specified sample of α. For these optimizations, an L-BFGS [32,33] algorithm that can utilize function and gradient information is used in this work, thereby reducing the cost of each optimization and ensuring excellent scaling in the number of variables with epistemic uncertainties.

Nonetheless, because of the expense of these optimizations, strategies to reduce the number of samples and thus the computational cost associated with sampling must be employed. For this work, a surrogate is created for f_{ext} as a function of the aleatory variables, which enables the extraction of a large number of samples in order to obtain accurate statistics for very low computational cost. Because the number of aleatory variables used here is relatively small, the required number of training points for an accurate surrogate is small, necessitating only a small amount of optimizations. Because the optimization results are viewed as general random variables, any surrogate can be used to represent the aleatory dependence of the variables. A Kriging surrogate model is employed in this work. The details of the construction of this particular Kriging model, which can utilize gradient and Hessian information and employ a dynamic training point selection, is described in previously published papers [34-37]. The center of the Kriging domain is prescribed by the mean value of α, D_α, and the boundary is taken to be two standard deviations σ_D away in all aleatory input dimensions. This implies that for the normally distributed input variables α more than 97% of all MC samples fall within the Kriging domain and the less accurate extrapolation capabilities of the Kriging surrogate model only need to be used for a small fraction of the samples. Since the purpose of this article is a robust design and not the accurate prediction of the tail statistics, this approach leads to very good results as demonstrated in section 'Robust design of a transonic airfoil'.

The deterministic optimization problem (1) can now be rewritten. The objective function can be written in terms of mean values of the functional outputs, $\bar{f}_{\text{ext}}$, and typically also becomes a function of the variances, $\text{Var}_{f_{\text{ext}}}$, for example, for robust design optimizations, objective functions are typically of the form

$$\mathcal{J} = w_1 \bar{f}_{\text{ext}} + w_2 \, \text{Var}_{f_{\text{ext}}}, \tag{4}$$

where w_i are some user-specified weights. The state equation residual equality constraint, R, is treated deterministically and thus needs to be satisfied for all values of α and β. The inequality constraints can be cast into a probabilistic statement such that the probability that the constraints are satisfied is greater than or equal to a desired or specified probability, P_k. This statement can be transformed [38] into a constraint involving mean values and standard deviations (also called moment matching formulation [39]) and the entire OUU problem can be expressed as [31,40]

$$\begin{aligned} \min_{\alpha,\beta} \quad & \mathcal{J} = \mathcal{J}(\bar{f}_{\text{ext}}, \text{Var}_{f_{\text{ext}}}, q, \alpha, \beta) \\ s.t. \quad & 0 = R(q, \alpha, \beta) \\ & 0 \leq g(\bar{f}_{\text{ext}}, q, \alpha, \beta) - k\sigma_g, \end{aligned} \tag{5}$$

where k is the number of standard deviations, σ_g, that the constraint g must be displaced in order to achieve P_k. The software package Ipopt (Interior Point Optimizer) [41] for large-scale nonlinear optimization with constraints is used for the solution of the OUU problem given by Equation (5). Ipopt also allows users to impose bound or box constraints on the design variables which can be very helpful in ensuring the stability of the flow analysis by preventing the exploration of too extreme regions of the design space.

Gradient evaluation

The gradient of the objective function, $\mathcal{J}$, given by Equation (4) with respect to design variables associated with aleatory uncertainties is given by

$$\frac{d\mathcal{J}}{dD_\alpha} = \frac{\partial \mathcal{J}}{\partial \bar{f}_{\text{ext}}} \frac{d\bar{f}_{\text{ext}}}{dD_\alpha} + \frac{\partial \mathcal{J}}{\partial \text{Var}_{f_{\text{ext}}}} \frac{d\text{Var}_{f_{\text{ext}}}}{dD_\alpha} = w_1 \frac{d\bar{f}_{\text{ext}}}{dD_\alpha} + w_2 \frac{d\text{Var}_{f_{\text{ext}}}}{dD_\alpha}. \tag{6}$$

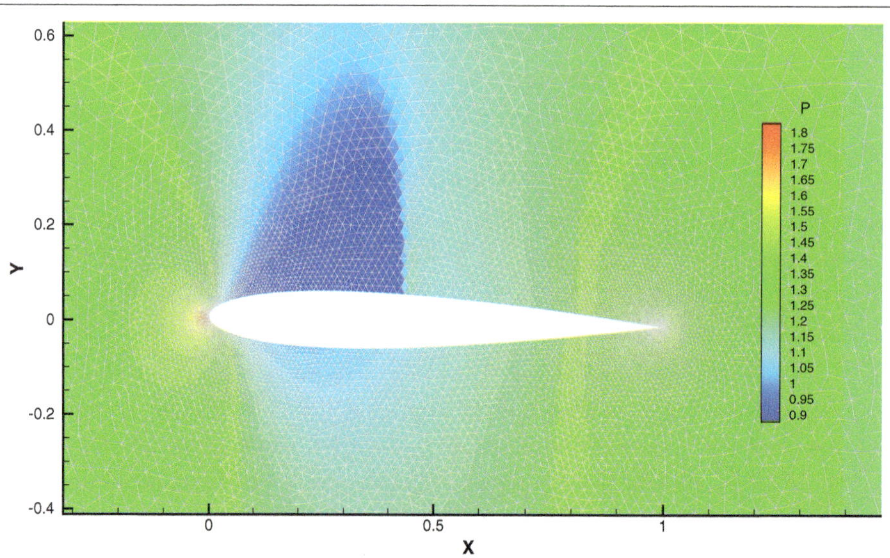

Figure 2 Pressure contours and mesh for angle of attack of 1.25° and a free-stream Mach number of 0.755.

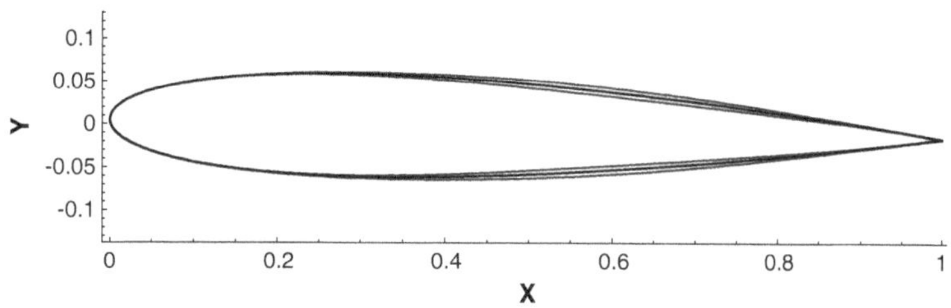

Figure 3 NACA 0012 airfoil (black) and airfoils resulting from two shape design variable values of ±0.005 (gray).

A Kriging surrogate is built to calculate $\bar{f}_{\text{ext}}$ and $\text{Var}_{f_{\text{ext}}}$ using N training points for each of which one has to calculate f_{ext} by solving an optimization problem as given by Equation (2) or (3). This Kriging surrogate is then sampled extensively $\tilde{N}$ times for inputs α^k, $k = 1, \ldots, \tilde{N}$ chosen based on their underlying probability distribution function. In this case, $\alpha \sim D_\alpha + \sigma_D Z$ with $Z \sim \mathcal{N}(0, 1)$ and the Kriging predictions are represented by $\hat{f}_{\text{ext}}(\alpha^k)$. The mean of the simulation output can then be approximated by

$$\bar{f}_{\text{ext}} \approx \frac{1}{\tilde{N}} \sum_{k=1}^{\tilde{N}} \hat{f}_{\text{ext}}(\alpha^k) \tag{7}$$

and the derivative can be approximated at the same time with little computational overhead *via* [42]

$$\frac{d\bar{f}_{\text{ext}}}{dD_\alpha} \approx \frac{1}{\tilde{N}} \sum_{k=1}^{\tilde{N}} \frac{d\hat{f}_{\text{ext}}(\alpha^k)}{d\alpha^k} \frac{d\alpha^k}{dD_\alpha} = \frac{1}{\tilde{N}} \sum_{k=1}^{\tilde{N}} \frac{d\hat{f}_{\text{ext}}(\alpha^k)}{d\alpha^k}, \tag{8}$$

where it is relatively straightforward to calculate $\frac{d\hat{f}_{\text{ext}}(\alpha^k)}{d\alpha^k}$ from the Kriging surrogate model [42,43]. This is especially true if the Kriging construction process can be gradient-enhanced since this derivative needs to be readily available for this.

Similarly, the variance and its derivative can be approximated as

$$\text{Var}_{f_{\text{ext}}} \approx \left(\frac{1}{\tilde{N}} \sum_{k=1}^{\tilde{N}} \hat{f}^2_{\text{ext}}(\alpha^k) \right) - \bar{f}^2_{\text{ext}} \tag{9}$$

$$\frac{d\text{Var}_{f_{\text{ext}}}}{dD_\alpha} \approx \left(\frac{2}{\tilde{N}} \sum_{k=1}^{\tilde{N}} \hat{f}_{\text{ext}}(\alpha^k) \frac{d\hat{f}_{\text{ext}}(\alpha^k)}{d\alpha^k} \right) - 2\bar{f}_{\text{ext}} \frac{d\bar{f}_{\text{ext}}}{dD_\alpha}. \tag{10}$$

The gradient of the objective function, $\mathcal{J}$, with respect to design variables associated with epistemic uncertainties is given by

$$\frac{d\mathcal{J}}{dD_\beta} = \frac{\partial \mathcal{J}}{\partial \bar{f}_{\text{ext}}} \frac{d\bar{f}_{\text{ext}}}{dD_\beta} + \frac{\partial \mathcal{J}}{\partial \text{Var}_{f_{\text{ext}}}} \frac{d\text{Var}_{f_{\text{ext}}}}{dD_\beta} = w_1 \frac{d\bar{f}_{\text{ext}}}{dD_\beta} + w_2 \frac{d\text{Var}_{f_{\text{ext}}}}{dD_\beta}. \tag{11}$$

Table 1 Validation of epistemic uncertainty propagation

Method	$c_{l_{\min}}$	$c_{l_{\max}}$	$c_{d_{\min}}$	$c_{d_{\max}}$
Optimization	0.195	0.344	3.56×10^{-3}	6.90×10^{-3}
LHS sampling	0.195	0.344	3.56×10^{-3}	6.90×10^{-3}

Table 2 Comparison of NLMC and Kriging aleatory uncertainty propagation

Method	$\bar{C}_l$	σ_{C_l}	$\bar{C}_d$	$\sigma^2_{C_d}$
NLMC ($\tilde{N} = 3,000$)	0.269	2.3×10^{-2}	5.54×10^{-3}	6.1×10^{-6}
Kriging ($N = 5$)	0.270	2.2×10^{-2}	5.65×10^{-3}	7.5×10^{-6}
Kriging ($N = 13$)	0.269	2.4×10^{-2}	5.53×10^{-3}	6.1×10^{-6}
Kriging ($N = 19$)	0.269	2.3×10^{-2}	5.54×10^{-3}	6.1×10^{-6}

However, it is not trivial to calculate $\frac{d\bar{f}_{\text{ext}}}{dD_\beta}$ and $\frac{d\text{Var}_{f_{\text{ext}}}}{dD_\beta}$, where D_β represents midpoints of epistemic uncertainty intervals since moving the midpoint will lead, in general, to different extrema for the training points and thus to a different Kriging surrogate which when sampled leads to different values of $\bar{f}_{\text{ext}}$ and $\text{Var}_{f_{\text{ext}}}$. In contrast, the aleatory gradient was easy to obtain since one only has to take into account how the sample points change while being able to reuse the same Kriging surrogate. The current workaround is to use the approximations

$$\frac{d\bar{f}_{\text{ext}}}{dD_\beta} \approx \left.\frac{df}{dD_\beta}\right|_{D_\alpha, D_\beta} \qquad \frac{d\text{Var}_{f_{\text{ext}}}}{dD_\beta} \approx 0 \tag{12}$$

that is the derivative of the mean value is approximated by the derivative of just f with respect to D_β at the mean values of the aleatory uncertainty variables α and midpoints of the intervals for the epistemic variables β. This derivative is, in general, non-zero since for the epistemic optimizations, the extreme value is typically encountered on the interval bound. The variances for the problems studied in this paper are much smaller than the mean values which allow the neglection of $\frac{d\text{Var}_{f_{\text{ext}}}}{dD_\beta}$. The following section will demonstrate that the presented approach can lead to successful robust designs.

Robust design of a transonic airfoil

The steady inviscid flow over a transonic NACA 0012 airfoil is considered as a flow example which is described in more detail in Mani and Mavriplis [44,45]. The computational mesh has about 20,000 triangular elements. The non-dimensionalized pressure contours for an angle of attack of 1.25° and a free-stream Mach number of 0.755 are shown in Figure 2 leading to a lift and drag coefficient of $C_l = 0.268$ and $C_d = 0.00521$, respectively.

In order to perform a robust lift-constrained drag minimization under mixed aleatory/epistemic uncertainty, one shape design variable on the upper surface and one on the lower surface which control the magnitude of Hicks-Henne sine bump functions [46] are allowed to vary. The resulting deformation of the mesh is calculated *via* a linear tension spring analogy [44,47]. Both shape design variables are assumed to have epistemic uncertainties due to manufacturing tolerances. A zero value corresponds to the original NACA 0012 airfoil and $s_{D_{u,l}}$ is taken to be 0.005. Figure 3 shows the original NACA 0012 airfoil and the airfoils resulting from design variable values of ±0.005.

The angle of attack and free-stream Mach number are assumed to have aleatory uncertainties which are both modeled with normal distributions. The mean values are given

Table 3 Comparison of NLMC and Kriging predictions for the initial guess with two shape design variables

	$\bar{C}_{l_{\min}}$	$\sigma_{C_{l_{\min}}}$	$\bar{C}_{d_{\max}}$	$\sigma^2_{C_{d_{\max}}}$
NLMC (3,000 optimizations)	0.195	2.1×10^{-2}	7.22×10^{-3}	7.9×10^{-6}
Kriging (13 optimizations)	0.195	2.1×10^{-2}	7.22×10^{-3}	7.9×10^{-6}

Table 4 Robust design results with two shape design variables

k	P_k	$\bar{C}_{d_{\max}}$	$\sigma^2_{C_{d\max}}$	$\bar{C}_{l_{\min}}$	$\sigma_{C_{l_{\min}}}$	D_u	D_l	D_{AoA}	D_M
0	0.5000	7.94×10^{-3}	8.1×10^{-6}	0.600	3.1×10^{-2}	2.50×10^{-2}	2.43×10^{-2}	1.85	0.711
1	0.8413	9.95×10^{-3}	1.1×10^{-5}	0.631	3.4×10^{-2}	2.50×10^{-2}	2.28×10^{-2}	1.85	0.730
2	0.9772	1.36×10^{-2}	1.0×10^{-5}	0.657	2.3×10^{-2}	2.49×10^{-2}	2.40×10^{-2}	1.84	0.736
3	0.9986	1.91×10^{-2}	1.8×10^{-5}	0.677	2.5×10^{-2}	2.49×10^{-2}	2.50×10^{-2}	1.85	0.745
4	0.9999	2.10×10^{-2}	2.7×10^{-5}	0.673	2.0×10^{-2}	2.44×10^{-2}	2.44×10^{-2}	1.85	0.750
Deterministic		1.36×10^{-3}	-	0.600	-	1.76×10^{-2}	2.06×10^{-2}	1.58	0.734

by the design variable values, D_{AoA} and D_M, and the standard deviations are prescribed as $\sigma_{D_{\mathrm{AoA}}} = 0.1$ and $\sigma_{D_M} = 0.01$, respectively. A robust design problem as given by Equation (5) can be posed by using

$$\mathcal{J} := \bar{C}_{d_{\max}} + \sigma^2_{C_{d\max}} \tag{13}$$

as objective function and

$$g := \bar{C}_{l_{\min}} - C_l^* \qquad \sigma_g := \sigma_{C_{l_{\min}}} \tag{14}$$

as inequality constraint to maintain a target lift coefficient of $C_l^* = 0.6$. Box constraints on all four design variables are used to prevent the generation of invalid geometries from the mesh movement algorithm and solver convergence issues:

$$D_{u,l} \in [-0.025, 0.025] \qquad D_{AoA} \in [0, 1.85] \qquad D_M \in [0.6, 0.78] \tag{15}$$

Even though one flow solve takes only about 10 s on 12 Intel Xeon processors with 3.33 GHz each it is still prohibitively expensive to obtain the mixed aleatory/epistemic optimization under uncertainty results through either nested sampling or exhaustive sampling of optimization results. In order to provide validation for the OUU framework with mixed aleatory/epistemic uncertainty the uncertainty propagations of aleatory and epistemic variables are validated only for the initial and optimized points and also only using 3,000 sample points. But before these combined results are shown, the uncertainty propagations of aleatory and epistemic variables are validated separately.

First, optimization is used to propagate the epistemic uncertainties within the problem. For this test, the aleatory variables are fixed at their mean value taken to be $D_{\mathrm{AoA}} = 1.25$ and $D_M = 0.755$, and optimization is performed over the epistemic variables $D_u = D_l = 0$ to determine the associated intervals for the output functions of interest. The interval produced through optimization is validated by performing Latin hypercube sampling (with 500 samples plus the corners of the domain) over the epistemic variables, again with the aleatory variables fixed at their mean values. The excellent agreement can be seen in Table 1. Note that the optimizations only took a few functions and adjoint gradient evaluation each.

Table 5 Comparison of NLMC and Kriging predictions for optimal design with two shape design variables obtained for $k = 1$

	$\bar{C}_{d_{\max}}$	$\sigma^2_{C_{d\max}}$	$\bar{C}_{l_{\min}}$	$\sigma_{C_{l_{\min}}}$
NLMC (3,000 optimizations)	9.95×10^{-3}	1.0×10^{-5}	0.636	4.0×10^{-2}
Kriging (13 optimizations) ($\tilde{N} = 3{,}000$)	1.02×10^{-2}	4.9×10^{-6}	0.634	3.3×10^{-2}
Kriging (13 optimizations) ($\tilde{N} = 10^5$)	9.95×10^{-3}	1.1×10^{-5}	0.631	3.4×10^{-2}

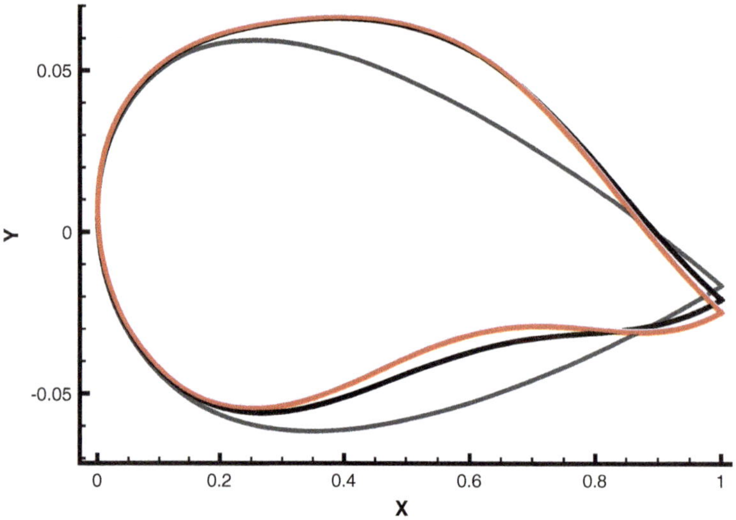

Figure 4 NACA 0012 at $\alpha = 1.25$ (gray), deterministically (black) and robustly ($k = 2$, red) optimized airfoils (two shape design variables).

With the optimization portion of the method validated, the ability of the Kriging surrogate model to capture the aleatory variation of the output functions of interest is tested next. For this test, the original NACA 0012 airfoil is used (i.e. no epistemic uncertainty), and sampling from Kriging surrogates (build from a varying number of training points, N) is performed over the aleatory variables $D_{\text{AoA}} = 1.25$ and $D_M = 0.755$, respectively. In order to provide validated data, full nonlinear MC (NLMC) sampling is performed over the aleatory variables, and both distributions are characterized by calculating statistics of interest using the same samples. For a reasonable trade-off between acquiring accurate statistics and computational cost for the NLMC, $3,000$ samples are used. Because the epistemic variables for this test are fixed, each training point for the Kriging or sample point for the NLMC requires only a single CFD simulation. A summary of these comparisons can be found in Table 2.

The Kriging model constructed from only 13 training points yields reasonable results for a fraction of the cost of a full NLMC simulation. Thus, all the required Kriging response surfaces for the actual robust design runs are constructed from thirteen training points and the sampling is performed using $\tilde{N} = 10^5$ Latin hypercube samples to keep the statistical error small. Lastly, in Table 3, a comparison of NLMC and Kriging predictions is presented using the same $3,000$ samples for the initial airfoil and flow conditions

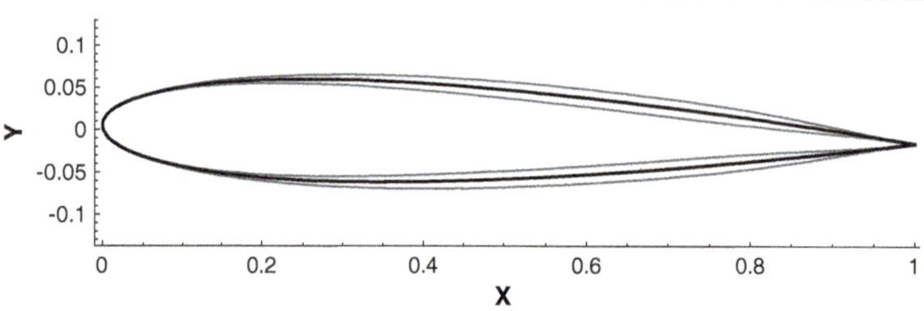

Figure 5 NACA 0012 airfoil (black) and airfoils resulting from six shape design variable values of ± 0.005 (in gray).

Table 6 Robust design results with six shape design variables

k	P_k	$\bar{C}_{d_{\max}}$	$\sigma^2_{C_{d_{\max}}}$	$\bar{C}_{l_{\min}}$	$\sigma_{C_{l_{\min}}}$	D_{AoA}	D_M
0	0.5000	2.75×10^{-3}	2.0×10^{-8}	0.600	1.8×10^{-2}	1.75	0.600
1	0.8413	2.87×10^{-3}	2.3×10^{-8}	0.618	1.8×10^{-2}	1.85	0.602
2	0.9772	3.28×10^{-3}	2.0×10^{-7}	0.640	2.0×10^{-2}	1.85	0.623
3	0.9986	5.60×10^{-3}	5.5×10^{-6}	0.666	2.2×10^{-2}	1.85	0.645
4	0.9999	1.31×10^{-2}	2.4×10^{-5}	0.700	2.5×10^{-2}	1.85	0.668
Deterministic		1.21×10^{-3}	-	0.600	-	1.85	0.600

($D_u = D_l = 0, D_{\text{AoA}} = 1.25$ and $D_M = 0.755$) which demonstrates the good quality of the predictions of the proposed approach for statistics of the lift and drag coefficients. Note that this time, each training point for the Kriging or sample point for the NLMC requires a solution of the optimization problem given by Equation (2) or (3).

Using the presented framework for the entire robust design gives the results presented in Table 4. The number of required optimization iterations for convergence (norm of gradient less than 10^{-4}) varies between 12 and 22 for all the presented cases. One can see that the average drag increases as the desired probability, P_k, of maintaining the target lift coefficient of $C_l^* = 0.6$ is increased. The principal mechanism of achieving this higher probability is to increase the mean Mach number. Note that a deterministic lift-constrained drag minimization yields a minimal drag of $C_d = 1.36 \times 10^{-3}$ at a Mach number of 0.734 and a lower angle of attack of 1.58°. In Table 5, a comparison of NLMC and Kriging predictions using the same $3,000$ samples for the optimal design with $k = 1$ is presented which demonstrates the quality of the predictions for statistics of the lift and drag coefficients. Also shown in the same table are the resulting statistics if the constructed Kriging is sampled $\tilde{N} = 10^5$ times. The original NACA 0012 as well as the deterministically and robustly ($k = 2$) optimized airfoils are all shown in Figure 4. One can see that the robustly optimized airfoil looks different from the deterministically optimized one especially along the lower surface.

The total number of CFD function equivalent evaluations is approximately:

Number of optimization iterations $\times$ 2 (one for minimum lift and one for maximum drag) $\times$ 13 (number of training points) $\times$ number of optimization iterations per epistemic optimization $\times$ 2 (one function and one gradient call) $\approx 1,600$.

Scalability of the framework

In order to demonstrate the scalability of the framework, the number of epistemic design variables is increased from two to six and later to fourteen. First, three shape design variables are placed on the upper surface and three on the lower surface (at 40%, 60% and 80% chords) and Figure 5 shows the original NACA 0012 airfoil and the airfoils resulting from

Table 7 Comparison of NLMC and Kriging predictions for the optimal design with six shape design variables obtained for $k = 2$

	$\bar{C}_{d_{\max}}$	$\sigma^2_{C_{d_{\max}}}$	$\bar{C}_{l_{\min}}$	$\sigma_{C_{l_{\min}}}$
NLMC (3,000 optimizations)	3.33×10^{-3}	2.3×10^{-7}	0.640	2.0×10^{-2}
Kriging (13 optimizations) ($\tilde{N} = 3,000$)	3.33×10^{-3}	1.8×10^{-7}	0.640	2.0×10^{-2}
Kriging (13 optimizations) ($\tilde{N} = 10^5$)	3.28×10^{-3}	2.0×10^{-7}	0.640	2.0×10^{-2}

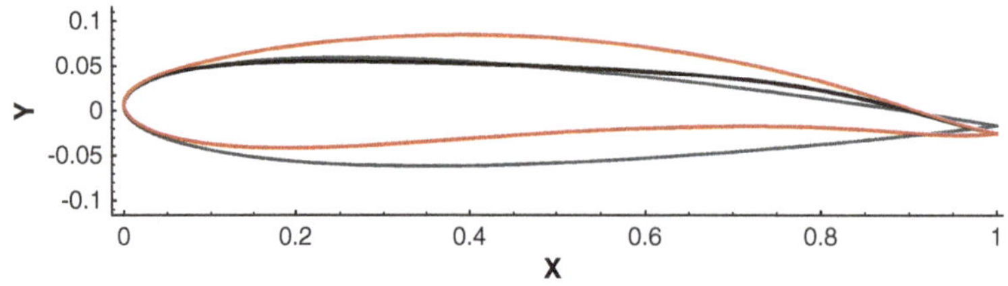

Figure 6 NACA 0012 at $\alpha = 1.25$ (gray), deterministically (black) and robustly ($k = 2$, red) optimized airfoils (six shape design variables).

perturbations of all six shape design variables of ± 0.005. The box constraints to prevent invalid meshes and flow convergence issues are as follows:

$$D_{1,6} \in [-0.01, 0.01] \qquad D_{2-5} \in [-0.02, 0.02] \qquad D_{\text{AoA}} \in [0, 1.85] \qquad D_M \in [0.6, 0.78]\,, \tag{16}$$

where $D_{1,6}$ are the shape design variables closest to the trailing edge on the lower and upper surfaces, respectively. The robust design results are presented in Table 6. The number of required optimization iterations for convergence (again norm of gradient less than 10^{-4}) varies between 9 and 27 for all the presented cases.

Again, the average drag and mean Mach number increase as the desired probability, P_k, of maintaining the target lift coefficient is increased. Compared with Table 4, one also gets lower drag coefficients since the required Mach number to maintain the same lift was reduced through shape modifications. In Table 7, a comparison of NLMC and Kriging predictions using the same 3,000 samples for the optimal design with $k = 2$ is presented which once again demonstrates the quality of the predictions for statistics of the lift and drag coefficients. The total number of CFD function equivalent evaluations has increased from the two epistemic variable case and is approximately as follows: Number of optimization iterations $\times\ 2 \times 13 \times$ number of optimization iterations per epistemic optimization $\times\ 2 \approx 1,900$. This increase is mostly due to the fact that Ipopt now requires a few more iterations to converge the outer optimization problem. However, this is only an increase of about 20% though the number of epistemic variables increased from two to six. The original NACA 0012 as well as the deterministically and robustly ($k = 2$) optimized airfoils are all shown in Figure 6. Once again, one can see that the robustly optimized airfoil looks different from the deterministically optimized one this time especially along the upper surface.

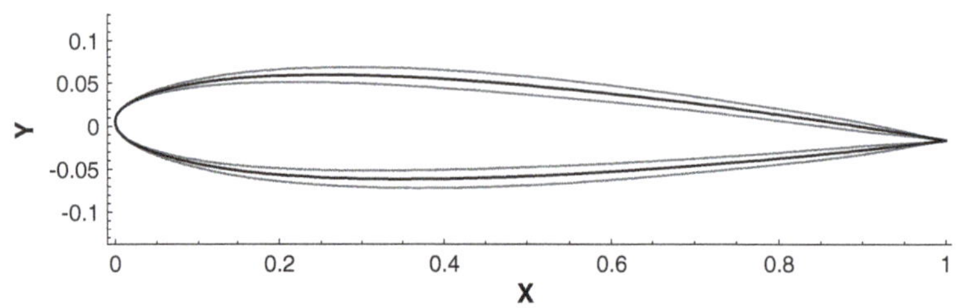

Figure 7 NACA 0012 airfoil (black) and airfoils resulting from fourteen shape design variable values of ± 0.0025 (in gray).

Table 8 Robust design results with fourteen shape design variables

k	P_k	$\bar{C}_{d\max}$	$\sigma^2_{C_{d\max}}$	$\bar{C}_{l\min}$	$\sigma_{C_{l\min}}$	D_{AoA}	D_M
0	0.5000	3.21×10^{-3}	1.2×10^{-8}	0.800	1.5×10^{-2}	4.00	0.461
1	0.8413	3.40×10^{-3}	1.5×10^{-8}	0.816	1.6×10^{-2}	4.00	0.485
2	0.9772	3.61×10^{-3}	1.8×10^{-8}	0.833	1.6×10^{-2}	4.00	0.508
3	0.9986	3.84×10^{-3}	2.3×10^{-8}	0.852	1.7×10^{-2}	4.00	0.531
4	0.9999	4.23×10^{-3}	6.3×10^{-8}	0.877	1.9×10^{-2}	4.00	0.562
Deterministic		1.66×10^{-3}	-	0.800	-	3.76	0.300

As a last demonstration, the number of epistemic design variables is increased from 6 to 14. Therefore, seven shape design variables are placed on the upper surface and seven on the lower surface (at 20%, 30%, 40%, 50%, 60%, 80%, and 90% chord) and Figure 7 shows the original NACA 0012 airfoil and the airfoils resulting from perturbations of all fourteen shape design variables of ± 0.0025. The box constraints to prevent invalid meshes and flow convergence issues are as follows:

$$D_{1,2,13,14} \in [-0.00125, 0.00125] \quad D_{3-12} \in [-0.01, 0.01] \quad D_{AoA} \in [0, 4] \quad D_M \in [0.3, 0.78], \tag{17}$$

where $D_{1,14}$ are the shape design variables closest to the trailing edge on the lower and upper surface, respectively. The target lift coefficient is increased to $C_l^* = 0.8$ to make this problem more difficult and the results are presented in Table 8. The number of required optimization iterations for convergence (again norm of gradient less than 10^{-4}) varies between 16 to 23 for all the presented cases. The average drag and mean Mach numbers increase as the desired probability of maintaining the target lift coefficient is increased. The total number of CFD function equivalent evaluations has increased again from the two and six epistemic variable case to approximately 2,000, which is only a modest increase considering how many more epistemic variables are used.

Conclusions

This article describes the use of gradient-based optimizations and Kriging surrogate models for the propagation of mixed aleatory/epistemic uncertainties for a robust lift-constrained drag minimization problem. Uncertainty due to epistemic variables is propagated via a box-constrained optimization approach, while the uncertainty due to aleatory variables is propagated *via* sampling of a Kriging surrogate model built with the optimization results. This statistics-of-intervals approach makes robust design under mixed aleatory/epistemic uncertainty possible while at the same time keeping the computational cost for these types of problems manageable.

Acknowledgments

This work was partially supported by the University of Dayton Research Council seed grants. I would also like to thank Karthik Mani for making his flow and adjoint solver available as well as Wataru Yamazaki for his Kriging model.

References

1. Luckring, JM, Hemsch, MJ, Morrison, JH: Uncertainty in computational aerodynamics. In the 41st AIAA Aerospace Sciences Meeting & Exhibit, Reno, 6–9 January 2003
2. Gumbert, CR, Newman, PA, Hou, GJ: Effect of random geometric uncertainty on the computational design of 3-D wing. In the 20th AIAA Applied Aerodynamics Conference, St. Louis, 24–26 June 2002
3. Ben-Tal, A, Ghaoui, LE, Nemirovski, A: Foreword: special issue on robust optimization. Math. Prog. **107**(1–2), 1–3 (2006)

4. Pilch, M, Trucano, TG, Helton, JC: Ideas underlying quantification of margins and uncertainties (QMU): a white paper. Tech. Rep. SAND2006-5001, Sandia National Laboratories, Albuquerque, NM (2006)
5. Helton, JC, Johnson, JD, Oberkampf, WL, Storlie, CB: A sampling-based computational strategy for the representation of epistemic uncertainty in model predictions with evidence theory. Tech. Rep. SAND2006-5557, Sandia National Laboratories, Albuquerque, NM (2006)
6. Diegert, K, Klenke, S, Novotny, G, Paulsen, R, Pilch, M, Trucano, T: Toward a more rigorous application of margins and uncertainties within the nuclear weapons life cycle - a Sandia perspective. Tech. Rep. SAND2007-6219, Sandia National Laboratories, Albuquerque, NM (2007)
7. Oberkampf, WL, Barone, MF: Measures of agreement between computation and experiment: validation metrics. J. Comput. Phys. **217**, 5–36 (2006)
8. Roy, CJ, Oberkampf, WL: A comprehensive framework for verification, validation, and uncertainty quantification in scientific computing. Comput. Methods Appl. Mech. Eng. **200**(25–28), 2131–2144 (2011)
9. Eldred, MS, Swiler, LP: Efficient Algorithms for Mixed Aleatory-Epistemic Uncertainty Quantification with Application to Radiation-Hardened Electronics. Tech. Rep. SAND2009-5805, Sandia National Laboratories, Albuquerque, NM (2009)
10. Metropolis, N, Ulam, S: The Monte Carlo method. J. Am. Stat. Assoc. **44**, 335–341 (1949)
11. Bettis, BR, Hosder, S: Uncertainty quantification in hypersonic reentry flows due to aleatory and epistemic uncertainties. In the 49th AIAA Aerospace SciencesMeeting including the New Horizons Forum and Aerospace Exposition, Orlando, FL, 4–7 January 2011
12. Lockwood, B, Anitescu, M, Mavriplis, DJ: Mixed aleatory/epistemic uncertainty quantification for hypersonic flows via gradient-based optimization and surrogate models. In the 50th AIAA Aerospace Sciences Meeting Including the New Horizons Forum and Aerospace Exposition, Nashville, Tennessee, 9–12 January 2012
13. Gu, X, Renaud, J, Batill, S, Brach, R, Budhiraja, A: Worst case propagated uncertainty of multidisciplinary systems in robust design optimization. Struct. Multidisciplinary Optimization **20**(3), 190–213 (2000)
14. Pironneau, O: On optimum design in fluid mechanics. J. Fluid Mech. **64**(1), 97–110 (1974)
15. Errico, RM: What is an adjoint model? Bull. Am. Meteorological Soc. **8**(11), 2577–2591 (1997)
16. Beale, EML: On minimizing a convex function subject to linear inequalities. J. R. Stat. Soc. **17B**, 173–184 (1955)
17. Dantzig, GB: Linear programming under uncertainty. Manage. Sci. **1**, 197–206 (1955)
18. Tintner, G: Stochastic linear programming with applications to agricultural economics. In: Antosiewicz, HA (ed.) Proceedings of the Second Symposium in Linear Programming, Washington, DC, January 27–29 1955
19. Sahinidis, NV: Optimization under uncertainty: state-of-the-art and opportunities. Comput. Chem. Eng. **28**, 971–983 (2004)
20. Dantzig, GB, Infanger, G: Stochastic Programming: The State of the Art in Honor of George B. Dantzig. Springer, New York (2010)
21. Beyera, HG, Sendhoff, B: Robust optimization - a comprehensive survey. Comput. Methods Appl. Mech. Eng. **196**(33–34), 3190–3218 (2007)
22. Giunta, AA, Eldred, MS, Swiler, LP, Trucano, TG, Wojtkiewicz, SF: Perspectives on Optimization under Uncertainty: Algorithms and Applications. In the 10th AIAA/ISSMO Multidisciplinary Analysis and Optimization Conference, Albany, NY, 30 August - 1 September 2004
23. Li, M: Robust Optimization and sensitivity analysis with multi-objective genetic algorithms: single- and multi-disciplinary applications. PhD thesis, University of Maryland (2007)
24. Kouvelis, P, Yu, G: Robust Discrete Optimization and Its Applications. Kluwer, Boston (1997)
25. Ben-Tal, A, Ghaoui, LE, Nemirovski, A: Robust Optimization. Princeton Series in Applied Mathematics. Princeton University Press, Princeton (2009)
26. Zang, C, Friswell, MI, Mottershead, JE: A review of robust optimal design and its application in dynamics. Comput. Struct. **83**, 315–326 (2005)
27. Chen, W, Allen, J, Tsui, K, Mistree, F: Procedure for robust design: minimizing variations caused by noise factors and control factors. J. Mech. Design **118**(4), 478–485 (1996)
28. Chen, W, Du, X: Towards a better understanding of modeling feasibility robustness in engineering design. J. Mech. Design **122**(4), 385–394 (1999)
29. Mourelatos, Z, Liang, J: A methodology for trading-off performance and robustness under uncertainty. J. Mech. Design **128**, 856 (2006)
30. Zaman, K, McDonald, M, Mahadevan, S, Green, L: Robustness-based design optimization under data uncertainty. Struct. Multidisciplinary Optimization **44**(2), 183–197 (2011)
31. Putko, MM, Newmann, PA, Taylor III, AC, Green, LL: Approach for uncertainty propagation and robust design in CFD using sensitivity derivatives. In the 15th AIAA Computational Fluid Dynamics Conference, Anaheim, CA, 11–14 June 2001
32. Byrd, RH, Lu, P, Nocedal, J, Zhu, C: A limited memory algorithm for bound constrained optimization. SIAM J. Sci. Comput. **16**(5), 1190–1208 (1995)
33. Zhu, C, Byrd, RH, Lu, P, Nocedal, J: L-BFGS-B: A Limited Memory FORTRAN Code for Solving Bound Constrained Optimization Problems. Tech. Rep. NAM-11, Department of Electrical Engineering and Computer Science, Northwestern University, Evanston, Illinois, USA (1994)
34. Yamazaki, W, Mouton, S, Carrier, G: Efficient design optimization by physics-based direct manipulation free-form deformation. In the 12th AIAA/ISSMO Multidisciplinary Analysis and Optimization Conference, Victoria, British Columbia, 10–12 September 2008
35. Rumpfkeil, MP, Yamazaki, W, Mavriplis, DJ: Uncertainty analysis utilizing gradient and Hessian information. In: Sixth International Conference on Computational Fluid Dynamics, ICCFD6, St. Petersburg, Russia, July 2010
36. Yamazaki, W, Rumpfkeil, MP, Mavriplis, DJ: Design Optimization Utilizing Gradient/Hessian Enhanced Surrogate Model. In the 28th AIAA Applied Aerodynamics Conference, Chicago, Illinois, 28 June - 1 July 2010
37. Rumpfkeil, MP, Yamazaki, W, Mavriplis, DJ: A Dynamic Sampling Method for Kriging and Cokriging Surrogate Models. In the 49th AIAA Aerospace Sciences meeting including the New Horizons Forum and Aerospace Exposition, Orlando, FL, 4-7 January 2011

38. Du, X, Chen, W: Methodology for managing the effect of uncertainty in simulation-based design. AIAA J. **38**(8), 1471–1478 (2000)
39. Parkinson, A, Sorensen, C, Pourhassan, N: A general approach for robust optimal design. Trans. ASME **115**, 74–80 (1993)
40. Putko, MM, Taylor III, AC, Newmann, PA, Green, LL: Approach for input uncertainty propagation and robust design in CFD using sensitivity derivatives. J. Fluids Eng. **124**(1), 60–69 (2002)
41. Waechter, A, Biegler, LT: On the implementation of a primal-dual interior point filter line search algorithm for large-scale nonlinear programming. Math. Prog. **106**(1), 25–57 (2006)
42. Rumpfkeil, MP: Optimization under uncertainty using gradients, Hessians, and surrogate models. AIAA J. **51**(2), 444–451 (2013)
43. Han, ZH, Zimmermann, R, Goertz, S: On improving efficiency and accuracy of variable-fidelity surrogate modeling in aero-data for loads context. In: CEAS 2009 European Air and Space Conference, Manchester, 26–29 October 2009
44. Mani, K, Mavriplis, DJ: Unsteady discrete adjoint formulation for two-dimensional flow problems with deforming meshes. AIAA J. **46**(6), 1351–1364 (2008)
45. Mani, K, Mavriplis, DJ: Adjoint-based sensitivity formulation for fully coupled unsteady aeroelasticity problems. AIAA J. **47**(8), 1902–1915 (2009)
46. Hicks, R, Henne, P: Wing design by numerical optimization. J Aircraft **15**(7), 407–412 (1978)
47. Batina, JT: Unsteady euler airfoil solutions using unstructured dynamic meshes. AIAA J. **28**(8), 1381–1388 (1990)

4

Liu process and uncertain calculus

Xiaowei Chen[1] and Dan A Ralescu[2*]

*Correspondence:
ralescd@ucmail.uc.edu
[2]Department of Mathematical Sciences, University of Cincinnati, Cincinnati, OH 45221-0025, USA
Full list of author information is available at the end of the article

Abstract

Uncertain calculus is a branch of mathematics that deals with the integral and differential of functions of uncertain processes. This paper first introduces the Liu process as an uncertain process defined by the Liu integral. Some properties of Liu processes are investigated such as sample continuity property, finite variation property, and the fact that a continuously differentiable function of the Liu process is another Liu process, among others. Based on the Liu process, the uncertain integral is extended. Furthermore, some mathematical properties are proved, including the fundamental theorem, change of variable theorem, and integration by parts theorem. Finally, uncertain calculus with respect to multiple Liu processes is discussed.

Keywords: Uncertainty theory; Uncertain process; Uncertain integral

Introduction

A stochastic process, a sequence of random variables indexed by time, is a useful tool to deal with dynamical random phenomena. A very important such process is called the Wiener process and was defined by Wiener [1]. The Wiener process, also known as the Brownian motion, is a stochastic process with stationary independent increments, and the increments are normal random variables. Based on the Wiener process, stochastic calculus was developed by Ito [2]. It is a branch of mathematical theory dealing with the integration and differentiation of functions of a stochastic process. Stochastic calculus with respect to the Wiener process is also called the Ito calculus. It has important applications in asset pricing theory. In 1967, the stochastic integral was extended by Kunita and Watanabe [3] to square integrable martingales. Furthermore, stochastic integral with respect to a semimartingale was introduced by Doléans-Dade and Meyer [4]. Besides, stochastic calculus with respect to a local martingale, the Poisson process, and the Lévy process have been studied (see [5,6]).

Stochastic processes are defined based on probability theory. When we use it, a large sample size is needed to estimate probability distribution based on long-run frequency. However, Liu [7] pointed out that the sample size is often too small (even no sample) in practice and the degree of belief usually has much larger variance than the long-run frequency. Thus, we should deal with it by using uncertainty theory. These facts promoted Liu [8] to found an uncertainty theory. That is a branch of mathematics dealing with human uncertainty. In order to describe dynamic uncertain systems, an uncertain process was introduced by Liu [9] as a sequence of uncertain variables indexed by time. In addition, Zhang et al. [10] proposed a delayed renewal process. Chen [11] investigated some properties of uncertain stationary independent increments.

Based on the Liu canonical process, a type of uncertain process with stationary independent increments which follow normal uncertainty distribution, Liu [12] invented an uncertain calculus in 2009. This type of uncertain integral is called the Liu integral. It was extended to multiple Liu canonical processes [13]. It is used to deal with the integration and differentiation of uncertain processes. The theory of integration and differentiation of uncertain processes with respect to the Liu process is called the Liu calculus. In order to study an uncertain integral with respect to an uncertain process admitting jumps, an uncertain integral with respect to a renewal process was introduced by Yao [14]. We call this type of uncertain integral the Yao integral. The theory of integration and differentiation of uncertain processes with respect to a renewal process is called the Yao calculus. Chen [15] introduced an uncertain integral with respect to a finite variation process.

In addition, the uncertain differential equation driven by the Liu canonical process was introduced by Liu [9]. After that, Chen and Liu [16] proved the existence and uniqueness theorem for uncertain differential equations. Yao and Gao [17] studied stability theorems for uncertain differential equations. Chen and Liu [16] proposed solution methods for linear uncertain differential equations. Liu [18] and Yao [19] gave a method to solve nonlinear uncertain differential equations. Besides, Yao and Chen [20] introduced a numerical method for uncertain differential equations. Meanwhile, an uncertain differential equation has been applied to uncertain optimal control by Zhu [21], American option pricing by Chen [22], and other option pricing models by Peng and Yao [23]. Liu [24] discussed some basic concepts of uncertain finance. For the latest development of uncertainty theory, please see [25].

In this paper, we generalize the Liu process by the Liu integral. Our goal is to extend an uncertain integral with respect to the Liu process. This uncertain integral has the properties of continuity and linearity. In the framework of the uncertain integral, the fundamental theorem of differentiation of function of uncertain processes is proved. In addition, the integration by parts formula and the formula for change of variables are derived. The rest of the paper is organized as follows: some preliminary concepts of the uncertain process including the Liu calculus, Yao calculus, and uncertain calculus with respect to an uncertain finite variation process are recalled in the 'Preliminary' section. The uncertain integral with respect to the general Liu process will be defined in the 'Liu process' section. The uncertain differential will be discussed in the 'Uncertain integral with respect to the Liu process' section. At last, a brief summary is given in the 'Multifactor Liu process' section.

Preliminary

The uncertain measure $\mathcal{M}$ is a real-valued set function on a σ-algebra $\mathcal{L}$ over a nonempty set Γ satisfying normality, duality, subadditivity, and product axioms. The triplet $(\Gamma, \mathcal{L}, \mathcal{M})$ is called an uncertainty space.

Definition 1. ([8]) An uncertain variable is a measurable function from an uncertainty space $(\Gamma, \mathcal{L}, \mathcal{M})$ to the set of real numbers, i.e., for any Borel set B of real numbers, the set

$$\{\xi \in B\} = \{\gamma \in \Gamma | \xi(\gamma) \in B\}$$

is an event.

The uncertainty distribution function $\Phi : \Re \to [0,1]$ of an uncertain variable ξ is defined as $\Phi(x) = \mathcal{M}\{\xi \le x\}$. The expected value of an uncertain variable is defined as follows.

Definition 2. ([8]) Let ξ be an uncertain variable. Then the expected value of ξ is defined by

$$E[\xi] = \int_0^{+\infty} \mathcal{M}\{\xi \ge r\}dr - \int_{-\infty}^0 \mathcal{M}\{\xi \le r\}dr$$

provided that at least one of the two integrals is finite.

The expected value can also be written as

$$E[\xi] = \int_0^{+\infty} (1 - \Phi(r))dr - \int_{-\infty}^0 \Phi(r)dr$$

where $\Phi(r)$ is the uncertainty distribution of ξ. If ξ is an uncertain variable with finite expected value e, then the variance of ξ is defined as $\text{Var}[\xi] = E[(\xi - e)^2]$.

Let $\xi_1, \xi_2, \cdots, \xi_n$ be independent uncertain variables with uncertainty distributions $\Phi_1, \Phi_2, \cdots, \Phi_n$, respectively. Liu [26] showed that if the function $f(x_1, x_2, \cdots, x_n)$ is strictly increasing with respect to $x_1, x_2, \cdots, x_m$ and strictly decreasing with respect to $x_{m+1}, x_{m+2}, \cdots, x_n$, then

$$\xi = f(\xi_1, \xi_2, \cdots, \xi_n)$$

is an uncertain variable with inverse uncertainty distribution

$$\Psi^{-1}(\alpha) = f(\Phi_1^{-1}(\alpha), \cdots, \Phi_m^{-1}(\alpha), \Phi_{m+1}^{-1}(1-\alpha), \cdots, \Phi_n^{-1}(1-\alpha)). \tag{1}$$

Furthermore, Liu and Ha [27] proved that it has an expected value

$$E[\xi] = \int_0^1 f(\Phi_1^{-1}(\alpha), \cdots, \Phi_m^{-1}(\alpha), \Phi_{m+1}^{-1}(1-\alpha), \cdots, \Phi_n^{-1}(1-\alpha))d\alpha. \tag{2}$$

Definition 3. ([9]) Let T be an index set and let $(\Gamma, \mathcal{L}, \mathcal{M})$ be an uncertainty space. An uncertain process is a measurable function from $T \times (\Gamma, \mathcal{L}, \mathcal{M})$ to the set of real numbers, i.e., for each $t \in T$ and any Borel set B of real numbers, the set

$$\{X_t \in B\} = \{\gamma \in \Gamma | X_t(\gamma) \in B\}$$

is an event.

Definition 4. ([9]) Let $\xi_1, \xi_2, \cdots$ be iid positive uncertain variables. Define

$$S_0 = 0 \text{ and } S_n = \xi_1 + \xi_2 + \cdots + \xi_n$$

for $n \ge 1$. Then the uncertain process

$$N_t = \max_{n \ge 0}\{n | S_n \le t\}$$

is called a renewal process.

An uncertain process X_t is said to have independent increments if

$$X_{t_0}, X_{t_1} - X_{t_0}, X_{t_2} - X_{t_1}, \cdots, X_{t_k} - X_{t_{k-1}}$$

are independent uncertain variables where t_0 is the initial time and $t_1, t_2, \cdots, t_k$ are any times with $t_0 < t_1 < \cdots < t_k$. An uncertain process X_t is said to have stationary increments if, for any given $t > 0$, the increments $X_{s+t} - X_s$ are identically distributed uncertain variables for all $s > 0$. An uncertain process S_t is said to be a stationary independent increment process if it has stationary and independent increments. Liu [26] proved that the expected value of stationary independent increment process S_t is $E[S_t] = a + bt$.

Definition 5. ([12]) An uncertain process C_t is said to be a canonical Liu process if:

1. $C_0 = 0$ and almost all sample paths are Lipschitz continuous.
2. C_t has stationary and independent increments.
3. Every increment $C_{s+t} - C_s$ is a normal uncertain variable with expected value 0 and variance t^2, whose uncertainty distribution is

$$\Phi(x) = \left(1 + \exp\left(\frac{-\pi x}{\sqrt{3}t}\right)\right)^{-1}, \quad x \in \Re.$$

Liu integral

In order to deal with the integration and differentiation of uncertain processes, Liu [12] proposed an uncertain integral with respect to the Liu process.

Definition 6. ([12]) Let X_t be an uncertain process and C_t be a canonical Liu process. For any partition of closed interval $[a, b]$ with $a = t_1 < t_2 < \cdots < t_{k+1} = b$, the mesh is written as

$$\Delta = \max_{1 \le i \le k} |t_{i+1} - t_i|.$$

Then the Liu integral of X_t is defined as

$$\int_a^b X_t \mathrm{d}C_t = \lim_{\Delta \to 0} \sum_{i=1}^{k} X_{t_i} \cdot (C_{t_{i+1}} - C_{t_i})$$

provided that the limit exists almost surely and is finite. In this case, the uncertain process X_t is said to be Liu integrable.

Liu [12] verified the fundamental theorem of uncertain calculus, i.e., for a canonical Liu process C_t and a continuously differentiable function $h(t, c)$, the uncertain process $Z_t = h(t, C_t)$ has a differential

$$\mathrm{d}Z_t = \frac{\partial h}{\partial t}(t, C_t)\mathrm{d}t + \frac{\partial h}{\partial c}(t, C_t)\mathrm{d}C_t.$$

Yao integral

In order to study an uncertain integral with respect to an uncertain process admitting jumps, an uncertain integral with respect to a renewal process was introduced by Yao [14].

Definition 7. ([14]) Let X_t be an uncertain process and N_t be an uncertain renewal process. For any partition of closed interval $[a, b]$ with $a = t_1 < t_2 < \cdots < t_{k+1} = b$, the mesh is written as

$$\Delta = \max_{1 \le i \le k} |t_{i+1} - t_i|.$$

Then the Yao integral of X_t is defined as

$$\int_a^b X_t \mathrm{d}N_t = \lim_{\Delta\to 0} \sum_{i=1}^{k} X_{t_i} \cdot (N_{t_{i+1}} - N_{t_i})$$

provided that the limit exists almost surely and is finite. In this case, the uncertain process X_t is said to be Yao integrable.

Yao [14] verified the fundamental theorem of uncertain calculus, i.e., for a renewal process N_t and a continuously differentiable function $h(t,n)$, the uncertain process $Z_t = h(t,N_t)$ has a differential

$$\mathrm{d}Z_t = \frac{\partial h}{\partial t}(t,N_t)\mathrm{d}t + h(t,N_t) - h(t,N_{t^-}).$$

Uncertain integral with finite variation processes

Definition 8. ([15]) Let X_t be an uncertain process and let A_t be a finite variation process. For any partition of closed interval $[a,b]$ with $a = t_1 < t_2 < \cdots < t_{k+1} = b$, the mesh is written as

$$\Delta = \max_{1\le i\le k} |t_{i+1} - t_i|.$$

Then the uncertain integral of X_t with respect to A_t is

$$\int_a^b X_t \mathrm{d}A_t = \lim_{\Delta\to 0} \sum_{i=1}^{k} X_{t_i} \cdot (A_{t_{i+1}} - A_{t_i})$$

provided that the limit exists almost surely and is finite. In this case, the uncertain process X_t is said to be integrable with respect to the finite variation process A_t.

Suppose that A_t is a finite variation process and $h(t,s)$ a continuously differentiable function, the uncertain process $Z_t = h(t,A_t)$ has a differential

$$\begin{aligned}\mathrm{d}h(t,A_t) = {} & \frac{\partial h}{\partial t}(t,A_t)\mathrm{d}t + \frac{\partial h}{\partial x}(t,A_{t^-})\mathrm{d}A_t \\ & - \frac{\partial h}{\partial x}(t,A_{t^-})(A_t - A_t) + h(t,A_t) - h(t,A_{t^-}).\end{aligned}$$

Uncertain differential equation

Definition 9. ([9]) Suppose C_t is a canonical Liu process, and f and g are some given functions. Given an initial value X_0, then

$$\mathrm{d}X_t = f(t,X_t)\mathrm{d}t + g(t,X_t)\mathrm{d}C_t \tag{3}$$

is called an uncertain differential equation with an initial value X_0. A solution is an uncertain process X_t that satisfies (3) identically in t.

Theorem 1. (Existence and Uniqueness Theorem [16]) *The uncertain differential equation (3) has a unique solution if the coefficients $f(x,t)$ and $g(x,t)$ satisfy the Lipschitz condition*

$$|f(x,t) - f(y,t)| + |g(x,t) - g(y,t)| \le L|x-y|, \ \forall x,y \in \Re, \ t \ge 0$$

and linear growth condition

$$|f(x,t)| + |g(x,t)| \leq L(1+|x|), \ \forall x \in \Re, \ t \geq 0$$

for some constant L. Moreover, the solution is sample-continuous.

Definition 10. ([20]) The α-path ($0 < \alpha < 1$) of an uncertain differential equation

$$\mathrm{d}X_t = f(t, X_t)\mathrm{d}t + g(t, X_t)\mathrm{d}C_t$$

with initial value X_0 is a deterministic function X_t^α with respect to t that solves the corresponding ordinary differential equation

$$\mathrm{d}X_t^\alpha = f(t, X_t^\alpha)\mathrm{d}t + |g(t, X_t^\alpha)|\Phi^{-1}(\alpha)\mathrm{d}t$$

where $\Phi^{-1}(\alpha)$ is the inverse uncertainty distribution of the standard normal uncertain variable, i.e.,

$$\Phi^{-1}(\alpha) = \frac{\sqrt{3}}{\pi}\ln\frac{\alpha}{1-\alpha}, \quad 0 < \alpha < 1.$$

Theorem 2. (Yao-Chen Formula [20]) *Let X_t and X_t^α be the solution and α-path of the uncertain differential equation*

$$\mathrm{d}X_t = f(t, X_t)\mathrm{d}t + g(t, X_t)\mathrm{d}C_t,$$

respectively. Then

$$\mathcal{M}\{X_t \leq X_t^\alpha, \forall t\} = \alpha,$$
$$\mathcal{M}\{X_t > X_t^\alpha, \forall t\} = 1 - \alpha.$$

The Yao-Chen formula relates an uncertain differential equation and a family of ordinary differential equations just like the Feyman-Kac formula relates a stochastic differential equation and a partial differential equation. Besides, Yao [28] studied the integral of solution to uncertain differential equations.

Liu process

Definition 11. Let C_t be a canonical Liu process, and μ_s and σ_s be two uncertain processes. Then the uncertain process

$$X_t = X_0 + \int_0^t \mu_s \mathrm{d}s + \int_0^t \sigma_s \mathrm{d}C_s \tag{4}$$

is called a Liu process with drift μ_t and diffusion σ_t. A Liu process X_t may also be written in differential form:

$$\mathrm{d}X_t = \mu_t \mathrm{d}t + \sigma_t \mathrm{d}C_t.$$

Example 1. A canonical Liu process C_t is a Liu process with initial value 0, drift 0, and diffusion 1.

Example 2. An arithmetic Liu process $X_t = et + \sigma C_t$ is a Liu process with initial value 0, drift e, and diffusion σ.

Example 3. A geometric Liu process $X_t = \exp(et + \sigma C_t)$ is a Liu process with initial value 1, drift eX_t, and diffusion σX_t.

Example 4. The uncertain process $X_t = t^2 + C_t^3$ is a Liu process with initial value 0, drift $2t$, and diffusion $3C_t^2$.

Example 5. The solution to an uncertain differential equation

$$\mathrm{d}X_t = f(t, X_t)\mathrm{d}t + g(t, X_t)\mathrm{d}C_t \tag{5}$$

is a Liu process with drift $f(t, X_t)$ and diffusion $g(t, X_t)$.

Example 6. Let $h(t, c)$ be a continuously differentiable function. Then $h(t, C_t)$ is a Liu process with drift $\frac{\partial h}{\partial t}(t, C_t)$ and diffusion $\frac{\partial h}{\partial c}(t, C_t)$.

Next, we will discuss the properties of the path for the Liu process. Let $\Pi : a_0 = t_0 < t_1 < \cdots < t_n = t$ be a partition of the closed interval $[0, t]$, and the mesh size of the partition Π is defined as $\|\Pi\|_t = \max\limits_{1 \le i \le n} |t_i - t_{i-1}|$. The total variation of the uncertain process X_s over the partition Π is defined as

$$V_t[\Pi] = \sum_{i=1}^{n} |X_{t_i} - X_{t_{i-1}}|.$$

Theorem 3. *Suppose that X_t is a Liu process. Let $\{\Pi_m\}_{m=0}^{\infty}$ be any sequence of partitions of closed interval $[0, t]$ with the mesh size $\lim\limits_{m \to +\infty} \|\Pi_m\|_t = 0$. Then the total variation of X_t satisfies*

$$\mathcal{M}\left\{\limsup_{m \to \infty} \sum_{i=1}^{n} |X_{t_{i+1}} - X_{t_i}| = +\infty\right\} = 0. \tag{6}$$

Proof. Note that

$$\begin{aligned}
\mathcal{M}\left\{\limsup_{m \to \infty} V_t[\Pi_m] = \infty\right\} &= \mathcal{M}\left\{\limsup_{m \to \infty} \sum_{i=1}^{n} \left|\int_{t_i}^{t_{i+1}} \mu_s \mathrm{d}s + \int_{t_i}^{t_{i+1}} \sigma_s \mathrm{d}C_s\right| = \infty\right\} \\
&\le \mathcal{M}\left\{\gamma \,|\, \limsup_{m \to \infty} \sum_{i=1}^{n} \int_{t_i}^{t_{i+1}} |\mu_s(\gamma)| \mathrm{d}s = \infty\right\} \\
&\quad + \mathcal{M}\left\{\gamma \,|\, \limsup_{m \to \infty} \sum_{i=1}^{n} \int_{t_i}^{t_{i+1}} |\sigma_s(\gamma)| \mathrm{d}|C_s(\gamma)| = \infty\right\} \\
&= \mathcal{M}\left\{\gamma \,|\, \int_0^t |\mu_s(\gamma)| \mathrm{d}s = \infty\right\} \\
&\quad + \mathcal{M}\left\{\gamma \,|\, \int_0^t |\sigma_s(\gamma)| \mathrm{d}|C_s(\gamma)| = \infty\right\}.
\end{aligned}$$

By the definition of the Liu integral, we know that

$$\int_0^t \mu_s \mathrm{d}s + \int_0^t \sigma_s \mathrm{d}C_s$$

exists almost surely. Then

$$\int_0^t |\mu_s| \mathrm{d}s < \infty \text{ and } \int_0^t |\sigma_s| \mathrm{d}|C_s| < \infty, \text{ almost surely.}$$

Thus, we get

$$\mathcal{M}\left\{\limsup_{\Delta\to 0}\sum_{i=1}^{k}|X_{t_{i+1}}-X_{t_i}|=+\infty\right\}=0. \tag{7}$$

The theorem is proved. Therefore, a Liu process is an uncertain finite variation process. □

Theorem 4. *The Liu process is a sample-continuous uncertain process.*

Proof. In order to prove this theorem, we only need to prove

$$X_{t+\Delta t}-X_t=\int_t^{t+\Delta t}\mu_s\mathrm{d}s+\int_t^{t+\Delta t}\sigma_s\mathrm{d}C_s\to 0,\ \text{as}\ \Delta t\to 0.$$

By the definition of the Liu integral, we know that the integral

$$\int_0^t\alpha_s(\gamma)\mathrm{d}s+\int_0^t\sigma_s(\gamma)\mathrm{d}C_s(\gamma)$$

is finite for almost all the sample path $\gamma\in\Gamma$. It follows from the continuity of the integral that $\int_t^{t+\Delta t}\mu_s(\gamma)\mathrm{d}s\to 0$ and $\int_t^{t+\Delta t}\sigma_s(\gamma)\mathrm{d}C_s(\gamma)\to 0$ as $\Delta t\to 0$ almost surely. Therefore, the theorem is proved. □

Uncertain integral with respect to the Liu process

Definition 12. Let X_t be a Liu process with drift μ_t and diffusion σ_t, and Y_t be an uncertain process. We define an uncertain integral with respect to the Liu process as

$$\int_0^t Y_s\mathrm{d}X_s=\int_0^t Y_s\mu_s\mathrm{d}s+\int_0^t Y_s\sigma_s\mathrm{d}C_s.$$

Theorem 5. *The uncertain integral of an uncertain process with respect to the Liu process is also a Liu process.*

Proof. For any integrable uncertain process Y_t and a Liu process X_t with the form

$$\mathrm{d}X_t=\mu_t\mathrm{d}t+\sigma_t\mathrm{d}C_t,$$

the uncertain integral $Z_t=\int_0^t Y_s\mathrm{d}X_s$ is

$$Z_t=\int_0^t Y_s\mu_s\mathrm{d}s+\int_0^t Y_s\sigma_s\mathrm{d}C_s.$$

It follows from the definition of the Liu process that Z_t is indeed a Liu process. □

Theorem 6. *Suppose that X_t is a Liu process and Y_t is an integrable uncertain process with respect to X_t on interval $[a,b]$. Then it is integrable on each subinterval of $[a,b]$. Moreover, if $c\in[a,b]$, then*

$$\int_a^b Y_t\mathrm{d}X_t=\int_a^c Y_t\mathrm{d}X_t+\int_c^b Y_t\mathrm{d}X_t.$$

Proof. Let $[a_0, b_0]$ be a subinterval of $[a, b]$. By the definition of the uncertain integral with respect to the Liu process, we get

$$\int_a^b Y_s \mathrm{d}X_s = \int_a^b Y_s \mu_s \mathrm{d}s + \int_a^b Y_s \sigma_s \mathrm{d}C_s.$$

Since Y_s is integrable on $[a, b]$, $Y_s\mu_s$ and $Y_s\sigma_s$ are integrable on the interval $[a, b]$ with respect to t and C_t, respectively. Then $Y_s\mu_s$ and $Y_s\sigma_s$ are integrable on any subinterval of $[a, b]$. Therefore, we obtain

$$\int_{a'}^{b'} Y_s \mathrm{d}X_s = \int_{a'}^{b'} Y_s \mu_s \mathrm{d}s + \int_{a'}^{b'} Y_s \sigma_s \mathrm{d}C_s.$$

Next,

$$\begin{aligned}
\int_a^b Y_s \mathrm{d}X_s &= \int_a^b Y_s \mu_s \mathrm{d}s + \int_a^b Y_s \sigma_s \mathrm{d}C_s \\
&= \int_a^c Y_s \mu_s \mathrm{d}s + \int_a^c Y_s \sigma_s \mathrm{d}C_s + \int_c^b Y_s \mu_s \mathrm{d}s + \int_c^b Y_s \sigma_s \mathrm{d}C_s \\
&= \int_a^c Y_s \mathrm{d}X_s + \int_c^b Y_s \mathrm{d}X_s.
\end{aligned}$$

Hence, the theorem is proved. □

Theorem 7. *Suppose that X_t is a Liu process, Y_t and Z_t are integrable uncertain processes with respect to the Liu process X_t on $[a, b]$, and α and β are real numbers. Then*

$$\int_a^b (\alpha Y_t + \beta Z_t) \mathrm{d}X_t = \alpha \int_a^b Y_t \mathrm{d}X_t + \beta \int_a^b Z_t \mathrm{d}X_t.$$

Proof. It follows from the definition of the uncertain integral with respect to the Liu process that

$$\begin{aligned}
\int_a^b (\alpha_t Y_t + \beta Z_t) \mathrm{d}X_t &= \int_a^b (\alpha_t Y_t + \beta Z_t)\mu_t \mathrm{d}t + \int_a^b (\alpha_t Y_t + \beta Z_t)\sigma_t \mathrm{d}C_t \\
&= \int_a^b \alpha_t Y_t \mu_t \mathrm{d}t + \int_a^b \beta Z_t \mu_t \mathrm{d}t \\
&\quad + \int_a^b \alpha_t Y_t \sigma_t \mathrm{d}C_t + \int_a^b \beta Z_t \sigma_t \mathrm{d}C_t \\
&= \alpha \left(\int_a^b Y_t \mu_t \mathrm{d}t + \int_a^b Y_t \sigma_t \mathrm{d}C_t \right) \\
&\quad + \beta \left(\int_a^b Z_t \mu_t \mathrm{d}t + \int_a^b Z_t \sigma_t \mathrm{d}C_t \right) \\
&= \alpha \int_a^b Y_t \mathrm{d}X_t + \beta \int_a^b Z_t \mathrm{d}X_t.
\end{aligned}$$

The theorem is proved. □

Theorem 8. (Fundamental Theorem) *Let X_t be a Liu process with drift μ_t and diffusion σ_t and $f(t,x)$ be a continuously differentiable function. Then $f(t,X_t)$ is a Liu process with*

$$\begin{aligned} \mathrm{d}f(t,X_t) &= \frac{\partial f}{\partial t}(t,X_t)\mathrm{d}t + \frac{\partial f}{\partial x}(t,X_t)\mathrm{d}X_t \\ &= \left(\frac{\partial f}{\partial t}(t,X_t) + \frac{\partial f}{\partial x}(t,X_t)\mu_t\right)\mathrm{d}t + \frac{\partial f}{\partial x}(t,X_t)\sigma_t\mathrm{d}C_t. \end{aligned} \tag{8}$$

Proof. Since the function f is continuously differentiable, by using the Taylor series expansion of $f(t,X_t)$ with respect to all its arguments, which in this case are t and X_t, we take the Taylor series expansion out to the first order of t and X_t. Then the infinitesimal increment of $f(t,X_t)$ has a first-order approximation

$$\Delta f(t,X_t) = \frac{\partial f}{\partial t}(t,X_t)\Delta t + \frac{\partial f}{\partial x}(t,X_t)\Delta X_t \tag{9}$$

$$= \left(\frac{\partial f}{\partial t}(t,X_t) + \frac{\partial f}{\partial x}(t,X_t)\mu_t\right)\Delta t + \frac{\partial f}{\partial x}(t,X_t)\sigma_t\Delta C_t. \tag{10}$$

In fact, the uncertain differential (8) is equivalent to the uncertain integral equation

$$f(t,X_t) = f(0,X_0) + \int_0^t \frac{\partial f}{\partial t}(s,X_s)\mathrm{d}s + \int_0^t \frac{\partial f}{\partial x}(s,X_s)\mathrm{d}X_s.$$

□

Theorem 9. (Change of Variables) *Let f and g be continuously differentiable functions. Then for any $s > 0$, we have*

$$\int_0^t f'(g(X_s))g'(X_s)\mathrm{d}X_s = f(g(X_t)) - f(g(X_0)).$$

Proof. Since f and g are continuously differentiable functions, it follows from the fundamental theorem that

$$\mathrm{d}f(g(X_t)) = f'(g(X_t))g'(X_t)\mathrm{d}X_t.$$

Then we get

$$f(g(X_t)) = f(g(X_0)) + \int_0^t f'(g(X_s))g'(X_s)\mathrm{d}X_s.$$

□

Theorem 10. (Integration by Parts) *Suppose that X_t is a Liu process and $F(x)$ is a continuously differentiable function. Then*

$$\int_0^t F(s)\mathrm{d}X_s = F(t)X_t - F(0)X_0 - \int_0^t X_s\mathrm{d}F(s).$$

Proof. It follows from the fundamental theorem that

$$\mathrm{d}(F(t)X_t) = F'(t)X_t\mathrm{d}t + F(t)\mathrm{d}X_t.$$

Thus,

$$\begin{aligned} \int_0^t F(s)\mathrm{d}X_s &= F(t)X_t - F(0)X_0 - \int_0^t F'(s)X_s\mathrm{d}s \\ &= F(t)X_t - F(0)X_0 - \int_0^t X_s\mathrm{d}F(s). \end{aligned}$$

Multifactor Liu process

□

Definition 13. Let $C_{1t}, C_{2t}, \cdots, C_{nt}$ be canonical Liu processes, and μ_t and $\sigma_{1t}, \sigma_{2t}, \cdots, \sigma_{nt}$ be uncertain processes. Then the uncertain process

$$X_t = X_0 + \int_0^t \mu_s \mathrm{d}s + \sum_{i=1}^{n} \int_0^t \sigma_{is} \mathrm{d}C_{is} \tag{11}$$

is called a multifactor Liu process with drift μ_t and diffusions $\sigma_{1t}, \sigma_{2t}, \cdots, \sigma_{nt}$. The Liu process X_t may also be written in differential form:

$$\mathrm{d}X_t = \mu_t \mathrm{d}t + \sum_{i=1}^{n} \sigma_{it} \mathrm{d}C_{it}. \tag{12}$$

Definition 14. Let X_t be a multifactor Liu process with drift μ_t and diffusions $\sigma_{1t}, \sigma_{2t}, \cdots, \sigma_{nt}$ and Y_t be an uncertain process. We define an uncertain integral with respect to the multifactor Liu process as

$$\int_0^t Y_s \mathrm{d}X_s = \int_0^t Y_s \mu_s \mathrm{d}s + \sum_{i=1}^{n} \int_0^t Y_s \sigma_{is} \mathrm{d}C_{is}.$$

Theorem 11. (Multifactor Fundamental Theorem) *Let $Y_{1t}, Y_{2t}, \cdots, Y_{nt}$ be Liu processes. If $f(t, y_1, y_2, \cdots, y_n)$ is a continuously differentiable function, then the uncertain process $Z_t = h(t, Y_{1t}, Y_{2t}, \cdots, Y_{nt})$ is a Liu process with*

$$\mathrm{d}Z_t = \frac{\partial f}{\partial t}(t, Y_{1t}, Y_{2t}, \cdots, Y_{nt})\mathrm{d}t + \sum_{i=1}^{n} \frac{\partial f}{\partial y_i}(t, Y_{1t}, Y_{2t}, \cdots, Y_{nt})\mathrm{d}Y_{it}.$$

Proof. Since the function f is continuously differentiable, by using the Taylor series expansion, the infinitesimal increment of Z_t has a first-order approximation

$$\Delta f(t, X_t) = \frac{\partial f}{\partial t}(t, Y_{1t}, Y_{2t}, \cdots, Y_{nt})\Delta t \tag{13}$$

$$+ \sum_{i=1}^{n} \frac{\partial f}{\partial y_i}(t, Y_{1t}, Y_{2t}, \cdots, Y_{nt})\Delta Y_{it}. \tag{14}$$

□

Example 7. Suppose that X_t and Y_t are two Liu processes. Since

$$\frac{\partial(xy)}{\partial t} = 0, \quad \frac{\partial(xy)}{\partial x} = y, \text{ and } \frac{\partial(xy)}{\partial y} = x,$$

the uncertain process $X_t Y_t$ is a Liu process with

$$\mathrm{d}X_t Y_t = X_t \mathrm{d}Y_t + Y_t \mathrm{d}X_t.$$

Conclusions

This paper introduced the concept of the Liu process which is defined by the Liu integral. Based on the Liu process, we extended the Liu integral on such a process. Some basic properties of this integral were discussed. Furthermore, the uncertain differential was introduced, and the fundamental theorem of uncertain calculus was derived. The integration by parts theorem was also discussed. Finally, we have studied multifactor Liu processes.

Acknowledgements

This work was supported by the National Natural Science Foundation of China (grant no. 61273044), Nankai University Project Funds for Young Teachers (no. NKQ1118), and Tianjin Municipal Research Program of Application Foundation and Advanced Technology of China (grant no. 10JCYBJC07300).

Author details

[1]Department of Risk Management and Insurance, Nankai University, Tianjin 300071, China. [2]Department of Mathematical Sciences, University of Cincinnati, Cincinnati, OH 45221-0025, USA.

References

1. Wiener, N: Differential space. J. Math. Phys. **2**, 131–174 (1923)
2. Ito, K: Stochastic integral. Proc. Jpn. Acad. **20**, 519–524 (1944)
3. Kunita, H, Watanabe, S: On square integrable martingales. Nagoya Math. **30**, 209–245 (1967)
4. Doléans-Dade, C, Meyer, P: Universté de strasbourg séminaire de probabilités. Lect. Notes Math. **5**, 143–173 (1970)
5. Jacod, J: Calculus stocastique et problemes de martingales. Lect. Notes Math. **714**, 66–112 (1979)
6. Kallenberg, O: Foundations of Modern Probability. 2nd edn. Springer, New York (1997)
7. Liu, B: Why is there a need for uncertainty theory? J. Uncertain Syst. **6**(1), 3–10 (2012)
8. Liu, B: Uncertainty Theory. 2nd edn. Springer, Berlin (2007)
9. Liu, B: Fuzzy process, hybrid process and uncertain process. J. Uncertain Syst. **2**(1), 3–16 (2008)
10. Zhang, X, Ning, Y, Meng, G: Delayed renewal process with uncertain interarrival times. Fuzzy Optimization Decis. Mak. **12**(1), 79–87 (2013)
11. Chen, X: Variation analysis of uncertain stationary independent increment processes. Eur. J. Oper. Res. **222**(2), 312–316 (2012)
12. Liu, B: Some research problems in uncertainty theory. J. Uncertain Syst. **3**(1), 3–10 (2009)
13. Liu, B, Yao, K: Uncertain integral with respect to multiple canonical processes. J. Uncertain Syst. **6**(4), 250–255 (2012)
14. Yao, K: Uncertain calculus with renewal process. Fuzzy Optimization Decis. Mak. **11**(3), 285–297 (2012)
15. Chen, X: Uncertain calculus with respected to finite variation processes. Technical Report. (2013). http://orsc.edu.cn/online/110303.pdf
16. Chen, X, Liu, B: Existence and uniqueness theorem for uncertain differential equations. Fuzzy Optimization Decis. Mak. **9**(1), 69–81 (2010)
17. Yao, K, Gao, J, Gao, Y: Some stability theorems of uncertain differential equation. Fuzzy Optimization Decis. Mak. **12**(1), 3–13 (2013)
18. Liu, Y: An analytic method for solving uncertain differential equations. J. Uncertain Syst. **6**(4), 244–249 (2012)
19. Yao, K: A type of nonlinear uncertain differential equations with analytic solution. (2013). http://orsc.edu.cn/online/110928.pdf
20. Yao, K, Chen, X: A numerical method for solving uncertain differential equations. J. Intell. Fuzzy Syst (2013). in press
21. Zhu, Y: Uncertain optimal control with application to a portfolio selection model. Cybern. Syst. **41**(7), 535–547 (2010)
22. Chen, X: American option pricing formula for uncertain financial market. Int. J. Oper. Res. **8**(2), 32–37 (2011)
23. Peng, J, Yao, K: A new option pricing model for stocks in uncertainty markets. Int. J. Oper. Res. **7**(4), 213–224 (2010)
24. Liu, B: Toward uncertain finance theory. J. Uncertainty Anal. Appl. **1**, 1 (2013)
25. Liu, B: Uncertainty Theory, 4th edn. (2013). http//:orsc.edu.cn/liu/ut.pdf
26. Liu, B: Uncertainty Theory: A Branch of, Mathematics for Modeling Human Uncertainty. Springer, Berlin (2010)
27. Liu, Y, Ha, M: Expected value of function of uncertain variables. J. Uncertain Syst. **4**(3), 181–186 (2010)
28. Yao, K: Extreme values and integral of solution of uncertain differential equation. J. Uncertainty Anal. Appl. **1**, 3 (2013)

5

Estimation of a linear model with two-parameter symmetric platykurtic distributed errors

Asrat Atsedeweyn Andargie* and K Srinivasa Rao

* Correspondence: asrat07@gmail.com
Department of Statistics, Andhra University, Visakhapatnam 530003, India

Abstract

Purpose: A linear regression model with Gaussian-distributed error terms is the most widely used method to describe the possible relationship between outcome and predictor variables. However, there are some drawbacks of Gaussian errors such as the distribution being mesokurtic. In many practical situations, the variables under study may not be mesokurtic but are platykurtic. Hence, to analyze this sort of platykurtic variables, a multiple regression model with symmetric platykurtic (SP) distributed errors is needed. In this paper, we introduce and develop a multiple linear regression model with symmetric platykurtic distributed errors for the first time.

Methods: We used the methods of ordinary least squares (OLS) and maximum likelihood (ML) to estimate the model parameters. The properties of the ML estimators with respect to the symmetric platykurtic distributed errors are discussed. The model selection criteria such as Akaike information criteria (AIC) and Bayesian information criteria (BIC) for the models are used. The utility of the proposed model is demonstrated with both simulation and real-time data.

Results: A comparative study of symmetric platykurtic linear regression model with the Gaussian model revealed that the former gives good fit to some data sets. The results also revealed that ML estimators are more efficient than OLS estimators in terms of the relative efficiency of the one-step-ahead forecast mean square error.

Conclusions: The study shows that the symmetric platykurtic distribution serves as an alternative to the normal distribution. The developed model is useful for analyzing data sets arising from agricultural experiments, portfolio management, space experiments, and a wide range of other practical problems.

Keywords: Maximum likelihood; Multiple linear regression model; Simulation; Symmetric platykurtic distribution

Introduction

Regression analysis is one of the most commonly used statistical methodologies in many branches of science and engineering used for discovering functional relationships between variables. The most typical example of regression analysis is multiple linear regression modeling, which is used for predicting values of one or more response variables from any factor of interest, the independent variables. It has received applications in almost every area of science, engineering, and medicine. Comprehensive accounts of the theory and applications of the linear regression model are discussed in Seber [1], Montgomery et al. [2], Grob [3], Sengupta and Jammalamadaka [4], Seber and Lee [5], Weisberg [6], and Yan and Su [7]. This technique is usually based on a statistical model in which the error terms are assumed to be independent and identically distributed random variables, whose distribution is considered to be multivariate normal with a zero mean vector and a positive definite covariance matrix [8]. However, in many disciplines, scientific research based on empirical studies or theoretical reasoning provided support for the presence of skewness or heavy tails in the distribution of the error terms. The departures from normality may be caused also by the presence of outlying values in the responses. Examples can be found, amongst others, in Fama [9] and Sutton [10]. For these reasons, several researchers proposed to perform multivariate regression analysis using a model that assumes a different parametric distribution family for the error terms.

Zeckhauser and Thompson [11] studied on a linear regression model with power distributions. Zellner [12] and Sutradhar and Ali [13] studied on a regression model with a multivariate t error variable. Tiku et al. [14-16] investigated a linear regression model with symmetric innovations, discussed a first-order autoregressive model with symmetric innovations, and presented a linear model with t distribution, respectively. Sengupta and Jammalamadaka [17] studied on linear models. Liu and Bozdogan [18] studied on power exponential (PE) multiple regression. Wong and Bian [19,20] studied on multiple regression coefficients in a linear model with errors being Student's t distribution and a linear regression model with underlying distribution being a generalized logistic distribution, respectively. Liu and Bozdogan [21] studied on multivariate regression models with PE random errors under various assumptions. Soffritti and Galimberti [22] discussed a multivariate linear regression model under the assumption that the error terms follow a finite mixture of normal distributions. Jafari and Hashemi [23] studied on linear regression with the error term of skew-normal distribution. Jahan and Khan [24] investigated the g-and-k distribution as the underlying assumption for the distribution of error in a simple linear regression model. Bian et al. [25] studied a multiple linear regression model with underlying Student's t distribution.

No serious attempt is made to develop and analyze multiple regression models with symmetric platykurtic (SP) errors. For this reason, to achieve more flexibility in statistical modeling and model selection, and to robustify many multiple statistical procedures, the purpose of this paper is to introduce and develop a multivariate linear regression model for conditions in which the distribution of error terms is assumed to be independent and identically distributed SP random errors with mean 0 and constant variance σ^2.

Model description

The multiple regression model assumes a linear (in parameters) relationship between a dependent variable $Y = (y_1, y_2, ..., y_n)'$ and a set of independent variables $X_i' = (x_{i0}, x_{i1}, x_{i2}, ..., x_{ik})$, where the first regressor $x_{i0} = 1$ is a constant and $i = 1, 2,..., n$. The model of interest is the standard linear regression model of the following form:

$$y_i = \beta_0 + \beta_1 x_{i1} + \cdots + \beta_k x_{ik} + \varepsilon_i \ (i = 1, 2, ..., n), \tag{1}$$

where y_i is an observed dependent variable, x_{ik} are observed independent variables, β_0, β_1,..., β_k are unknown regression coefficients to be estimated, and ε_i are independently and identically distributed. Using linear algebra notation, model (1) may be written alternatively in matrix form as

$$\mathbf{y} = \mathbf{X}\beta + \in, \in \sim SP\left(0, \sigma^2 I\right). \tag{2}$$

In this model,

$$\underset{nx1}{\mathbf{y}} = \begin{pmatrix} y_1 \\ y_2 \\ \cdot \\ \cdot \\ \cdot \\ y_n \end{pmatrix}, \underset{nx(k+1)}{\mathbf{X}} = \begin{pmatrix} x_{10} & x_{11} & \cdot & \cdot & \cdot & x_{1k} \\ x_{20} & x_{21} & \cdot & \cdot & \cdot & x_{2k} \\ \cdot & \cdot & \cdot & & & \cdot \\ \cdot & \cdot & & \cdot & & \cdot \\ \cdot & \cdot & & & \cdot & \cdot \\ x_{n0} & x_{n1} & \cdot & \cdot & \cdot & x_{nk} \end{pmatrix}, \underset{(k+1)x1}{\beta} = \begin{pmatrix} \beta_0 \\ \beta_1 \\ \cdot \\ \cdot \\ \cdot \\ \beta_k \end{pmatrix}, \underset{nx1}{\in} = \begin{pmatrix} \varepsilon_1 \\ \varepsilon_2 \\ \cdot \\ \cdot \\ \cdot \\ \varepsilon_n \end{pmatrix}, \tag{3}$$

where $\mathbf{y}$ is a column vector of n elements, $\mathbf{X}$ is an $nx(k+1)(k+1<n)$ nonrandom design matrix of covariates (with its first column having all elements equal to 1, the second column being filled by the observed values of x_1, the $(k+1)$th column being filled by the observed values of x_k), β is a column vector of the $(k+1)$ elements, σ is an unknown scale parameter, and $\in$ is an $nx1$ column vector of error terms with zero mean and constant variance $\sigma^2 I$.

Model assumptions

In order to complete the description of the model, some assumptions about the nature of the errors are necessary. It is assumed that the errors are independent and identically distributed (i.i.d.) random variables whose distribution is assumed to be two-parameter SP rather than normal, with zero mean and a positive definite variance-covariance matrix $\sigma^2 I$ of dimension nxn, that is

$$\varepsilon_i \sim MSP\left(0, \sigma^2 I\right). \tag{4}$$

These assumptions are summarized in the matrix vector form as

$$E(\in) = 0, Cov(\in) = E(\in\in') = \sigma^2 \underset{nxn}{I}, \tag{5}$$

where the notation E stands for the expected value and Cov represents an nxn variance-covariance matrix. The vector $\mathbf{0}$ is a column vector with n zero elements, and I is an identity matrix of order nxn. The parameter σ^2 is unspecified, along with the vector parameter β. The elements of β are real-valued, while σ is positive. The covariates are either nonrandom or are independent of the errors. $E(y) = X\beta$, and $Cov(y) = \sigma^2 I$. We shall use the triplet $(y, X\beta, \sigma^2 I)$ for linear model (2).

Article headings

The manuscript is organized in eight sections. The 'Introduction' section frames the objective of the paper and reviews related literatures. In the 'Properties of the two parameter symmetric platykurtic distribution' section, we introduced the two-parameter SP distribution, in notation SP(μ,σ). We derived the maximum likelihood (ML) estimators in the 'Maximum likelihood estimation of the model parameters' section. In order to obtain numerical solutions to the ML estimate problem, the Newton–Raphson (NR) iterative method has been used. In the 'Properties of the estimators through simulation study' section, we show the asymptotic properties of the estimators. In the 'Least squares estimation of the model parameters' section, OLS estimation for the model parameters is studied. Comparison of MLE with OLS estimators and that of the proposed model with the Gaussian model are done in the 'Comparative study of the model' section. The 'Application of the model' section demonstrates the usefulness of the present model on real data. Finally, the 'Summary and conclusions' section concludes the paper.

Properties of the two-parameter symmetric platykurtic distribution

Even though the sample frequency curve has a symmetric and bell shape, it often has fat tails than normal and the almost universally used Gaussian distribution may badly fit the fat tails. The family of symmetric platykurtic distributions can best model these features. The origin and genesis of this distribution is given by Srinivasa Rao et al. [26] for analyzing statistical data that arise from biological, sociological, agricultural, and environmental experiments. It became popular thereafter and has received considerable attention in the present time in modeling image segmentation and economic and financial data as a generalization of normal distribution (e.g., [27]). The probability density function (pdf) of such a family of distributions is

$$f_r(y/\mu,\sigma,\gamma) = \frac{\left[2\gamma + \left(\frac{y-\mu}{\sigma}\right)^2\right]^{\gamma} e^{-\frac{1}{2}\left(\frac{y-\mu}{\sigma}\right)^2}}{\sum_{j=0}^{\gamma}\binom{\gamma}{j}(2\gamma)^{\gamma-j}2^{j+\frac{1}{2}}\Gamma\left(j+\frac{1}{2}\right)\sigma}. \tag{6}$$

The distribution depends on three parameters μ, σ, and γ. These parameters can be interpreted as follows:

- μ is a real number and may be thought of as a location measure.
- σ is positive and measures the dispersion or the scale of the distribution.
- γ is a kurtosis parameter which determines the shape of the distribution, taking values $\gamma = 0, 1, 2, \ldots, n$. If $\gamma = 0$, we retrieve the normal distribution $N(\mu,\sigma^2)$. If we take $\gamma = 1$, we get a two-parameter symmetric platykurtic distribution.

In this section, we briefly present a two-parameter symmetric platykurtic distribution which is appropriate for data having kurtosis $\beta = 2.52$. A random variable Y is said to follow a two-parameter SP distribution if the density function of Y is

$$f(y/\mu,\sigma) = \frac{\left[2 + \left(\frac{y-\mu}{\sigma}\right)^2\right] e^{-\frac{1}{2}\left(\frac{y-\mu}{\sigma}\right)^2}}{3\sigma\sqrt{2\pi}}, y \in \mathfrak{R}, \tag{7}$$

where $-\infty < \mu < \infty$ and $\sigma > 0$ are location and scale parameters, respectively. We denote this by $Y \sim \mathrm{SP}(\mu,\sigma^2)$. The various shapes of the frequency curves are shown in Figure 1.

Some important properties of the random variable Y in a univariate context and those which are needed for the simulation of the variate are as follows:

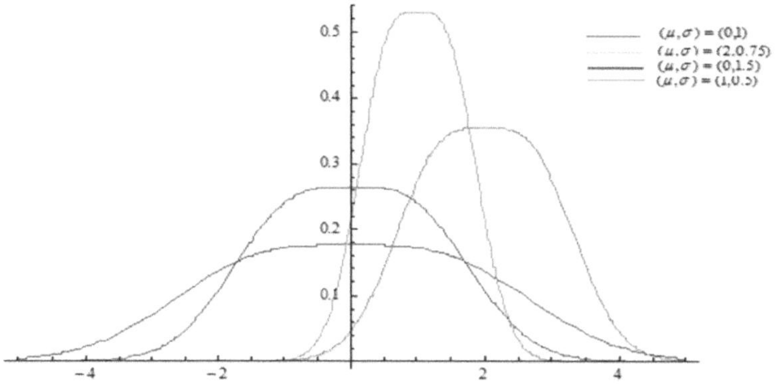

Figure 1 The probability density function curve for different values of μ and σ. $(\mu,\sigma) = (1,0.5)$, $(0,1.5)$, $(0,1)$, and $(2,0.75)$.

1. The distribution function of the random variable Y specified by the probability density function (7) is given by

$$F_Y(y) = \int_{-\infty}^{y} \frac{\left[2 + \left(\frac{t-\mu}{\sigma}\right)^2\right] e^{-\frac{1}{2}\left(\frac{t-\mu}{\sigma}\right)^2}}{3\sigma\sqrt{2\pi}} dt$$

which, on simplification, reduces to

$$F_Y(y) = \frac{1}{\sigma\sqrt{2\pi}} \int_{-\infty}^{y} e^{-\frac{1}{2}\left(\frac{t-\mu}{\sigma}\right)^2} dt - \frac{(y-\mu)e^{-\frac{1}{2}\left(\frac{y-\mu}{\sigma}\right)^2}}{3\sigma\sqrt{2\pi}} = F_0(y;\mu,\sigma) - F_1(y;\mu,\sigma), \tag{8}$$

where $F_0(y;\mu,\sigma) = \frac{1}{\sigma\sqrt{2\pi}} \int_{-\infty}^{y} e^{-\left(\frac{t-\mu}{\sigma}\right)^2} dt \ y \in \mathfrak{R}$, is the distribution function of the normal random variable with mean μ and variance σ, while $F_1(y) = \frac{(y-\mu)}{3\sigma\sqrt{2\pi}} e^{-\frac{1}{2}\left(\frac{y-\mu}{\sigma}\right)^2}$, $y \in \mathfrak{R}$ is the nondistribution function (cannot be a cumulative density function) since it is negative for $Y > \mu$.

2. Numerical approximations for the two-parameter symmetric platykurtic cumulative distribution function (CDF): Following Marsaglia's [28] approximation for standard normal distribution who suggested a simple algorithm based on the Taylor series expansion, we have also approximated the values $F(y;\mu,\sigma)$ as follows. For standard normal distribution with arbitrary precision, $\Phi(y) = \frac{1}{2} + \phi(y)\left(y + \frac{y^3}{3} + \frac{y^5}{3.5} + \frac{y^7}{3.5.7} + \ldots\right)$ where ϕ and Φ is the pdf and CDF of the normal distribution with mean μ and variance σ. Accordingly, after little algebra, the standard symmetric platykurtic CDF, $F(y)$, is approximated by

$$F(y) = \frac{1}{2} + f(y)\left(\frac{-y}{3} + \sum_{i=1}^{n} \frac{y^i}{i!!}\right) \tag{9}$$

where $n = 1, 3, 5, \ldots, n$ and $n!!$ denotes the double factorial that is the product of every odd number from 1 to n.

3. The cumulant-generating function is the logarithm of the moment-generating function:

$$g(t;\mu,\sigma^2) = \mu t + \frac{1}{2}\sigma^2 t^2 + \ln\left[1 + \frac{(\sigma t)^2}{3}\right]. \tag{10}$$

The cumulants k_n are extracted from the cumulant-generating function via differentiation (at zero) of $g(t)$. That is, the cumulants appear as the coefficients in the Maclaurin series of $g(t)$:

$$k_1 = g'(0) = \mu, k_2 = g''(0) = \frac{5\sigma^2}{3}, k_3 = g'''(0) = \cdots = k_n = g^{(n)}(0) = 0 \tag{11}$$

That is, the first two cumulants are equal to the mean μ and the variance $\frac{5\sigma^2}{3}$ of the two-parameter symmetric platykurtic distribution, respectively, whereas all higher-order cumulants are equal to zero.

4. Hazard rate function of the distribution: The hazard function $h(y;\mu,\sigma)$ of the two-parameter symmetric platykurtic distribution used in this paper is utilized to characterize life phenomena and can be written as

$$h(y;\mu,\sigma) = \frac{\left(2 + \left(\frac{y-\mu}{\sigma}\right)^2\right)\phi\left(\frac{y-\mu}{\sigma}\right)}{3\sigma Q(y) + (y-\mu)\phi\left(\frac{y-\mu}{\sigma}\right)}, \tag{12}$$

where ϕ is the pdf of the normal distribution with mean μ and variance σ, whereas the Q-function $Q(y)$ is the complement of the standard normal CDF, $Q(y) = 1 - \Phi(y)$.

Recently, it was observed by Gupta and Gupta [29] that the reversed hazard function plays an important role in the reliability analysis. The reversed hazard function of the two-parameter SP(μ,σ) is

$$r(y;\mu,\sigma) = \frac{f(y;\mu,\sigma)}{\text{SP}(y;\mu,\sigma)} = \frac{\left(2 + \left(\frac{y-\mu}{\sigma}\right)^2\right)\phi\left(\frac{y-\mu}{\sigma}\right)}{3\sigma\Phi\left(\frac{y-\mu}{\sigma}\right) - (y-\mu)\phi\left(\frac{y-\mu}{\sigma}\right)}. \tag{13}$$

It is well known that the hazard function or the reversed hazard function uniquely determines the corresponding probability density function.

5. Entropy: The entropy for a two-parameter symmetric platykurtic distribution random variable y with probability density function $f(y)$ on the real line is defined by

$$h(y) = \frac{-529}{1,500} + \ln\left(3\sigma\sqrt{2\pi}\right). \tag{14}$$

It can be recalled that the entropy for normal distribution is $\frac{1}{2}\ln(2\pi e\sigma^2)$. If f and g are the probability distributions of symmetric platykurtic and normal distributions, respectively, then the relative entropy $D(f||g)$ from f to g is

$$D(f||g) = \int_{-\infty}^{\infty} f(y) \log \frac{f(y)}{g(y)} dy = 0.7088. \tag{15}$$

This gives us a measure of something like the distance between the two probability distributions, in the sense that the relative entropy is always positive, is zero if and only if the two distributions are the same, and increases as the distributions diverge. Some of the more important properties of the SP distribution are summarized in the Appendix.

Maximum likelihood estimation of the model parameters

In this section, we consider the homoscedastic regression model $(Y, X\beta, \sigma^2 I)$ with $\varepsilon \sim$ SP $(0, \sigma^2 I)$. The unknown parameters of this model are the coefficient vector β and the error variance σ^2. We deal with the problem of ML estimation of these parameters, which requires some fairly intricate mathematics, from the observables Y and X. The MLEs of β and σ^2 are the parameter values that maximize the likelihood function:

$$L(y;\beta,\sigma) = \Pi_{i=1}^{n} f\left(y_i/X\beta,\sigma^2 I\right) = \prod_{i=1}^{n} \frac{\left[2+\left(\frac{y_i - x_i'\beta}{\sigma}\right)^2\right] e^{-\frac{1}{2}\left(\frac{y_i - x_i'\beta}{\sigma}\right)^2}}{3\sigma\sqrt{2\pi}}$$

$$= \left(3\sigma\sqrt{2\pi}\right)^{-n} \prod_{i=1}^{n} \left[\left(2+\left(\frac{y_i - x_i'\beta}{\sigma}\right)^2\right) e^{-\frac{1}{2}\left(\frac{y_i - x_i'\beta}{\sigma}\right)^2}\right]. \tag{16}$$

The log-likelihood function in the i.i.d. case, ignoring the additive constants, equals

$$\ell = \ln L(y;\beta,\sigma) = -\frac{n}{2}\ln\sigma^2 + \sum_{i=1}^{n} \ln h\left(\frac{y_i - x_i'\beta}{\sigma}\right)$$

$$\text{where } h\left(\frac{y_i - x_i'\beta}{\sigma}\right) = \left[2+\left(\frac{y_i - x_i'\beta}{\sigma}\right)^2\right] e^{-\frac{1}{2}\left(\frac{y_i - x_i'\beta}{\sigma}\right)^2}. \tag{17}$$

The unknown parameters of this model are the coefficient vector β and the error variance σ^2. The MLE is that $\hat{\theta}_{\text{MLE}} = \left(\hat{\beta}, \hat{\sigma}^2\right)$ which maximizes the log-likelihood. Taking the partial derivatives of the log of the likelihood with respect to the $(k+1)x1$ vector β and nxn matrix of $\sigma^2 I$ and setting the result equal to zero will produce (18) and (19). The MLEs are the solutions of the equations:

$$\frac{\partial \ell}{\partial \beta} = -\frac{1}{\sigma}\sum_{i=1}^{n} x_i \left(\frac{\partial \ln h(u)}{\partial u}\right)\Bigg|_{u=\frac{y_i - x_i'\beta}{\sigma}} = 0 \quad = \frac{1}{\sigma}\sum_{i=1}^{n} x_i \left(\frac{u^3}{2+u^2}\right)\Bigg|_{u=\frac{y_i - x_i'\beta}{\sigma}} = 0 \tag{18}$$

and

$$\frac{\partial \ell}{\partial \sigma^2} = -\frac{n}{2\sigma^2} - \frac{1}{2\sigma^2}\sum_{i=1}^{n}\left(u\frac{\partial \ln h(u)}{\partial u}\right)\Bigg|_{u=\frac{y_i - x_i'\beta}{\sigma}} = 0 \quad = -\frac{n}{2\sigma^2} + \frac{1}{2\sigma^2}\sum_{i=1}^{n}\left(\frac{u^4}{2+u^2}\right)\Bigg|_{u=\frac{y_i - x_i'\beta}{\sigma}} = 0. \tag{19}$$

Since there are no closed-form solutions to the likelihood equations, numerical methods such as the Fisher scoring or Newton–Raphson iterative method can be used to obtain the MLEs. The usual or standard procedure for implementing this solution is to use the Newton–Raphson iteration method given by

$$\theta^{(n+1)} = \theta^{(n)} - \left(H^{(n)}\right)^{-1} S^{(n)}, \theta = (\beta, \sigma^2)', \tag{20}$$

where H is the matrix of the second derivative and S is the vector of the first derivative of the log-likelihood function both evaluated at the current values of the parameter vector θ.

Here we begin with some starting value, say $\theta^{(0)}$, and improve it by finding some better approximation $\theta^{(1)}$ to the required root. This procedure can be iterated to go from a current approximation $\theta^{(n)}$ to a better approximation $\theta^{(n+1)}$.

Variance-covariance matrix

To obtain the asymptotic variances and covariance of the estimates, it is required to construct the Hessian matrix of the log-likelihood. The negative expected values of the second-order partial derivatives of the log-likelihood equations (18) and (19) can be used to estimate the asymptotic covariance matrix of parameter estimates and can be found as follows. Each summand in the right-hand side of (19) has zero expectation. Therefore,

$$\begin{aligned} E\left[\left(\frac{\partial \ell}{\partial \beta}\right)\left(\frac{\partial \ell}{\partial \beta}\right)'\right] &= \frac{1}{\sigma^2}\sum_{i=1}^{n} x_i x_i' \int_{-\infty}^{\infty}\left(\frac{\partial \ln h(u)}{\partial u}\right)^2\Bigg|_{u=\frac{y_i - x_i'\beta}{\sigma}} \cdot \frac{1}{\sigma} h\left(\frac{y_i - x_i'\beta}{\sigma}\right) dy_i \\ &= \frac{1}{\sigma^2}\sum_{i=1}^{n} x_i x_i' \int_{-\infty}^{\infty}\left(\frac{\partial \ln h(u)}{\partial u}\right)^2 h(u)du \\ &= I_\mu X'X, \end{aligned} \tag{21}$$

$$\text{where } I_\mu = \frac{1}{\sigma^2}\int_{-\infty}^{\infty}\left(\frac{\partial \ln h(u)}{\partial u}\right)^2 h(u)du = \frac{1}{\sigma^2}\int_{-\infty}^{\infty}\frac{u^6}{u^2+2}e^{-\frac{1}{2}u^2}du = \frac{4.934}{\sigma^2}. \tag{22}$$

Further,

$$\begin{aligned} E\left[\left(\frac{\partial \ell}{\partial \beta}\right)\left(\frac{\partial \ell}{\partial \sigma^2}\right)\right] &= \frac{1}{2\sigma^3}\sum_{i=1}^{n} x_i \int_{-\infty}^{\infty} u\left(\frac{\partial \ln h(u)}{\partial u}\right)^2\Bigg|_{u=\frac{y_i - x_i'\beta}{\sigma}} \cdot \frac{1}{\sigma} h\left(\frac{y_i - x_i'\beta}{\sigma}\right) dy_i \\ &= \frac{1}{2\sigma^3}\sum_{i=1}^{n} x_i \int_{-\infty}^{\infty} u\left(\frac{\partial \ln h(u)}{\partial u}\right)^2 h(u)du = 0 \end{aligned} \tag{23}$$

as the integrand in the last expression is an odd function of u, that is, making use of the fact that the integrand of the off-diagonal term is an odd function of u.

Finally,

$$E\left[\left(\frac{\partial \ell}{\partial \sigma^2}\right)^2\right] = \frac{n^2}{4\sigma^4} + \frac{n}{2\sigma^4}\sum_{i=1}^{n} E\left[\left(u\frac{\partial \ln h(u)}{\partial u}\right)\Bigg|_{u=\frac{y_i - x_i'\beta}{\sigma}}\right]$$

$$+\frac{1}{4\sigma^4}\sum_{i=1}^{n}\sum_{j=1}^{n} E\left[\left(u\frac{\partial \ln h(u)}{\partial u}\right)\Bigg|_{u=\frac{y_i - x_i'\beta}{\sigma}} \cdot \left(u\frac{\partial \ln h(u)}{\partial u}\right)\Bigg|_{u=\frac{y_i - x_i'\beta}{\sigma}}\right]$$

$$= \frac{n^2}{4\sigma^4} + \frac{n^2}{2\sigma^4}\int_{-\infty}^{\infty} u\frac{\partial h(u)}{\partial u}du + \frac{n^2 - n}{4\sigma^4}\left(\int_{-\infty}^{\infty} u\frac{\partial h(u)}{\partial u}du\right)^2$$

$$+\frac{n}{4\sigma^4}\int_{-\infty}^{\infty}\left(u\frac{\partial \ln h(u)}{\partial u}\right)^2 .h(u)du.$$

Simplification of the expression of this term is aided by the identity

$$\int_{-\infty}^{\infty} u\frac{\partial h(u)}{\partial u}du = -1 = \frac{n^2}{4\sigma^4} + \frac{n^2}{2\sigma^4}(-1) + \frac{n^2 - n}{4\sigma^4}(-1)^2 + \frac{n}{4\sigma^4}\int_{-\infty}^{\infty}\left(u\frac{\partial \ln h(u)}{\partial u}\right)^2 .h(u)du = nI_{\sigma^2} \tag{24}$$

$$\text{where } I_{\sigma^2} = \frac{1}{4\sigma^4}\left[\int_{-\infty}^{\infty}\left(u\frac{\partial \ln h(u)}{\partial u}\right)^2 .h(u)du - 1\right] = \frac{1}{4\sigma^4}\left[\int_{-\infty}^{\infty}\frac{u^8}{u^2+2}e^{-\frac{1}{2}u^2}du - 1\right] \tag{25}$$

$$= \frac{1}{4\sigma^4}[27.731 - 1] = \frac{26.731}{4\sigma^4} = \frac{6.68275}{\sigma^4}.$$

Therefore, the information matrix for $\theta = (\beta, \sigma^2)'$ is

$$I(\theta) = \begin{pmatrix} I_\mu X'X & 0 \\ 0 & nI_{\sigma^2} \end{pmatrix}. \tag{26}$$

The information for β is $I_\mu X'X$. Therefore, the design issues can be addressed with reference to the matrix $X'X$ - just as in the normal case. The scalar I_μ is equal to $\frac{4.934}{\sigma^2}$ (greater than σ^{-2} when the components of **y** are normally distributed), whereas the scalar I_{σ^2} is equal to $\frac{6.68275}{\sigma^4}$ (greater than σ^{-4} when the components of **y** are normally distributed). The large sample variances and covariance of the estimates can be approximated by inverting the usual symmetric information matrix (26):

$$Var(\theta) = [I(\theta)]^{-1} = \begin{pmatrix} (I_\mu X'X)^{-1} & 0 \\ 0 & (nI_{\sigma^2})^{-1} \end{pmatrix} \tag{27}$$

Thus, the square root of the elements on the diagonal of this matrix will give us the standard errors associated with the coefficients.

Simulation and results

The proposed approach was evaluated through Monte Carlo experiments in which artificial data sets were generated from model (2) using Wolfram Mathematica 9. To facilitate exposition of the method of estimation, a multiple data set with two independent variables and one dependent variable are simulated from a model with prespecified parameters for various sample sizes n = 100, 1,000, 3,000, 5,000, 10,000. (The sample size in this case, 10,000, is relatively large, and so finite sample bias is less of an issue.) The dependent variable (Y) is simulated from the symmetric platykurtic distribution with mean 0 and variance 1 while the two predictor x_1 and x_2 variables are generated from normal and lognormal distributions, respectively, with the prespecified mean and variance using the following simulation protocol:

$$\begin{aligned} x_1 &= \text{RandomReal}[\text{NormalDistribution}[10, 5]]; \\ x_2 &= \text{RandomReal}[\text{LogNormalDistribution}[0, 3]] \end{aligned} \tag{28}$$

and use as explanatory variables for the regression model. Without loss of generality, we considered the values of the model parameters as

$$\begin{pmatrix} \beta_0 \\ \beta_1 \\ \beta_2 \end{pmatrix} = \begin{pmatrix} 4 \\ 2 \\ 3 \end{pmatrix} \text{and} \begin{pmatrix} \mu \\ \sigma \end{pmatrix} = \begin{pmatrix} 0 \\ 1 \end{pmatrix}. \tag{29}$$

Summary statistics of estimations of the regression model with symmetric platykurtic distributed errors using the ML procedure are presented in Table 1.

The numerical results in Table 1 suggest that as the sample size increases, the estimates of the parameters become more precise. The ML method provides good estimates of the underlying model not only of the regression coefficients but also of the

Table 1 Summary of ML estimation of the regression model for the simulated data

Sample size (n)	Parameter	Estimate	Standard error	Wald 95% confidence limits		Chi-square	Pr > chi-square
				Lower	Upper		
100	β_0	4.0747	0.2515	3.5817	4.5677	262.46	<.0001
	β_1	1.9914	0.0216	1.9491	2.0337	8,510.03	<.0001
	β_2	2.9998	0.0010	2.9978	3.0019	8,263,598	<.0001
	σ	1.0668	0.0754	0.9288	1.2254		
1,000	β_0	3.8958	0.0696	3.7593	4.0323	3,129.59	<.0001
	β_1	2.0062	0.0065	1.9935	2.0189	96,198.0	<.0001
	β_2	2.9999	0.0001	2.9997	3.0000	1.408E9	<.0001
	σ	1.0242	0.0229	0.9803	1.0701		
3,000	β_0	4.0504	0.0407	3.9707	4.1301	9,916.51	<.0001
	β_1	1.9938	0.0036	1.9867	2.0009	303,055	<.0001
	β_2	2.9999	0.0000	2.9999	3.0000	3.731E9	<.0001
	σ	1.0037	0.0130	0.9786	1.0294		
5,000	β_0	4.0012	0.0315	3.9394	4.0630	16,095.1	<.0001
	β_1	2.0013	0.0028	1.9957	2.0068	506,637	<.0001
	β_2	3.0000	0.0000	3.0000	3.0000	6.34E10	<.0001
	σ	0.9988	0.0100	0.9794	1.0186		
10,000	β_0	4.0052	0.0220	3.9620	4.0483	33,060.4	<.0001
	β_1	1.9981	0.0020	1.9942	2.0019	1,031,236	<.0001
	β_2	3.0000	0.0000	3.0000	3.0000	9.55E11	<.0001
	σ	0.9850	0.0070	0.9714	0.9987		

correlation matrix. The fitted linear regression model with symmetric platykurtic distributed error terms to the simulated data, based on the sample of size 10,000, is

$$\hat{Y} = 4.0052 + 1.9981X_1 + 3.000X_2. \quad (30)$$

Standard errors of the estimates were estimated by the square root of the diagonal elements of the inverse of the Hessian of the log-likelihood function. Thus, the estimated standard errors are

$$\text{s.e.}\left(\hat{\beta}_0\right) = 0.0220, \text{s.e.}\left(\hat{\beta}_1\right) = 0.0020, \text{s.e.}\left(\hat{\beta}_2\right) = 0.000, \text{and s.e.}(\hat{\sigma}) = 0.0070. \quad (31)$$

Properties of the estimators through simulation study

If certain regularity conditions of the density are met, the MLEs are most attractive because they possess many asymptotic or large sample properties. Derivations of the asymptotic properties require some fairly intricate mathematics. The three properties of the regular densities (moments of the derivatives of the log-likelihood) are used in establishing the properties of MLEs. The properties of the ML estimators are as follows.

Consistency

One of the basic properties of a good estimator is that it differs from a true value by a very small amount as n becomes large. This implies that we can reach the exact value of θ by indefinitely increasing the sample size. Mathematically, it is expressed as

$$p\lim\hat{\theta} = \theta. \quad (32)$$

More specifically, a consistent estimator should not only be unbiased, but it should also have a variance which is as small as possible. This leads to two definitions:

$$\begin{aligned} &E\left(\hat{\theta}\right) \rightarrow^{n\rightarrow\infty} \theta \text{ and} \\ &V\left(\hat{\theta}\right) \rightarrow^{n\rightarrow\infty} 0 \end{aligned} \quad (33)$$

From (27), it is clear that the variance tends to zero as $n \rightarrow \infty$ in each case, so we conclude that the estimators are consistent since they are composed of i.i.d. observations.

Asymptotic normality

Greene's derivation of the asymptotic normality of the MLE applies here. The first derivative of the log-likelihood evaluated at the MLE equals zero. So

$$S\left(\hat{\theta}\right) = 0. \quad (34)$$

Expand this set of equations in a Taylor series around the true parameters θ_0 using the mean value theorem to truncate the Taylor series at the second term:

$$S\left(\hat{\theta}\right) = S(\theta_0) + H(\bar{\theta})\left(\hat{\theta}\text{-}\theta_0\right) = 0. \quad (35)$$

The Hessian is evaluated at a point $\bar{\theta}$ that is between $\hat{\theta}$ and θ_0 $\left[\bar{\theta} = w\hat{\theta} + (1-w)\theta_0\right.$ for some $\left.0 < w < 1\right]$. We then rearrange this function and multiply the result by $\sqrt{n}$ to obtain

$$\sqrt{n}\left(\hat{\theta}-\theta_0\right) = \left[-H(\bar{\theta})\right]^{-1}\left[\sqrt{n}S(\theta_0)\right] \tag{36}$$

Using (32), the probability limits of $\sqrt{n}\left(\hat{\theta}-\theta\right)$ and $\sqrt{n}\left(\hat{\theta}-\theta\right)$ go to zero because of the consistency of $\hat{\theta}$ (i.e., $p\lim\left(\hat{\theta}-\theta\right) = 0$ and $p\lim\left(\hat{\theta}-\bar{\theta}\right) = 0$). The second derivatives are continuous functions. Therefore, if the limiting distribution exists, then

$$\sqrt{n}\left(\hat{\theta}-\theta_0\right) \xrightarrow{d} \left[-H(\theta_0)\right]^{-1}\left[\sqrt{n}S(\theta_0)\right]. \tag{37}$$

By dividing $H(\theta_0)$ and $S(\theta_0)$ by n, we obtain

$$\sqrt{n}\left(\hat{\theta}-\theta_0\right) \xrightarrow{d} \left[-\frac{1}{n}H(\theta_0)\right]^{-1}\left[\sqrt{n}\bar{S}(\theta_0)\right] \tag{38}$$

We may apply the Lindeberg-Levy central limit theorem to $[\sqrt{n}\bar{S}(\theta_0)]$ because it is $\sqrt{n}$ times the mean of a random sample. By virtue of $V[S_i(\theta)] = -E\left[\frac{1}{n}H(\theta)\right]$, the limiting variance of $[\sqrt{n}\bar{S}(\theta)]$ is $-E\left[\frac{1}{n}H(\theta)\right]$, so

$$\sqrt{n}\bar{S}(\theta) \xrightarrow{d} N\left\{0, -E\left[\frac{1}{n}H(\theta)\right]\right\} \tag{39}$$

By virtue of $E[S_i(\theta)] = 0, p\lim\left[-\frac{1}{n}H(\theta)\right] = -E\left[\frac{1}{n}H(\theta)\right]$. This result is a constant matrix, so we can combine results to obtain

$$\left[-\frac{1}{n}H(\theta)\right]^{-1}\sqrt{n}\bar{S}(\theta) \rightarrow^{d} N\left[0, \left\{-E\left[\frac{1}{n}H(\theta)\right]\right\}^{-1}\left\{-E\left[\frac{1}{n}H(\theta)\right]\right\}\left\{-E\left[\frac{1}{n}H(\theta)\right]\right\}^{-1}\right] \text{ or}$$
$$\sqrt{n}\left(\hat{\theta}-\theta\right) \rightarrow^{d} N\left[0, \left\{-E\left[\frac{1}{n}H(\theta)\right]\right\}^{-1}\right]. \tag{40}$$

It follows that the MLEs are asymptotically normal with asymptotic distribution:

$$\hat{\theta}\sim N\left[\theta, \{I(\theta)\}^{-1}\right]. \tag{41}$$

Asymptotic efficiency

The information matrix forms a tool of interest to verify efficiency, *viz.* the attainment of the information limit to the variance of the estimator. An estimator whose variance is as small as the Cramer-Rao lower bound when the sample size tends to infinity is

Table 2 Nonlinear OLS summary of residual errors

Equation	DF model	DF error	SSE	MSE	Root MSE	*R*-square	Adjusted *R*-square
y	3	9996	16650.3	1.6657	1.2906	1.0000	1.0000

DF degrees of freedom, *SSE* sum of the squared errors.

called asymptotically efficient. This means that an estimator which reaches 100% efficiency only in the $n \to \infty$ limit is called asymptotically efficient. It can be shown that the Cramer-Rao lower bound for $\hat{\theta} = \left(\hat{\beta}_0, \hat{\beta}_1, \hat{\beta}_2\right)'$ and $\hat{\sigma}^2$, respectively, are

$$\begin{aligned} \mathrm{Var}\left(\hat{\theta}\right) &\geq \frac{1}{-nE\left(\dfrac{\partial^2 \ln f(y)}{\partial \theta^2}\right)} = \left(I_\mu X'X\right)^{-1} \text{ and} \\ \mathrm{Var}(\hat{\sigma}^2) &\geq \frac{1}{-nE\left(\dfrac{\partial^2 \ln f}{(\partial \sigma^2)^2}\right)} = \left(nI_{\sigma^2}\right)^{-1}, \end{aligned} \tag{42}$$

where I_μ and I_{σ^2} are as defined in (22) and (25), respectively.

This means that any unbiased estimator that achieves this lower bound is efficient and no better unbiased estimator is possible. Now look back at the variance-covariance matrix (27). It is interesting to note that the variances of the estimators in the variance-covariance matrix do asymptotically coincide with the Cramer-Rao lower bound (42). This means that our MLEs are 100% asymptotically efficient. The asymptotic variance of the MLE is, in fact, equal to the Cramer-Rao lower bound for the variance of a consistent and asymptotically normally distributed estimator [30].

Invariance

Last, the invariance property is a mathematical result of the method of computing MLEs; it is not a statistical result as such. If it is desired to analyze a continuous and continuously differentiable function of an MLE, then the function of $\hat{\theta}$ will, itself, be the MLE since the MLE is invariant to one-to-one transformations of θ.

These four properties explain the prevalence of the ML technique. The second is a particularly powerful result. The third greatly facilitates hypothesis testing and the construction of interval estimates. The MLE has the minimum variance achievable by a consistent and asymptotically normally distributed estimator.

Least squares estimation of the model parameters

The most widely used technique for estimating the unknown regression coefficients in a standard linear regression model is undeniably the method of ordinary least squares (OLS). The least squares estimates of β_0, β_1, and β_2 are the values which minimize

Table 3 Nonlinear OLS parameter estimates

Parameter	Estimate	Approximate standard error	*t* value	Approximate Pr > |*t*|
β_0	3.9598	0.0287	138.16	<.0001
β_1	2.003412	0.00256	782.76	<.0001
β_2	2.99998	0.000015	202,020	<.0001

Number of observations: used = 10,000, missing = 0; statistics for the system: objective = 1.6652; objective × *N* = 1665.2.

Table 4 Comparison of the OLS with ML estimators of the SP regression model

		OLS	MLE
$\hat{\beta}_0$	Bias	−0.04016	0.0109
	MSE	0.002436516	0.000914
$\hat{\beta}_1$	Bias	0.003412	−0.0022
	MSE	1.81953E − 05	1.11E − 05
$\hat{\beta}_2$	Bias	−1.7E − 05	0.0001
	MSE	5.14E − 10	2E − 08

$$\mathrm{SS} = \sum_{i=1}^{n} \left(y_i - \beta_0 - \beta_1 x_{1i} - \beta_2 x_{2i}\right)^2. \tag{43}$$

This leads to a closed-form expression for the estimated value of the unknown parametric vector β:

$$\hat{\beta} = \left(X^T X\right)^{-1} X^T y = \left(\frac{1}{n}\sum_{1}^{n} x_i x_i^T\right)^{-1} \left(\frac{1}{n}\sum_{1}^{n} x_i y_i\right). \tag{44}$$

Tables 2 and 3 summarize the results of the OLS estimation for a sample of size 10,000 using the same simulated observations obtained in the 'Simulation and results' section.

From Table 3, it is observed that the OLS estimates differ significantly from the ML estimates and the ML estimators are closer to the true values of the parameters compared to the OLS estimators.

Comparative study of the model

Comparison of estimators of the linear regression model

In this section, ML and OLS estimators are compared in fitting the multivariate linear model with two-parameter symmetric platykurtic error terms. One-step-ahead forecasting is commonly used to compare the performance of different models [31,32]. For

Table 5 Information criteria and model diagnostics of ML estimators of the parameters

Model	Sample size (n)	Information criterion		Model diagnostics
		AIC	BIC	RMSE
Symmetric platykurtic	100	−362.2424	−360.0588	0.16106
	1,000	−3,097.2222	−3,095.2041	0.21222
	3,000	−9,716.3881	−9,714.3821	0.19792
	5,000	−15,524.364	−15,522.361	0.21167
	10,000	−33,554.338	−33,552.336	0.18674
Normal	100	−3.9620	−1.7784	0.96601
	1,000	17.3923	19.4103	1.00722
	3,000	−58.9903	−56.9843	0.98972
	5,000	−73.1305	−71.1269	0.99242
	10,000	20.9714	22.9732	1.00090

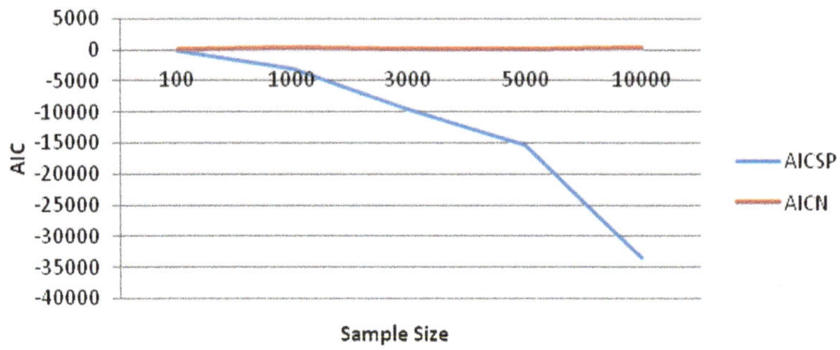

Figure 2 Comparison of the SP linear model versus the N linear model using AIC.

each estimation technique, bias and mean square error (MSE) are calculated for the sample size of 10,000 where

$$\mathrm{MSE}\left(\hat{\beta}\right) = \mathrm{var}\left(\hat{\beta}\right) + \left[\mathrm{bias}\left(\hat{\beta}\right)\right]^2. \tag{45}$$

The computational result is presented in Table 4.

As we expect, the results reported in Table 4 show that ML estimators have both smaller one-step-ahead forecast bias and less MSE than OLS estimators. This reveals that ML estimators exhibit superior performance to OLS estimators. This confirms the fact that deviations from normality cause OLS estimators to be poor estimators.

Comparison of the SP-LRM with the N-LRM

The linear regression model with symmetric platykurtic errors (SP-LRM) and linear regression model with normal errors (N-LRM) were applied to the simulated data sets. This produced parameter estimates $(\hat{\beta}_0, \hat{\beta}_1, \hat{\beta}_2, \hat{\sigma})$ for the model with SP errors and parameter estimates $(\hat{\beta}_0, \hat{\beta}_1, \hat{\beta}_2, \hat{\sigma})$ for the model with N errors. The simulated multivariate data are used to compare the performance of the linear regression model with symmetric platykurtic error terms with that of the linear regression model with normal error terms. In order to determine the best linear model among the fitted ones, we computed the Akaike information criteria (AIC) and Bayesian information criteria (BIC) with model diagnostics root-mean-square error (RMSE). Needless to say, the proposed model would be chosen as the best model according to the minimum of AIC, BIC, and RMSE. The output of the simulation study using various sample sizes is presented in Table 5.

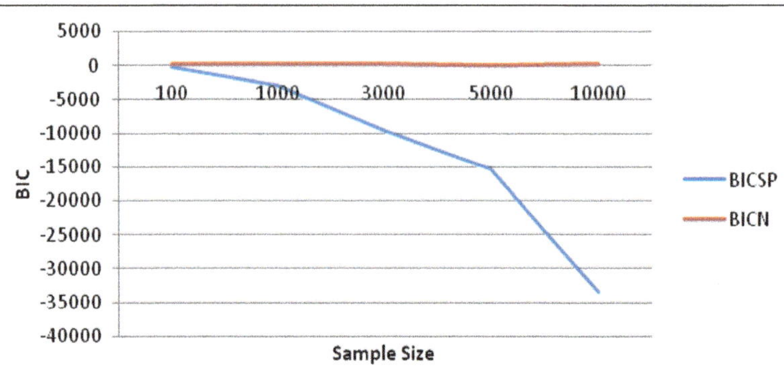

Figure 3 Comparison of the SP linear model versus the N linear model using BIC.

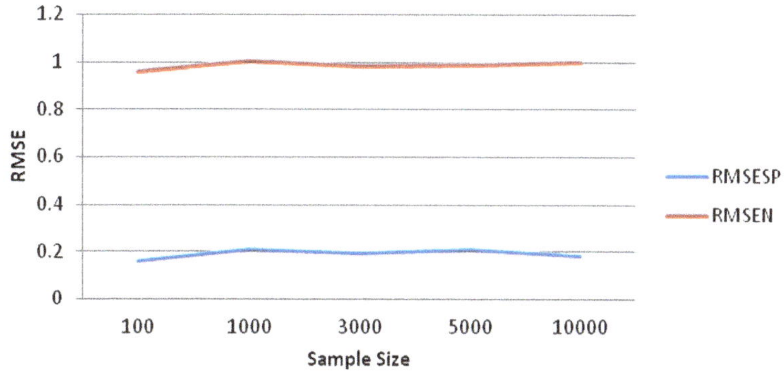

Figure 4 Comparison of the SP linear model versus the N linear model using RMSE.

Models with small values for the criterion are potential candidate models where it can be seen that the SP distribution provides the best fit to the data. That is, both information criteria and model diagnostics indicate that the linear model with symmetric platykurtic distributed error terms consistently performed best across all the sample sizes of the simulation. The fact that SP-LRM is superior to N-LRM is also consistently noticed from Figures 2, 3, and 4.

Application of the model

As an illustration of the proposed methodology, we considered a real data set concerning 202 athletes collected at the Australian Institute of Sport, courtesy of Richard Telford and Ross Cunningham [33]. It is also available within the package *sn* in *R*. The variables examined are body mass index (BMI), red cell count (RCC), white cell count (WCC), and plasma ferritine concentration (PFC). The first is a biometrical variable, while the remaining three concern blood composition. They are summarized in Table 6. We studied the linear dependence of the biometrical variable on the blood composition variables.

We compute the skewness, kurtosis, and Jarque-Bera statistic to test the normality hypothesis for the body mass index. The results are shown in Table 7. Several other statistics could be used to test normality, such as the modified Shapiro-Wilk statistic, Anderson-Darling test, and Kolmogorov-Smirnov test. However, as the Jarque-Bera statistic is one of the most powerful tests of normality, and the results of the other statistics are similar, we report only the results of the Jarque-Bera statistic and its corresponding skewness and kurtosis. The 1% level of significance shown in the table leads to rejection of the null hypothesis of normality for the body mass index.

Table 8 shows the results of fitting multiple regression models with symmetric platykurtic distributed errors and Gaussian errors (SP-LRM and N-LRM) to the Australian Institute of Sport data set using the maximum likelihood method of estimation. The standard errors are the asymptotic standard errors based on the observed information matrix given in (26).

Table 6 Variables of the Australian Institute of Sport data frame with 202 observations

	Variables
Response variables	$Y = BMI$
	$X1 = RCC$
Independent variables	$X2 = WCC$
	$X3 = PFC$

Table 7 Tests for departure from normality of the response variable (BMI)

Response variable	Skewness	Kurtosis	Jarque-Bera statistic
BMI	1.174474881	2.65370385	36.85416*

*$p < 1\%$.

From Table 8, it is observed that the estimates differ slightly between the two models. Following Lachos et al. [34], we propose selecting the best fit between N-LRM and SP-LRM by inspection of information criteria such as AIC and BIC (the preferred model is the one with the smallest value of the criterion). The AIC and BIC values shown at the bottom of Table 8 indicate that the SP-LRM outperforms the N-LRM. Therefore, the fitted model for BMI is

$$\hat{Y} = 4.7799 + 3.1472X_1 + 0.0899X_2 + 0.1988X_3 \tag{46}$$

with the estimated standard errors

$$\begin{aligned} &\text{s.e.}\left(\hat{\beta}_0\right) = 2.3617, \text{s.e.}\left(\hat{\beta}_1\right) = 0.4575, \text{s.e.}\left(\hat{\beta}_2\right) = 0.1018, \\ &\text{s.e.}\left(\hat{\beta}_3\right) = 0.0337, \text{and s.e.}(\hat{\sigma}) = 0.1918. \end{aligned} \tag{47}$$

Summary and conclusions

A multiple linear regression model generalizes the simple linear regression model by allowing the response variable to depend on more than one explanatory variable. In this paper, we have explored the idea of using a symmetric platykurtic distribution for analyzing nonnormal errors in the multivariate linear regression model. The symmetric platykurtic distribution serves as an alternative to the normal distribution with platykurtic nature. The maximum likelihood estimators of the model parameters are derived and we found them feasible. Through simulation studies, the properties of these estimators are studied. Traditional OLS estimation is carried out in parallel and the results are compared. The simulated results reveal that the ML estimators are more efficient than the OLS estimators in terms of the relative efficiency of one-step-ahead forecast mean square error. A comparative study of the developed regression model with the Gaussian model revealed that this model gives good fit to some data sets. The asymptotic properties of the maximum likelihood estimators are studied, and the large sample theory with respect to regression coefficients is also presented. The utility of the proposed model is demonstrated with real-time data. This regression model is much more useful for analyzing data sets arising from agricultural experiments, portfolio management, space experiments, and a wide range other practical problems. The calculations

Table 8 ML estimates of model parameters calculated from the real data set

Parameter	N-LRM		SP-LRM	
	Estimate	SE	Estimate	SE
$\hat{\beta}_0$	4.7799	2.3617	4.7800	2.3382
$\hat{\beta}_1$	3.1472	0.4575	3.1472	0.4530
$\hat{\beta}_2$	0.08992	0.1018	0.0899	0.1008
$\hat{\beta}_3$	0.1988	0.0337	0.1989	0.0333
$\hat{\sigma}$	2.5133	0.1918	2.4883	0.1238
AIC (smaller is better)	407.0530		376.2918	
BIC (smaller is better)	408.2227		378.4526	

in this paper make considerable use of a combination of three popular statistical packages: Mathematica 9.0, Matlab R2012b, and SAS 9.0.

Appendix

Summary of properties of two-parameter symmetric platykurtic distribution

1. Notation — $NS(\mu, \sigma^2)$

2. Parameters — $\mu \in \Re$–Mean (location); $\sigma^2 > 0$–Variance (squared scale)

3. Support — $y \in \Re$

4. PDF — $\frac{\left[2+\left(\frac{y-\mu}{\sigma}\right)^2\right]e^{-\frac{1}{2}\left(\frac{y-\mu}{\sigma}\right)^2}}{3\sigma\sqrt{2\pi}}$

5. CDF — $\frac{1}{\sigma\sqrt{2\pi}}\int_{-\infty}^{y} e^{-\frac{1}{2}\left(\frac{t-\mu}{\sigma}\right)^2} dt - \frac{(y-\mu)e^{-\frac{1}{2}\left(\frac{y-\mu}{\sigma}\right)^2}}{3\sigma\sqrt{2\pi}}$

6. Mean — μ

7. Median — μ

8. Mode — μ

9. Variance — $\frac{5}{3}\sigma^2$

10. Skewness — 0

11. Kurtosis — 2.52

12. Entropy — $\frac{-529}{1500} + ln\left(3\sigma\sqrt{2\pi}\right)$

13. MGF — $e^{\mu t+\frac{t^2\sigma^2}{2}}\left[1+\frac{(\sigma t)^2}{3}\right]$

14. CF — $e^{\mu it-\frac{1}{2}\sigma^2 t^2}\left\{1+\frac{(\sigma it)^2}{3}\right\}$

15. CGF — $\mu t + \frac{1}{2}\sigma^2 t^2 + ln\left[1+\frac{(\sigma t)^2}{3}\right]$

16. Central moments — $\mu_{2n} = \left[\frac{\left(n+\frac{3}{2}\right)\Gamma\left(n+\frac{1}{2}\right)}{3\sqrt{\pi}}\right]2^{n+1}\sigma^{2n}$, even central moments; $\mu_{2n+1} = 0$, odd central moments

17. Fisher information — $\begin{pmatrix} I_\mu X'X & 0 \\ 0 & nI_{\sigma^2} \end{pmatrix}$; $I_\mu = \frac{4.934}{\sigma^2}$, $I_{\sigma^2} = \frac{6.68275}{\sigma^4}$

Competing interests
The authors declare that they have no competing interests.

Acknowledgements
The authors are grateful to the Editor of JUAA, two anonymous referees, and SpringerOpen Copyediting Management for the helpful comments and suggestions on the earlier version of this article. The present version of the paper owes much to their precise and kind remarks.

References

1. Seber, GAF: Linear Regression Analysis. Wiley, New York (1977)
2. Montgomery, DC, Peck, EA, Vining, GG: Introduction to Linear Regression Analysis, 3rd edn. Wiley, New York (2001)

3. Grob, J: Linear Regression. In: Lecture Notes in Statistics, vol. 175. Springer, Berlin (2003)
4. Sengupta, D, Jammalamadaka, SR: Estimation in the linear model. In: Linear Models: An Integrated Approach, pp. 93–131. World Scientific, River Edge (2003)
5. Seber, GAF, Lee, AJ: Linear Regression Analysis, 2nd edn. Wiley, New York (2003)
6. Weisberg, S: Applied Linear Regression, 3rd edn. Wiley, New York (2005)
7. Yan, X, Su, XG: Linear Regression Analysis: Theory and Computing. World Scientific, Hackensack (2009)
8. Srivastava, MS: Methods of Multivariate Statistics. Wiley, New York (2002)
9. Fama, EF: The behaviour of stock market prices. J. Bus. **38**, 34–105 (1965)
10. Sutton, J: Gibrat's legacy. J. Econ. Lit. **35**, 40–59 (1997)
11. Zeckhauser, R, Thompson, M: Linear regression with non-normal error terms. Rev. Econ. Stat. **52**(3), 280–286 (1970)
12. Zellner, A: Bayesian and non-Bayesian analysis of the regression model with multivariate Student-t error terms. J. Am. Stat. Assoc. **71**(354), 400–405 (1976)
13. Sutradhar, BC, Ali, MM: Estimation of the parameters of a regression model with a multivariate t error variable. Commun. Stat. Theory **15**, 429–450 (1986)
14. Tiku, ML, Wong, WK, Bian, G: Estimating parameters in autoregressive models in non-normal situations: symmetric innovations. Commun. Stat. Theory Methods **28**(2), 315–341
15. Tiku, ML, Wong, WK, Vaughan, DC, Bian, G: Time series models with non-normal situations: symmetric innovations. J. Time Ser. Anal. **2**(5), 571–596 (2000)
16. Tiku, ML, Islam, MQ, Selcuk, AS: Non-normal regression, II: symmetric distributions. Commun. Stat. Theory Methods **30**(6), 1021–1045 (2001)
17. Sengupta, D, Jammalamadaka, SR: The symmetric non-normal case. In: Linear Models: An Integrated Approach, pp. 131–133. World Scientific, River Edge (2003)
18. Liu, M, Bozdogan, H: Power exponential multiple regression model selection with ICOMP and genetic algorithms. Working paper, Springer, Tokyo (2004)
19. Wong, WK, Bian, G: Estimation of parameters in autoregressive models with asymmetric innovations. Stat. Prob. Lett. **71**(1), 61–70 (2005)
20. Wong, WK, Bian, G: Robust estimation of multiple regression model with asymmetric innovations and its applicability on asset pricing model. Euras. Rev. Econ. Financ. **1**(4), 7 (2005)
21. Liu, M, Bozdogan, H: Multivariate regression models with power exponential random errors and subset selection using genetic algorithms with information complexity. Eur. J. Pure Appl. Math. **1**(1), 4–37 (2008)
22. Soffritti, G, Galimberti, G: Multivariate linear regression with non-normal errors: a solution based on mixture models. Stat. Comput. **21**(4), 523–536 (2011)
23. Jafari, H, Hashemi, R: Optimal designs in a simple linear regression with skew-normal distribution for error term. J. Appl. Math. **1**(2), 65–68 (2011)
24. Jahan, S, Khan, A: Power of t-test for simple linear regression model with non-normal error distribution: a quantile function distribution approach. J. Sci. Res. **4**(3), 609–622 (2012)
25. Bian, G, McAleer, M, Wong, WK: Robust estimation and forecasting of the capital asset pricing model. Ann. Financ. Econ. (2013). in press
26. Srinivasa Rao, K, Vijay Kumar, CVSR, Lakshmi Narayana, J: On a new symmetrical distribution. J. Indian Soc. Agric. Stat. **50**(1), 95–102 (1997)
27. Seshashayee, M, Srinivas Rao, K, Satyanarayana, CH, Srinivasa Rao, P: Image segmentation based on a finite generalized symmetric platykurtic mixture model with K-means. Int. J. Comput. Sci. Issu. **8**(3), 2 (2011)
28. Marsaglia, G: Evaluating the normal distribution. J. Stat. Softw. **11**(4), 1–7 (2004)
29. Gupta, RC, Gupta, RD: Proportional reversed hazard rate model and its applications. J. Stat. Plann. Inference **137**(11), 3525–3536 (2007)
30. Greene, W: Econometric Analysis, 5th edn. Prentice-Hall, Upper Saddle River (2003)
31. Clements, MP, Hendry, DF: An empirical study of seasonal unit roots in forecasting. Int. J. Forecast. **13**(3), 341–355 (1997)
32. Chiang, TC, Qiao, Z, Wong, WK: New evidence on the relation between return volatility and trading volume. J. Forecast. **29**(5), 502–515 (2010)
33. Ferreira, JTAS, Steel, MF: Bayesian multivariate regression analysis with a new class of skewed distributions. Statistics Research Report, 419. Department of Statistics, University of Warwick (2004)
34. Lachos, VH, Bolfarine, H, Arellano-Valle, RB, Montenegro, LC: Likelihood based inference for multivariate skew-normal regression models. Commun. Stat. Theory Methods **36**, 1769–1786 (2007)

6

Bi-matrix game in bifuzzy environment

Sankar Kumar Roy[1*] and Prasanta Mula[2]

*Correspondence:
sankroy2006@gmail.com
[1] Department of Applied Mathematics with Oceanology and Computer Programming, Vidyasagar University, Midnapore, West Bengal 721102, India
Full list of author information is available at the end of the article

Abstract

In this paper, a new concept of bifuzzy bi-matrix game is introduced where all elements of the payoff matrices are characterized by bifuzzy variables. The uncertainties of entries of payoff matrices (bifuzzy variables) are measured by bifuzzy measure known as Chance measure. Combining the bifuzzy set theory and bi-matrix game theory, the solution concept of bifuzzy bi-matrix game theory is introduced. The quadratic programming problem plays the major role to solve bifuzzy bi-matrix game. In order to show the applicability and feasibility of our proposed method, a real-life bi-matrix game problem is considered and solved.

Keywords: Bifuzzy variable; Expected value operator; Chance measure; Quadratic programming; Nash equilibrium strategy

Introduction

Game theory [1] is a bag of analytical tools designed to help us understand the phenomena that we observed when decision-makers interact. The models of game theory are highly abstract representations of classes of real-life situations. The basic assumption is that the decision-makers pursued well-defined exogenous objectives and take into account their knowledge and expectations of other decision-maker's behavior.

A game is a formal description of a strategic situation. Game theory is applied whenever the actions of several agents are interdependent. These agents may be individuals, groups, firms, or any combination of these. The concept of game theory provides a language to formulate, structure, analyze, and understand the strategic scenario. A payoff is a number, also called utility, that reflects the desirability of an outcome to a player, for whatever reason. When the outcomes are uncertain variables, payoffs are usually weighted with their uncertainty. The expected payoff incorporates the player's attitude toward risk. A game in strategic form, also called normal form, is a compact representation of a game in which players simultaneously choose their strategies. The resulting payoffs are presented in a table with a cell for each strategy combination. In a game in strategic form, a strategy is one of the given possible actions of a player. In an extensive game, a strategy is a complete plan of choices, one for each decision point of the player.

In game theory, the players have to choose appropriate strategy to optimize their gain. There have been different methodologies in bi-matrix game theory to optimize the solution, a well-known technique is Wolfe's modified simplex method. In traditional game theory, the elements of payoff matrix are considered as real numbers, and this basically depends upon the gain of the players. In recent years, attempts have been made to extend the results of crisp game theory to the fuzzy games [2-7]. In credibilistic bi-matrix game [8], the payoff elements are fuzzy variables. To find the optimum solution, applying the fuzzy measurable function known as credibility defined by Liu [9] in 2002. Using credibility theory and game theory, the credibilistic bi-matrix game is converted into crisp quadratic programming problem which depends upon the confidence level of the players' payoff matrix. In bifuzzy matrix game [10], the elements of the payoff matrix are bifuzzy variables, and its uncertainty is measured by bifuzzy measurable function known as Chance measure introduced by Liu [9]. Zhou and Liu [11] introduced the concept of chance distribution for bifuzzy variable, and they also introduced the expected value operator and its linearity on bifuzzy variable.

In many cases, fuzzy variable is not sufficient to handle some type of real-life practical problems on bi-matrix game. As a result, a new type of variable is incorporated which known as bifuzzy variable of bifuzzy set theory. Due to this reason, the elements of payoff matrices (bi-matrix game) are considered as bifuzzy variables in this paper. In fuzzy variables, the bounds are real numbers whereas in bifuzzy variable, the bounds are fuzzy variables and the uncertainty of bifuzzy variable is measured by chance measure. Combining the bifuzzy set theory and bi-matrix game, bifuzzy bi-matrix game has been formulated. The optimum solution of bifuzzy bi-matrix game, depending upon their confidence levels, is discussed in this paper. Finally, a practical example is included to explain the proposed methodology, and then, solution procedure has been discussed.

Preliminaries

Given a universe $\Theta, P(\Theta)$ is the power set of Θ, and a set function *Pos* defines on $P(\Theta)$ which is called a possibility measure if it satisfies the following conditions

1. $\mathrm{Pos}(\phi) = 0$, ϕ is an empty set.
2. $\mathrm{Pos}(\Theta) = 1$.
3. $\mathrm{Pos}(\cup_{i\in I} A_i) = \sup_{i\in I} \mathrm{Pos}(A_i)$ for any subclass $\{A_i | i \in I\}$ of $P(\Theta)$.

The necessity measure of a set A is defined as the impossibility of the opposite set A^c. Then, the necessity measure of A is defined by

$$\mathrm{Nec}(A) = 1 - \mathrm{Pos}(A^c).$$

The triplet $(\Theta, P(\Theta), \mathrm{Pos})$ is usually called a possibility space, which is also called a pattern space. In addition, a self dual set function is called credibility measure [9,12] and defined as follows:

$$\mathrm{Cr}(A) = \frac{1}{2}\left[\mathrm{Pos}(A) + 1 - \mathrm{Pos}(A^c)\right]$$

for any $A \in P(\Theta)$ where A^c is the complement of A.

A fuzzy variable ξ is defined as a function from a credibility space $(\Theta, P(\Theta), \mathrm{Cr})$ to the set of real numbers. Based on credibility measure, the expected value of fuzzy variable ξ is defined as:

$$E[\xi] = \int_0^\infty \mathrm{Cr}\{\xi \geq r\}dr - \int_{-\infty}^0 \mathrm{Cr}\{\xi \leq r\}dr,$$

provided that at least one of the two integrals is finite.

Example 1. *(Liu and Lui [13]) Let $\xi = (a,b,c,d)$ be a trapezoidal fuzzy variable with $c \le a < b \le d$. Then, we have*

$$E[\xi] = \frac{1}{4}(a+b+c+d)$$

Given a credibility space $(\Theta, P(\Theta), \mathrm{Cr})$ which is complete, we obtain the definition of bifuzzy variable as follows

Definition 1. *(Liu [9]) A bifuzzy variable is a function from the credibility space $(\Theta, P(\Theta), Cr)$ to the set of fuzzy variables.*

Definition 2. *(Liu [9]) Let $(\Theta, P(\Theta), Cr)$ be a credibility space. A map $\zeta = (\zeta_1, \zeta_2, ..., \zeta_n)^T : \Theta \to F_\alpha{}^n$ is said to be an n-array bifuzzy vector if for any Borel subset $\mathbf{B}$ of $\overline{\Re}^n$, the function*

$$Cr\{\theta \in \Theta \mid \zeta_\theta(\theta') \in \mathbf{B}\}$$

is treated as measurable with respect to θ, as $n = 1$, where ζ is called a bifuzzy variable.

Definition 3. *(Liu [9]) An n-dimensional bifuzzy vector is a function from the credibility space $(\Theta, P(\Theta), Cr)$ to the set of n-dimensional fuzzy vector.*

Theorem 1. *(Liu [9]) Assume that ζ is a bifuzzy variable. Then for any set $\mathbf{B}$ of $\Re$, we have*

(a) The possibility $Pos\{\zeta(\theta) \in \mathbf{B}\}$ is a fuzzy variable.
(b) The necessity $Nec\{\zeta(\theta) \in \mathbf{B}\}$ is a fuzzy variable.
(c) The credibility $Cr\{\zeta(\theta) \in \mathbf{B}\}$ is a fuzzy variable.

Definition 4. *(Liu [9]) Suppose ζ is a bifuzzy variable, the expected value of ζ is a real number defined as,*

$$E[\zeta] = \int_0^{\infty} Cr\{\theta \in \Theta | E[\zeta(\theta)] \ge r\}dr - \int_{-\infty}^{0} Cr\{\theta \in \Theta | E[\zeta(\theta)] \le r\}dr$$

provided that at least one of the two integrals is finite.

Theorem 2. *(Liu [9]) Let ζ be a bifuzzy variable. If the expected value $E[\zeta(\theta)]$ is finite for each θ, then $E[\zeta(\theta)]$ is a fuzzy variable.*

Definition 5. *(Zhou and Liu [11]) Assume that ζ and ρ are bifuzzy variables with finite expected values. If for each $\theta \in \Theta$, the fuzzy variables $\zeta(\theta)$ and $\rho(\theta)$ are independent, and $E[\zeta(\theta)]$ and $E[\rho(\theta)]$ are independent fuzzy variables, then for any real numbers a and b, we have*

$$E[a\zeta + b\rho] = aE[\zeta] + bE[\rho]$$

Example 2. *Let $\zeta = (\xi_1, \xi_2, \xi_3, \xi_4)$ be a bifuzzy variable with independent trapezoidal fuzzy variable $\xi_i = (a_i, b_i, c_i, d_i)$ for $i = 1, 2, 3, 4$. Then we have*

$$\begin{aligned}
E[\zeta] &= \frac{1}{4}E[\xi_1 + \xi_2 + \xi_3 + \xi_4] = \frac{1}{4}[E[\xi_1] + E[\xi_2] + E[\xi_3] + E[\xi_4]] \\
&= \frac{1}{4}[\frac{1}{4}(a_1 + b_1 + c_1 + d_1) + \frac{1}{4}(a_2 + b_2 + c_2 + d_2) + \frac{1}{4}(a_3 + b_3 + c_3 + d_3) \\
&\quad + \frac{1}{4}(a_4 + b_4 + c_4 + d_4)] \\
&= \frac{1}{16}(a_1 + b_1 + c_1 + d_1 + a_2 + b_2 + c_2 + d_2 + a_3 + b_3 + c_3 + d_3 + a_4 + b_4 + c_4 + d_4)
\end{aligned}$$

where the fuzzy variable $\xi_i = (a_i, b_i, c_i, d_i)$ such that $c_i \le a_i < b_i \le d_i$ for $i = 1, 2, 3, 4$.

Definition 6. *(Liu [9]) Let ζ be a bifuzzy variable and* **B** *be a set of real number $\Re$. Then, the chance measure denoted by Ch of bifuzzy event $\zeta \in$* **B** *is a function from* $(0, 1]$ *to* $[0, 1]$ *defined as,*

$$Ch\{\zeta \in B\}(\alpha) = sup_C r\{A\} \geq \alpha inf_{\theta \in A} Cr\{\zeta(\theta) \in \mathbf{B}\}$$

Theorem 3. *(Zhou and Liu [11]) Let ζ be a bifuzzy variable and* **B** *be a Borel set of $\Re$. For any given $\alpha^* > 0.5$, we write $\delta^* = Ch\{\zeta \in \mathbf{B}\}(\alpha^*)$. Then, we have*

$$Cr\{\theta \in \Theta | Cr\{\zeta(\theta) \in \mathbf{B}\} \geq \delta^*\} \geq \alpha^*.$$

Definition 7. *(Liu [9]) Let ζ be a bifuzzy variable and $\alpha, \delta \in (0, 1]$. Then*

$$\zeta_{sup}(\alpha, \delta) = sup\{w | Ch\{\zeta \geq w\}(\alpha) \geq \delta\}$$

is called the $(\alpha, \delta)-$optimistic value to ζ and

$$\zeta_{inf}(\alpha, \delta) = inf\{w | Ch\{\zeta \leq w\}(\alpha) \geq \delta\}$$

is called the $(\alpha, \delta)-$pessimistic value to ζ

Theorem 4. *(Zhou and Liu [11]) Let ζ be a bifuzzy variable. Assume that $\zeta_{sup}(\alpha, \delta)$ is the $(\alpha, \delta)-$optimistic value of ζ and $\zeta_{inf}(\alpha, \delta)$ is the $(\alpha, \delta)-$pessimistic value to ζ. If $\alpha > 0.5$ and $\delta > 0.5$ then we have,*

$$Ch\{\zeta \leq \zeta_{inf}(\alpha, \delta)\}(\alpha) \geq \delta, \quad Ch\{\zeta \geq \zeta_{sup}(\alpha, \delta)\}(\alpha) \geq \delta$$

Example 3. *Let us consider the trapezoidal fuzzy variable $\xi = (a, b, c, d)$ with the crisp relation $a \leq b < c \leq d$. Then, the fuzzy measures namely possibility, necessity, and credibility are defined by Liu [9] as follows:*

$$\text{Pos}\{\xi \leq 0\} = \begin{cases} 1 & if\, b \leq 0, \\ \frac{a}{a-b} & if\, a \leq 0 \leq b, \\ 0 & if\, otherwise. \end{cases}$$

$$\text{Nec}\{\xi \leq 0\} = \begin{cases} 1 & if\, d \leq 0, \\ \frac{c}{c-d} & if\, c \leq 0 \leq d, \\ 0 & if\, otherwise. \end{cases}$$

$$\text{Cr}\{\xi \leq 0\} = \begin{cases} 1 & if\, d \leq 0, \\ \frac{2c-d}{2(c-d)} & if\, c \leq 0 \leq d, \\ \frac{1}{2} & if\, b \leq 0 \leq c, \\ \frac{a}{2(a-b)} & if\, a \leq 0 \leq b, \\ 0 & if\, otherwise. \end{cases}$$

Theorem 5. *(Liu [9]) Let us consider the trapezoidal fuzzy variable $\xi = (a, b, c, d)$ and confidence level $\alpha \in (0, 1]$, we have*

(1) $\text{Pos}\{\xi \leq 0\} \geq \alpha$, *if and only if* $(1 - \alpha)a + \alpha b \leq 0$
(2) $\text{Nec}\{\xi \leq 0\} \geq \alpha$, *if and only if* $(1 - \alpha)c + \alpha d \leq 0$
(3) when $\alpha \leq \frac{1}{2}$, $\text{Cr}\{\xi \leq 0\} \geq \alpha$, *if and only if* $(1 - 2\alpha)a + 2\alpha b \leq 0$
(4) when $\alpha > \frac{1}{2}$, $\text{Cr}\{\xi \leq 0\} \geq \alpha$, *if and only if* $(2 - 2\alpha)c + (2\alpha - 1)d \leq 0$

Theorem 6. *(Liu [9]) Let $\xi_k = (a_k, b_k, c_k, d_k)$ be trapezoidal fuzzy variables for $k = 1, 2, \cdots, n$ and a function $g(x, \xi)$ can be written as,*

$$g(x,\xi) = h_1(x)\xi_1 + h_2(x)\xi_2 + \cdots + h_n(x)\xi_n + h_0(x)$$

If two functions are defined as $h_k^+(x) = h_k(x) \vee 0$ and $h_k^-(x) = -h_k(x) \wedge 0$ for $k = 1, 2, \cdots, n$, Then, there exists a confidence level $\alpha \in (0, 1]$,

(1) when $\alpha < 1/2$, $Cr\{g(x,\xi) \le 0\} \ge \alpha$, *if and only if,*

$$(1-2\alpha)\sum_{k=1}^{n}\left[a_k h_k^+(x) - d_k h_k^-(x)\right] + 2\alpha\sum_{k=1}^{n}\left[b_k h_k^+(x) - c_k h_k^-(x)\right] + h_0(x) \le 0$$

(2) when $\alpha \ge 1/2$, $Cr\{g(x,\xi) \le 0\} \ge \alpha$, *if and only if,*

$$(2-2\alpha)\sum_{k=1}^{n}\left[c_k h_k^+(x) - b_k h_k^-(x)\right] + (2\alpha-1)\sum_{k=1}^{n}\left[d_k h_k^+(x) - a_k h_k^-(x)\right] + h_0(x) \le 0$$

Theorem 7. *Let $\zeta_k = (\xi_{1k}, \xi_{2k}, \xi_{3k}, \xi_{4k})$ be bifuzzy variables for $k = 1, 2, \cdots, n$ with trapezoidal fuzzy variable $\xi_{mk} = (a_{mk}, b_{mk}, c_{mk}, d_{mk})$ for $m = 1, 2, 3, 4$ and a function $g(x, \zeta)$ can be written as,*

$$g(x,\zeta) = h_1(x)\zeta_1 + h_2(x)\zeta_2 + \cdots + h_n(x)\zeta_n + h_0(x)$$

If two functions are defined as $h_k^+(x) = h_k(x) \vee 0$ and $h_k^-(x) = -h_k(x) \wedge 0$ for $k = 1, 2, \cdots, n$. Then, there exists a confidence level $(\alpha, \delta) \in (0, 1]$,

(1) when $\delta < 1/2$, $Ch\{g(x,\zeta) \le 0\}(\alpha) \ge \delta$, *if and only if,*

$$(1-2\delta)P_{11} + 2\delta P_{12} + h_0(x) \le 0$$

(2) when $\delta \ge 1/2$, $Ch\{g(x,\zeta) \le 0\}(\alpha) \ge \delta$, *if and only if,*

$$(2-2\delta)P_{21} + (2\delta-1)P_{22} + h_0(x) \le 0$$

where P_{11} defined as,

(1) when $\alpha < 1/2$, *if and only if,*

$$P_{11} = (1-2\alpha)\sum_{k=1}^{n}\left[a_{1k} h_k^+(x) - d_{4k} h_k^-(x)\right] + 2\alpha\sum_{k=1}^{n}\left[b_{1k} h_k^+(x) - c_{4k} h_k^-(x)\right]$$

(2) when $\alpha \ge 1/2$, *if and only if,*

$$P_{11} = (2-2\alpha)\sum_{k=1}^{n}\left[c_{1k} h_k^+(x) - b_{4k} h_k^-(x)\right] + (2\alpha-1)\sum_{k=1}^{n}\left[d_{1k} h_k^+(x) - a_{4k} h_k^-(x)\right]$$

where P_{12} defined as,

(1) when $\alpha < 1/2$, *if and only if,*

$$P_{12} = (1-2\alpha)\sum_{k=1}^{n}\left[a_{2k} h_k^+(x) - d_{3k} h_k^-(x)\right] + 2\alpha\sum_{k=1}^{n}\left[b_{2k} h_k^+(x) - c_{3k} h_k^-(x)\right]$$

(2) when $\alpha \geq 1/2$, if and only if,

$$P_{12} = (2-2\alpha)\sum_{k=1}^{n}\left[c_{2k}h_k^+(x) - b_{3k}h_k^-(x)\right] + (2\alpha-1)\sum_{k=1}^{n}\left[d_{2k}h_k^+(x) - a_{3k}h_k^-(x)\right]$$

where P_{21} defined as,

(1) when $\alpha < 1/2$, if and only if,

$$P_{21} = (1-2\alpha)\sum_{k=1}^{n}\left[a_{3k}h_k^+(x) - d_{2k}h_k^-(x)\right] + 2\alpha\sum_{k=1}^{n}\left[b_{3k}h_k^+(x) - c_{2k}h_k^-(x)\right]$$

(2) when $\alpha \geq 1/2$, if and only if,

$$P_{21} = (2-2\alpha)\sum_{k=1}^{n}\left[c_{3k}h_k^+(x) - b_{2k}h_k^-(x)\right] + (2\alpha-1)\sum_{k=1}^{n}\left[d_{3k}h_k^+(x) - a_{2k}h_k^-(x)\right]$$

where P_{22} defined as,

(1) when $\alpha < 1/2$, if and only if,

$$P_{22} = (1-2\alpha)\sum_{k=1}^{n}\left[a_{4k}h_k^+(x) - d_{1k}h_k^-(x)\right] + 2\alpha\sum_{k=1}^{n}\left[b_{4k}h_k^+(x) - c_{1k}h_k^-(x)\right]$$

(2) when $\alpha \geq 1/2$, if and only if,

$$P_{22} = (2-2\alpha)\sum_{k=1}^{n}\left[c_{4k}h_k^+(x) - b_{1k}h_k^-(x)\right] + (2\alpha-1)\sum_{k=1}^{n}\left[d_{4k}h_k^+(x) - a_{1k}h_k^-(x)\right]$$

Proof. Since $h_k^+(x)$ and $h_k^-(x)$ both are nonnegative functions and $h_k(x) = h_k^+(x) - h_k^-(x)$, thus we have

$$\begin{aligned} g(x,\zeta) &= \sum_{k=1}^{n} h_k(x)\zeta_k + h_0(x) \\ &= \sum_{k=1}^{n}\left[h_k^+(x) - h_k^-(x)\right]\zeta_k + h_0(x) \\ &= \sum_{k=1}^{n}\left[h_k^+(x)\zeta_k + h_k^-(x)\zeta_k'\right] + h_0(x) \end{aligned}$$

where $\zeta_k' = (-\xi_{4k}, -\xi_{3k}, -\xi_{2k}, -\xi_{1k})$ for $k = 1, 2, \cdots, n$ with $-\xi_{mk} = (-d_{mk}, -c_{mk}, -b_{mk}, -a_{mk})$ for $m = 1, 2, 3, 4$. In the function $g(x,\zeta)$, the addition and multiplication on bifuzzy variables is determined by quadruple as follows,

$$g(x,\zeta) = \begin{pmatrix} \sum_{k=1}^{n}\left[\xi_{1k}h_k^+(x) - \xi_{4k}h_k^-(x)\right] + h_0(x) \\ \sum_{k=1}^{n}\left[\xi_{2k}h_k^+(x) - \xi_{3k}h_k^-(x)\right] + h_0(x) \\ \sum_{k=1}^{n}\left[\xi_{3k}h_k^+(x) - \xi_{2k}h_k^-(x)\right] + h_0(x) \\ \sum_{k=1}^{n}\left[\xi_{4k}h_k^+(x) - \xi_{1k}h_k^-(x)\right] + h_0(x) \end{pmatrix}^T$$

i.e.,

$$g(x,\zeta)=\begin{pmatrix} \sum_{k=1}^{n}\left[a_{1k}h_k^+(x)-d_{4k}h_k^-(x)\right]+h_0(x)\\ \sum_{k=1}^{n}\left[b_{1k}h_k^+(x)-c_{4k}h_k^-(x)\right]+h_0(x)\\ \sum_{k=1}^{n}\left[c_{1k}h_k^+(x)-b_{4k}h_k^-(x)\right]+h_0(x)\\ \sum_{k=1}^{n}\left[d_{1k}h_k^+(x)-a_{4k}h_k^-(x)\right]+h_0(x)\\ \sum_{k=1}^{n}\left[a_{2k}h_k^+(x)-d_{3k}h_k^-(x)\right]+h_0(x)\\ \sum_{k=1}^{n}\left[b_{2k}h_k^+(x)-c_{3k}h_k^-(x)\right]+h_0(x)\\ \sum_{k=1}^{n}\left[c_{2k}h_k^+(x)-b_{3k}h_k^-(x)\right]+h_0(x)\\ \sum_{k=1}^{n}\left[d_{2k}h_k^+(x)-a_{3k}h_k^-(x)\right]+h_0(x)\\ \sum_{k=1}^{n}\left[a_{3k}h_k^+(x)-d_{2k}h_k^-(x)\right]+h_0(x)\\ \sum_{k=1}^{n}\left[b_{3k}h_k^+(x)-c_{2k}h_k^-(x)\right]+h_0(x)\\ \sum_{k=1}^{n}\left[c_{3k}h_k^+(x)-b_{2k}h_k^-(x)\right]+h_0(x)\\ \sum_{k=1}^{n}\left[d_{3k}h_k^+(x)-a_{2k}h_k^-(x)\right]+h_0(x)\\ \sum_{k=1}^{n}\left[a_{4k}h_k^+(x)-d_{1k}h_k^-(x)\right]+h_0(x)\\ \sum_{k=1}^{n}\left[b_{4k}h_k^+(x)-c_{1k}h_k^-(x)\right]+h_0(x)\\ \sum_{k=1}^{n}\left[c_{4k}h_k^+(x)-b_{1k}h_k^-(x)\right]+h_0(x)\\ \sum_{k=1}^{n}\left[d_{4k}h_k^+(x)-a_{1k}h_k^-(x)\right]+h_0(x) \end{pmatrix}^T$$

Depending upon the confidence level of (δ,α), we can separate the above constraints using **Theorem 6**. □

Bi-matrix game

In this subsection, let us consider the bi-matrix game whose payoff elements are characterized by real numbers. Let $X\equiv\{1,2,\cdots,m\}$ be a set of strategies for the player I and $Y\equiv\{1,2,\cdots,n\}$ be a set of strategies for player II. Let R^n be the n-dimensional Euclidean space and R_+^n be its non-negative orthant. Here, e^T be a vector of element '1' whose dimension is specified as per specific context. Mixed strategies of players I and II are represented by $S^X=\{x\in R_+^m, e^Tx=1\}$ and $S^Y=\{y\in R_+^n, e^Ty=1\}$, respectively.

By considering real numbers a_{ij} and b_{ij} as the expected reward for the players I and II corresponding to the proposed strategies i and j, respectively. Then, the bi-matrix game can be defined as follows:

$$(A,B)=\begin{pmatrix} (a_{11},b_{11}) & (a_{12},b_{12}) & \cdots & (a_{1n},b_{1n})\\ (a_{21},b_{21}) & (a_{22},b_{22}) & \cdots & (a_{2n},b_{2n})\\ \cdot & \cdot & \cdots & \cdot\\ \cdot & \cdot & \cdots & \cdot\\ \cdot & \cdot & \cdots & \cdot\\ (a_{m1},b_{m1}) & (a_{m2},b_{m2}) & \cdots & (a_{mn},b_{mn}) \end{pmatrix}$$

Definition 8. *(Basar and Olsder [14]) A pair $(x^*,y^*)\in S^X\times S^Y$ is said to be Nash equilibrium strategy of the bi-matrix game $BG=(S^X,S^Y,A,B)$ if,*

$$\begin{aligned} &x^TAy^*\le x^{*T}Ay^*,\ \forall x\in S^X\\ \textit{and}\quad &x^{*T}By\le x^{*T}By^*,\ \forall y\in S^Y \end{aligned}$$

Theorem 8. *(Basar and Olsder [14]) Let $BG = (S^X, S^Y, A, B)$ be given bi-matrix game. A necessary and sufficient condition that (x^*, y^*) be an equilibrium strategy of BG is that it is a solution of the following quadratic programming problem,*

$$\begin{aligned} \max\ x^T(A+B)y - v - w& \\ \textit{subject to,}\quad Ay - ve &\le 0 \\ B^T x - we &\le 0 \\ e^T x - 1 &= 0 \\ e^T y - 1 &= 0 \\ v, w &\in \Re \\ x, y &\ge 0 \end{aligned}$$

Lemma 1. *If v^* be the value of the game for player I where v^* is given by*

$$v^* = {x^*}^T A y^* = \max_x \left\{ x^T A y^* : e^T x = 1, x \ge 0 \right\}$$

and w^* is the value of the game for player II where w^* is given by

$$w^* = {x^*}^T B y^* = \max_y \left\{ {x^*}^T B y : e^T y = 1, y \ge 0 \right\}$$

*And if we assume that $x_i{}' = \frac{x_i}{w}$ and $y_j{}' = \frac{y_j}{v}$ (for $i = 1, 2, \cdots, m; j = 1, 2, \cdots, n$). Then, the above **Theorem 8** reduces into the following quadratic programming problem,*

$$\begin{aligned} \max\ e^T x' + e^T y' &+ x'^T (A+B) y' \\ \textit{subject to,}\quad \sum_{j=1}^{n} a_{ij} y'_j &\le 1 \ \textit{for}\ i = 1, 2, \cdots, m \\ \sum_{i=1}^{m} b_{ij} x'_i &\le 1 \ \textit{for}\ j = 1, 2, \cdots, n \\ x', y' &\ge 0 \end{aligned}$$

where (x^*, y^*) and (v^*, w^*) are the Nash equilibrium strategy and outcome of the bi-matrix game BG.

Proof. Since $(x^*, y^*) \in S^X \times S^Y$ be the Nash equilibrium strategy of the bi-matrix game BG if and only if x^* and y^* are simultaneously solutions of the following two problems,

$$\begin{cases} \qquad \max\ x^T A y^* \\ \text{subject to,} \quad e^T x = 1 \\ \qquad x \ge 0. \end{cases} \tag{1}$$

and

$$\begin{cases} \qquad \max\ {x^*}^T B y \\ \text{subject to,} \quad e^T y = 1 \\ \qquad y \ge 0. \end{cases} \tag{2}$$

Here, $(x^*, y^*) \in S^X \times S^Y$ be an optimal strategy of BG that satisfy the conditions of (1) and (2). Also $e^T x' = \frac{1}{w}$ and $e^T y' = \frac{1}{v}$ that is, $e^T x'^* = \frac{1}{w^*}$ and $e^T y'^* = \frac{1}{v^*}$. So the constraints of

Theorem 8 can be written into the following form,

$$Ay^* \le v^*e \qquad \text{because} \quad x^TAy^* \le v^*x^Te = v^* \tag{3}$$

$$\text{Hence,} \quad Ay \le ve \quad \text{or} \quad Ay' \le e \tag{4}$$

$$i.e, \quad \sum_{j=1}^{n} a_{ij}y'_j \le 1 \;\text{ for } i = 1, 2, \cdots, m \tag{5}$$

and

$$x^{*T}B \le w^*e \qquad \text{because} \quad x^{*T}By \le w^*y^Te = w^* \tag{6}$$

$$\text{Hence,} \quad B^Tx \le we \quad \text{or} \quad B^Tx' \le e \tag{7}$$

$$i.e, \quad \sum_{i=1}^{m} b_{ij}x'_i \le 1 \;\text{ for } j = 1, 2, \cdots, n \tag{8}$$

Therefore, the objective function of the **Theorem 8** can be modified into the following form,

$$\max \; x^T(A+B)y \; - \; v - w \tag{9}$$

$$\text{or} \qquad \max \; (\frac{x}{w})^T (A+B)(\frac{y}{v}) \; + \; \frac{1}{w} + \frac{1}{v} \tag{10}$$

$$\text{Hence,} \qquad \max \; x'^T(A+B)y' \; + \; e^Tx' + e^Ty' \tag{11}$$

Now recall the Equations (5), (8), and (11) with $x', y' \ge 0$, the theorem is obvious. □

Bifuzzy bi-matrix game

In the crisp scenario, there exists a beautiful relationship on bi-matrix game. It is therefore natural to ask if something similar holds in bifuzzy scenario as well. In many applied situation, the elements of the bi-matrix game are not fixed. So the elements are imprecise. Therefore, to introduce the bi-matrix game, we considered the elements as bifuzzy variables and then it is measured by Chance. Let the bifuzzy variable ζ_{ij} denote the payoff element that player I gains or the bifuzzy variable ρ_{ij} be the payoff element that player II gains when players I and II play the strategies i and j, respectively. Then, the bifuzzy bi-matrix game can be represented as follows:

$$(\zeta, \rho) = \begin{pmatrix} (\zeta_{11}, \rho_{11}) & (\zeta_{12}, \rho_{12}) & \cdots & (\zeta_{1n}, \rho_{1n}) \\ (\zeta_{21}, \rho_{21}) & (\zeta_{22}, \rho_{22}) & \cdots & (\zeta_{2n}, \rho_{2n}) \\ \cdot & \cdot & \cdots & \cdot \\ \cdot & \cdot & \cdots & \cdot \\ \cdot & \cdot & \cdots & \cdot \\ (\zeta_{m1}, \rho_{m1}) & (\zeta_{m2}, \rho_{m2}) & \cdots & (\zeta_{mn}, \rho_{mn}) \end{pmatrix}$$

Definition 9. *Let ζ_{ij} and ρ_{ij} $(i = 1, 2, ..., m; j = 1, 2, \cdots, n)$ be independent bifuzzy variables. Then, (x^*, y^*) is called an expected bifuzzy Nash equilibrium strategy to the bifuzzy bi-matrix game $\{S^X, S^Y, \zeta, \rho\}$ if,*

$$v^* = E\left[x^{*T}\zeta y^*\right] \ge E\left[x^T\zeta y^*\right], \forall x \in S^X$$
$$w^* = E\left[x^{*T}\rho y^*\right] \ge E\left[x^{*T}\rho y\right], \forall y \in S^Y$$

The pair (v^*, w^*) is called an optimum value of the bifuzzy bi-matrix game.

Definition 10. *Let ζ_{ij} and ρ_{ij} $(i = 1,2,...,m; j = 1,2,\cdots,n)$ are different independent bifuzzy variables, $(\alpha,\delta) \in (0,1]$ and $w,v \in \Re$ be predetermined level of the bifuzzy payoffs. Then (x^*,y^*) is called a (α,δ)-bifuzzy equilibrium strategy to bifuzzy bi-matrix game $\{S^X,S^Y,\zeta,\rho\}$ if,*

$$\max\{v|\ Ch\{x^T\zeta y^* \geq v\}(\alpha) \geq \delta\} \leq \max\{v|\ Ch\{x^{*T}\zeta y^* \geq v\}(\alpha) \geq \delta\}$$

$$\max\{w|\ Ch\{x^{*T}\rho y \geq w\}(\alpha) \geq \delta\} \leq \max\{w|\ Ch\{x^{*T}\rho y^* \geq w\}(\alpha) \geq \delta\}$$

Lemma 2. *Let bifuzzy bi-matrix game $BG = \{S^X,S^Y,\zeta,\rho\}$ and the value of the game v^* for player I is given by*

$$v^* = E[x^{*T}\zeta y^*] = \max_x\{x^T\zeta y^*\ : e^Tx = 1, x \geq 0\}$$

and w^* is the value of the game for player II is given by

$$w^* = E[x^{*T}\rho y^*] = \max_y\{x^{*T}\rho y\ : e^Ty = 1, y \geq 0\}$$

where the max *operator is defined as,*

$$\max_{y\in S^Y}\{x^T\zeta y\} = x^T\zeta y_{\max} \text{ such that } E[x^T\zeta y] \leq E[x^T\zeta y_{\max}]$$

*If we assume that $x'_i = \frac{x_i}{w}$ and $y'_i = \frac{y_i}{v}$ (for $i = 1,2,\cdots,m; j = 1,2,\cdots,n$) and the expected value operator (E) defined on the objective function and the uncertainty of the bifuzzy constraints is measured with bifuzzy measurable function Chance with confidence level $(\alpha,\delta) \in (0,1]$. Then the **Lemma (1)** reduces into the following quadratic programming problem (QPP):*

$$\max\ e^Tx' + e^Ty' + E[x'^T(\zeta+\rho)y']$$

$$\text{subject to, } Ch\{\sum_{j=1}^{n}\zeta_{ij}y'_j \leq 1\}(\alpha) \geq \delta \quad \text{for} \quad i = 1,2,\cdots,m$$

$$Ch\{\sum_{i=1}^{m}\rho_{ij}x'_i \leq 1\}(\alpha) \geq \delta \quad \text{for} \quad j = 1,2,\cdots,n$$

$$y' \geq 0, x' \geq 0$$

where (x^*,y^*) is the equilibrium strategy and (v^*,w^*) is the equilibrium outcome of the bifuzzy bi-matrix game.

Proof. Since $(x^*,y^*) \in S^X \times S^Y$ be an equilibrium strategy of the bifuzzy bi-matrix game if and only if x^* and y^* are simultaneously solutions of the following two problems,

$$\begin{cases} \max\ x^T\zeta y^* \\ \text{subject to, } e^Tx = 1 \\ x \geq 0. \end{cases} \tag{12}$$

and

$$\begin{cases} \max\ x^{*T}\rho y \\ \text{subject to, } e^Ty = 1 \\ y \geq 0. \end{cases} \tag{13}$$

Here, $(x^*,y^*) \in S^X \times S^Y$ be an arbitrary strategy of the bifuzzy bi-matrix game that satisfy the conditions (3.12) and (3.13). Also, $e^Tx' = \frac{1}{w}$ and $e^Ty' = \frac{1}{v}$, that is, $e^Tx'^* = \frac{1}{w^*}$ and $e^Ty'^* = \frac{1}{v^*}$. So the constraints of **Lemma (1)** are the elements of bifuzzy variables for bifuzzy bi-matrix game. So we cannot directly write the inequality for bifuzzy payoff. For the uncertainty of the payoff elements, applying the bifuzzy uncertain measurable function Chance with confidence level $(\alpha, \delta) \in (0, 1]$ as follows:

$$\begin{aligned}
&\text{Ch }\{\zeta\, y^* \leq v^* e\}(\alpha) \geq \delta \text{ because, } \text{Ch}\{x^T \zeta y^* \leq v^* x^T e = v^*\}(\alpha) \geq \delta \\
\text{Hence, } &\text{Ch }\{\zeta\, y \leq ve\}(\alpha) \geq \delta \text{ or, } \text{Ch}\{\zeta\, y' \leq e\}(\alpha) \geq \delta \\
&\text{Ch }\{\sum_{j=1}^{n} \zeta_{ij}\, y'_j \leq 1\}(\alpha) \geq \delta \quad \text{for} \quad i = 1, 2, \cdots, m
\end{aligned}$$

and

$$\begin{aligned}
&\text{Ch }\{x^{*T} \rho \leq w^* e\}(\alpha) \geq \delta \text{ because, } \text{Ch}\{x^{*T} \rho y \leq w^* y^T e = w^*\}(\alpha) \geq \delta \\
\text{Hence, } &\text{Ch }\{\rho^T x \leq we\}(\alpha) \geq \delta \text{ or, } \text{Ch}\{\rho^T\, x' \leq e\}(\alpha) \geq \delta \\
&\text{Ch }\{\sum_{i=1}^{m} \rho_{ij}\, x'_i \leq 1\}(\alpha) \geq \delta \quad \text{for} \quad j = 1, 2, \cdots, n.
\end{aligned}$$

□

Quadratic programming problem

To derive the solution of the bifuzzy bi-matrix game, we are to solve the following bifuzzy quadratic programming problem,

$$\max \quad e^T x' + e^T y' + x'^T (\zeta + \rho) y'$$

$$\begin{aligned}
\text{subject to, } &\sum_{j=1}^{n} \zeta_{ij}\, y'_j \leq 1 \quad \text{for} \quad i = 1, 2, \cdots, m \\
&\sum_{i=1}^{m} \rho_{ij}\, x'_i \leq 1 \quad \text{for} \quad j = 1, 2, \cdots, n \\
&y' \geq 0, x' \geq 0.
\end{aligned}$$

Since the bifuzzy variables are present in the above quadratic programming problem, so traditional method is not applicable. To find the solution of the above problem, we have to introduce the bifuzzy expected operator to the objective function and the constraints, bifuzzy measurable function named as Chance with confidence level $(\alpha, \delta) \in (0, 1]$. So the bifuzzy quadratic programming problem becomes

$$\begin{aligned}
\max \quad & E[e^T x' + e^T y' + x'^T (\zeta + \rho) y'] \\
\text{subject to, } \quad & \text{Ch}\{\sum_{j=1}^{n} \zeta_{ij}\, y'_j \leq 1\}(\alpha) \geq \delta \quad \text{for} \quad i = 1, 2, \cdots, m \\
& \text{Ch}\{\sum_{i=1}^{m} \rho_{ij}\, x'_i \leq 1\}(\alpha) \geq \delta \quad \text{for} \quad j = 1, 2, \cdots, n \\
& y' \geq 0, x' \geq 0.
\end{aligned}$$

With the help of **Theorem 7**, we can easily find that $h^{+}(x) = x', h^{+}(y) = y'$, and $h^{-}(x) = 0, h^{-}(y) = 0$ because $y' \geq 0, x' \geq 0$, and using the bifuzzy set theory, the above problem can be written into crisp quadratic programming problem which depends upon the confidence level $(\alpha, \delta) \in (0, 1]$, as follows:

$$\begin{aligned}\max \frac{1}{16}\sum_{i=1}^{m}\sum_{j=1}^{n}\{&a_{ij11}+a_{ij12}+a_{ij21}+a_{ij22}+b_{ij11}+b_{ij12}+b_{ij21}+b_{ij22}\\&+c_{ij11}+c_{ij12}+c_{ij21}+c_{ij22}+d_{ij11}+d_{ij12}+d_{ij21}+d_{ij22}\\&+x_{ij11}+x_{ij12}+x_{ij21}+x_{ij22}+y_{ij11}+y_{ij12}+y_{ij21}+y_{ij22}\\&+z_{ij11}+z_{ij12}+z_{ij21}+z_{ij22}+t_{ij11}+t_{ij12}+t_{ij21}+t_{ij22}\}x'_i y'_j\\&+\sum_{i=1}^{m}x'_i+\sum_{j=1}^{n}y'_j\end{aligned}$$

subject to,

If $\delta < 0.5, \alpha < 0.5$,

$$\sum_{j=1}^{n}\{(1-2\delta)\{(1-2\alpha)a_{ij11}+2\alpha b_{ij11}\}+2\delta\{(1-2\alpha)a_{ij12}+2\alpha b_{ij12}\}\}\,y'_j \leq 1$$

$$\text{for } i = 1, 2, \cdots, m$$

$$\sum_{i=1}^{m}\{(1-2\delta)\{(1-2\alpha)x_{ij11}+2\alpha y_{ij11}\}+2\delta\{(1-2\alpha)x_{ij12}+2\alpha y_{ij12}\}\}\,x'_i \leq 1$$

$$\text{for } j = 1, 2, \cdots, n$$

$$y' \geq 0, \ x' \geq 0$$

If $\delta < 0.5, \alpha \geq 0.5$,

$$\sum_{j=1}^{n}\{(1-2\delta)\{(2-2\alpha)c_{ij11}+(2\alpha-1)d_{ij11}\}+2\delta\{(2-2\alpha)c_{ij12}+(2\alpha-1)d_{ij12}\}\}\,y'_j \leq 1$$

$$\text{for } i = 1, 2, \cdots, m$$

$$\sum_{i=1}^{m}\{(1-2\delta)\{(2-2\alpha)z_{ij11}+(2\alpha-1)t_{ij11}\}+2\delta\{(2-2\alpha)z_{ij12}+(2\alpha-1)t_{ij12}\}\}\,x'_i \leq 1$$

$$\text{for } j = 1, 2, \cdots, n$$

$$y' \geq 0, \ x' \geq 0$$

If $\delta \geq 0.5, \alpha < 0.5$,

$$\sum_{j=1}^{n}\{(2-2\delta)\{(1-2\alpha)a_{ij21}+2\alpha b_{ij21}\}+(2\delta-1)\{(1-2\alpha)a_{ij22}+2\alpha b_{ij22}\}\}\,y'_j \leq 1$$

$$\text{for } i = 1, 2, \cdots, m$$

$$\sum_{i=1}^{m}\{(2-2\delta)\{(1-2\alpha)x_{ij21}+2\alpha y_{ij21}\}+(2\delta-1)\{(1-2\alpha)x_{ij22}+2\alpha y_{ij22}\}\}\,x'_i \leq 1$$

$$\text{for } j = 1, 2, \cdots, n$$

$$y' \geq 0, \ x' \geq 0$$

If $\delta \geq 0.5, \alpha \geq 0.5$,

$$\sum_{j=1}^{n} \{(2-2\delta)\{(2-2\alpha)c_{ij21}+(2\alpha-1)d_{ij21}\}+(2\delta-1)\{(2-2\alpha)c_{ij22}+ (2\alpha-1)d_{ij22}\}\} y'_j \leq 1 \quad \text{for } i=1,2,\cdots,m$$

$$\sum_{i=1}^{m} \{(2-2\delta)\{(2-2\alpha)z_{ij21}+(2\alpha-1)t_{ij21}\}+(2\delta-1)\{(2-2\alpha)z_{ij22}+ (2\alpha-1)t_{ij22}\}\} x'_i \leq 1 \quad \text{for } j=1,2,\cdots,n$$

$$y' \geq 0, \ x' \geq 0$$

where the bifuzzy variable for player I's payoff is $\zeta = (\zeta_{ija}, \zeta_{ijb}, \zeta_{ijc}, \zeta_{ijc})$ and for player II's payoff is $\rho = (\rho_{ijx}, \rho_{ijy}, \rho_{ijz}, \rho_{ijt})$ with fuzzy variables defined as, $\zeta_{ija} = (a_{ij11}, b_{ij11}, c_{ij11}, d_{ij11})$, $\zeta_{ijb} = (a_{ij12}, b_{ij12}, c_{ij12}, d_{ij12})$, $\zeta_{ijc} = (a_{ij21}, b_{ij21}, c_{ij21}, d_{ij21})$, $\zeta_{ijd} = (a_{ij22}, b_{ij22}, c_{ij22}, d_{ij22})$ and $\rho_{ijx} = (x_{ij11}, y_{ij11}, z_{ij11}, t_{ij11})$, $\rho_{ijy} = (x_{ij12}, y_{ij12}, z_{ij12}, t_{ij12})$, $\rho_{ijz} = (x_{ij21}, y_{ij21}, z_{ij21}, t_{ij21})$, $\rho_{ijt} = (x_{ij22}, y_{ij22}, z_{ij22}, t_{ij22})$
Depending upon confidence level $(\delta, \alpha) \in (0, 1]$, the proposed bifuzzy bi-matrix game becomes a standard quadratic programming problem with crisp constraints and obtained the values of bifuzzy bi-matrix game w and v to the corresponding strategies of the players, using Wolfe's modified simplex method.

A numerical example

A bifuzzy bi-matrix game problem has been formulated for two mobile network companies at West Bengal in India, depending upon their profits earned. It is considered that both the companies have four plans for rate of call charges. The plans are 1 paisa/s, 30 paisa/min, 50 paisa/2 min, and 3 rupees/10 min for their networks. Due to the availability of network and depending upon the plan, assuming that the people talk more in the first plan rather than the second plan. So, the profit occurs more in the first plan as compared to other plan. Suppose for the first company, the profit of the plan-a (1 paisa/s) is about 180 to 190 million per hour, but at night time, the profit arises up to 195 million and during morning it reduces to 175 million. For the plan-b (30 paisa/min), profit is about 156 to 158 million per hour, but at night time, the profit arises up to 160 million and during morning, it reduces to 150 million. For the plan-c (50 paisa/2 min), profit is about 90 to 95 million per hour, but at night time, the profit arises up to 100 million and during morning, it reduces to 80 million. For the plan-d (3 rupees/10 min), profit is about 120 to 130 million per hour, but at night time, the profit arises up to 140 million and during morning, it reduces to 100 million. Similarly, profit of the second company for plan-a (1 paisa/s) is about 165 to 170 million per hour, but at night time, the profit arises up to 175 million and during morning, it reduces to 160 million. For the plan-b (30 paisa/min), profit is about 145 to 148 million per hour, but at night time, the profit arises up to 150 million and during morning, it reduces to 140 million. For the plan-c (50 paisa/2 min), profit is about 75 to 78 million per hour, but at night time, the profit arises up to 82 million and during morning, it reduces to 70 million. For the plan-d (3 rupees/10 min), profit is about 100 to 110 million per hour but at night time, the profit arises up to 130 million and during morning, it reduces to 90 million. For this marketing strategy, the first company losses 2 million during at morning time and gains 2 million during at night time and other times gain varies between −1 and 1 million per hour. Similarly, the second company losses 3 million during morning and gains 2 million during at night time and other times gain varies between -2 and 1 million per hour.

To optimize their profits, a bifuzzy bi-matrix game is formulated whose payoff matrix elements are bifuzzy variables representing the profit of the company per hour. For the first company, bifuzzy variable $\zeta_{ij} = (\xi_{ij}-2, \xi_{ij}-1, \xi_{ij}+1, \xi_{ij}+2)$, and for second company, bifuzzy variable $\rho_{ij} = (\varrho_{ij}-3, \varrho_{ij}-2, \varrho_{ij}+1, \varrho_{ij}+2)$ with their fuzzy variables ξ_{ij} and ϱ_{ij}, respectively. For the first company, fuzzy variables for plan-a is $\xi_{11} = (175, 180, 190, 195)$, plan-b $\xi_{12} = (150, 156, 158, 160)$, plan-c $\xi_{21} = (80, 90, 95, 100)$, and plan-d $\xi_{22} = (100, 120, 130, 140)$. Similarly for the second company, fuzzy variables for the plan-a is $\varrho_{11} = (160, 165, 170, 175)$, plan-b $\varrho_{12} = (140, 145, 148, 150)$, plan-c $\varrho_{21} = (70, 75, 78, 82)$, and plan-d $\varrho_{22} = (90, 100, 110, 130)$. Here, one of the models is shown for particular case, where values of the confidence levels $\delta = \frac{3}{4}$ and $\alpha = \frac{1}{4}$ (choice of decision maker) are considered. Applying the proposed methodology discussed in earlier section, the equivalent crisp quadratic programming problem for the bifuzzy bi-matrix game is constructed as follows:

$$\begin{aligned}
\max\ & 309.625000x_1'y_1' + 264.375000x_1'y_2' + 147.625000x_2'y_1' + 202.125000x_2'y_2' \\
& + x_1' + x_2' + y_1' + y_2' \\
\text{subject to,}\ & 179.000000y_1' + 154.500000y_2' \leq 1 \\
& 86.500000y_1' + 111.500000y_2' \leq 1 \\
& 164.000000x_1' + 74.000000x_2' \leq 1 \\
& 144.000000x_1' + 96.500000x_2' \leq 1 \\
& y_1', y_2', x_1', x_2' \geq 0
\end{aligned}$$

The solution of the above non-linear programming problem is

$$y' = (0.0, 0.006472492) \text{ and } x' = (0.0, 0.002072539)$$

which leads to the following solution of the original problem

$$\begin{aligned}
y^* &= (0, 1) \text{ and } w = 482.499969 \\
x^* &= (0, 1) \text{ and } v = 154.500000.
\end{aligned}$$

Now for different values of confidence levels (Cofd. lev.), the values of the game (Val. game) and the strategies (Str.) of the players I and II are presented in Table 1.

From the above tabulated values, it is observed that the values of the game increase with increasing the values of confidence levels. For particular case, when $\delta = 1, \alpha = 1$, the values of the game are maximum, and consequently, it can be concluded that both the companies have maximum profit. Again for $\delta = 0.1, \alpha = 0.1$, the values of the game are minimum and so the companies have minimum profit. Also, the decision maker (here company) has the right to choose the confidence levels. So for the selection of proper network and time condition, the companies have chosen the confidence levels and obtained the values of the game, i.e., the profits of the companies are maximum and appropriate.

Table 1 Values of conf. levels, values of game, and strategies of the players

Cofd. lev. δ	Cofd. lev. α	Val. game II w	Val. game I v	Str. II $y1$	Str. II $y2$	Str. I $x1$	Str. I $x2$
0.100000	0.100000	102.199997	149.399994	0.000000	1.000000	0.000000	1.000000
0.100000	0.200000	106.200005	150.600006	0.000000	1.000000	0.000000	1.000000
0.100000	0.300000	110.199997	151.800018	0.000000	1.000000	0.000000	1.000000
0.100000	0.400000	114.199997	153.000000	0.000000	1.000000	0.000000	1.000000
0.100000	0.500000	128.199997	156.199982	0.000000	1.000000	0.000000	1.000000
0.100000	0.600000	130.199997	156.599991	0.000000	1.000000	0.000000	1.000000
0.100000	0.700000	132.199997	157.000000	0.000000	1.000000	0.000000	1.000000
0.100000	0.800000	146.478531	157.400009	0.000000	1.000000	0.529248	0.470752
0.100000	0.900000	147.676834	157.800018	0.000000	1.000000	0.531335	0.468665
0.100000	1.000000	148.866653	158.199982	0.000000	1.000000	0.533333	0.466667
0.200000	0.100000	102.400002	149.600006	0.000000	1.000000	0.000000	1.000000
0.200000	0.200000	106.400002	150.800003	0.000000	1.000000	0.000000	1.000000
0.200000	0.300000	110.400002	152.000015	0.000000	1.000000	0.000000	1.000000
0.200000	0.400000	114.400002	153.199997	0.000000	1.000000	0.000000	1.000000
0.200000	0.500000	128.399994	156.399994	0.000000	1.000000	0.000000	1.000000
0.200000	0.600000	130.399994	156.800003	0.000000	1.000000	0.000000	1.000000
0.200000	0.700000	132.399994	157.200012	0.000000	1.000000	0.000000	1.000000
0.200000	0.800000	146.678558	157.600021	0.000000	1.000000	0.529248	0.470752
0.200000	0.900000	147.876846	158.000000	0.000000	1.000000	0.531335	0.468665
0.200000	1.000000	149.066666	158.399979	0.000000	1.000000	0.533333	0.466667
0.300000	0.100000	102.599998	149.800003	0.000000	1.000000	0.000000	1.000000
0.300000	0.200000	106.599998	150.999985	0.000000	1.000000	0.000000	1.000000
0.300000	0.300000	110.600006	152.200012	0.000000	1.000000	0.000000	1.000000
0.300000	0.400000	110.600006	152.200012	0.000000	1.000000	0.000000	1.000000
0.300000	0.500000	128.600006	156.599991	0.000000	1.000000	0.000000	1.000000
0.300000	0.600000	130.600021	157.000000	0.000000	1.000000	0.000000	1.000000
0.300000	0.700000	132.600006	157.400009	0.000000	1.000000	0.000000	1.000000
0.300000	0.800000	146.878555	157.800018	0.000000	1.000000	0.529248	0.470752
0.300000	0.900000	148.076859	158.199982	0.000000	1.000000	0.531335	0.468665
0.300000	1.000000	149.266678	158.600006	0.000000	1.000000	0.533333	0.466667
0.400000	0.100000	102.800003	150.000000	0.000000	1.000000	0.000000	1.000000
0.400000	0.200000	106.799995	151.199997	0.000000	1.000000	0.000000	1.000000
0.400000	0.300000	110.799995	152.399994	0.000000	1.000000	0.000000	1.000000
0.400000	0.400000	114.800011	153.600021	0.000000	1.000000	0.000000	1.000000
0.400000	0.500000	128.800003	156.800003	0.000000	1.000000	0.000000	1.000000
0.400000	0.600000	130.800003	157.200012	0.000000	1.000000	0.000000	1.000000
0.400000	0.700000	132.800003	157.600021	0.000000	1.000000	0.000000	1.000000
0.400000	0.800000	147.078552	158.000000	0.000000	1.000000	0.529248	0.470752
0.400000	0.900000	148.276825	158.399979	0.000000	1.000000	0.531335	0.468665
0.400000	1.000000	149.466675	158.800003	0.000000	1.000000	0.533333	0.466667
0.500000	0.100000	104.999992	152.200012	0.000000	1.000000	0.000000	1.000000
0.500000	0.200000	108.999992	153.399994	0.000000	1.000000	0.000000	1.000000
0.500000	0.300000	112.999985	154.600006	0.000000	1.000000	0.000000	1.000000
0.500000	0.400000	116.999992	155.800003	0.000000	1.000000	0.000000	1.000000
0.500000	0.500000	131.000000	159.000000	0.000000	1.000000	0.000000	1.000000
0.500000	0.600000	133.000000	159.399994	0.000000	1.000000	0.000000	1.000000
0.500000	0.700000	135.000015	159.800018	0.000000	1.000000	0.000000	1.000000
0.500000	0.800000	149.278549	160.200012	0.000000	1.000000	0.529248	0.470752
0.500000	0.900000	150.476837	160.600006	0.000000	1.000000	0.531335	0.468665
0.500000	1.000000	151.666656	161.000015	0.000000	1.000000	0.533333	0.466667

Table 1 Values of conf. levels, values of game, and strategies of the players *(Continued)*

0.600000	0.100000	105.199989	152.399994	0.000000	1.000000	0.000000	1.000000
0.600000	0.200000	109.200005	153.600021	0.000000	1.000000	0.000000	1.000000
0.600000	0.300000	113.199997	154.800003	0.000000	1.000000	0.000000	1.000000
0.600000	0.400000	117.200012	156.000015	0.000000	1.000000	0.000000	1.000000
0.600000	0.500000	131.200012	159.199997	0.000000	1.000000	0.000000	1.000000
0.600000	0.600000	133.199982	159.600006	0.000000	1.000000	0.000000	1.000000
0.600000	0.700000	135.199997	160.000000	0.000000	1.000000	0.000000	1.000000
0.600000	0.800000	149.478561	160.399994	0.000000	1.000000	0.529248	0.470752
0.600000	0.900000	150.676849	160.800003	0.000000	1.000000	0.531335	0.468665
0.600000	1.000000	151.866684	161.199997	0.000000	1.000000	0.533333	0.466667
0.700000	0.100000	151.866684	161.199997	0.000000	1.000000	0.533333	0.466667
0.700000	0.200000	109.399994	153.800018	0.000000	1.000000	0.000000	1.000000
0.700000	0.300000	113.399994	155.000000	0.000000	1.000000	0.000000	1.000000
0.700000	0.400000	117.400002	156.199982	0.000000	1.000000	0.000000	1.000000
0.700000	0.500000	131.400009	159.399994	0.000000	1.000000	0.000000	1.000000
0.700000	0.600000	133.399994	159.800018	0.000000	1.000000	0.000000	1.000000
0.700000	0.700000	135.399994	160.200012	0.000000	1.000000	0.000000	1.000000
0.700000	0.800000	149.678543	160.600006	0.000000	1.000000	0.529248	0.470752
0.700000	0.900000	150.876831	161.000015	0.000000	1.000000	0.531335	0.468665
0.700000	1.000000	152.066666	161.399994	0.000000	1.000000	0.533333	0.466667
0.800000	0.100000	105.599998	152.800018	0.000000	1.000000	0.000000	1.000000
0.800000	0.200000	105.599998	152.800018	0.000000	1.000000	0.000000	1.000000
0.800000	0.300000	113.599998	155.199997	0.000000	1.000000	0.000000	1.000000
0.800000	0.400000	117.599998	156.399994	0.000000	1.000000	0.000000	1.000000
0.800000	0.500000	131.600006	159.600006	0.000000	1.000000	0.000000	1.000000
0.800000	0.600000	133.599991	160.000000	0.000000	1.000000	0.000000	1.000000
0.800000	0.700000	135.600006	160.399994	0.000000	1.000000	0.000000	1.000000
0.800000	0.800000	149.878540	160.800003	0.000000	1.000000	0.529248	0.470752
0.800000	0.900000	151.076843	161.199997	0.000000	1.000000	0.531335	0.468665
0.800000	1.000000	152.266678	161.600006	0.000000	1.000000	0.533333	0.466667
0.900000	0.100000	105.799995	153.000000	0.000000	1.000000	0.000000	1.000000
0.900000	0.200000	109.800003	154.200012	0.000000	1.000000	0.000000	1.000000
0.900000	0.300000	113.800003	155.399994	0.000000	1.000000	0.000000	1.000000
0.900000	0.400000	117.800003	156.599991	0.000000	1.000000	0.000000	1.000000
0.900000	0.500000	131.800003	159.800018	0.000000	1.000000	0.000000	1.000000
0.900000	0.600000	133.800018	160.200012	0.000000	1.000000	0.000000	1.000000
0.900000	0.700000	135.800003	160.600006	0.000000	1.000000	0.000000	1.000000
0.900000	0.800000	150.078537	161.000015	0.000000	1.000000	0.529248	0.470752
0.900000	0.900000	151.276840	161.399994	0.000000	1.000000	0.531335	0.468665
0.900000	1.000000	152.466690	161.800003	0.000000	1.000000	0.533333	0.466667
1.000000	0.100000	106.000000	153.199997	0.000000	1.000000	0.000000	1.000000
1.000000	0.200000	110.000000	154.399994	0.000000	1.000000	0.000000	1.000000
1.000000	0.300000	114.000000	155.600006	0.000000	1.000000	0.000000	1.000000
1.000000	0.400000	118.000000	156.800003	0.000000	1.000000	0.000000	1.000000
1.000000	0.500000	132.000000	160.000000	0.000000	1.000000	0.000000	1.000000
1.000000	0.600000	134.000000	160.399994	0.000000	1.000000	0.000000	1.000000
1.000000	0.700000	136.000000	160.800003	0.000000	1.000000	0.000000	1.000000
1.000000	0.800000	150.278564	161.199997	0.000000	1.000000	0.529248	0.470752
1.000000	0.900000	151.476852	161.600006	0.000000	1.000000	0.531335	0.468665
1.000000	1.000000	152.666641	161.999985	0.000000	1.000000	0.533333	0.466667

Conclusion

The properties of bifuzzy variable and bi-matrix game theory have been investigated in the paper. Using bifuzzy theory, the procedure to solve the bifuzzy bi-matrix game problem has been discussed. A real-life practical problem for bifuzzy bi-matrix game has been included and solved using proposed method. Various solutions have provided for different confidence levels. The players have freedom to choose the appropriate confidence levels, and for inexact confidence levels, the solution of the bifuzzy bi-matrix game problem may not be optimum. For appropriate choice of confidence levels, the players have optimum values with strategies of the bifuzzy bi-matrix game. The present method can be used as a powerful decision-making tool for decision maker to take right decision for competitive systems.

Acknowledgements

The authors are grateful to acknowledge the anonymous referees for their helpful suggestions and comments which have led to an improvement in the quality of the paper.

Author details

[1]Department of Applied Mathematics with Oceanology and Computer Programming, Vidyasagar University, Midnapore, West Bengal 721102, India. [2]ISRO Satellite Centre, Old Airport Road, Vimanapura Post, Bangalore 560017, India.

References

1. Owen, G: Game Theory, Academic, San Diego (1995)
2. Zadeh, LA: Fuzzy sets. Inf. Control. **8**, 338–353 (1965)
3. Bector, CR, Chandra, S: Fuzzy Mathematical Programming and Fuzzy Matrix Games. Springer, Berlin, Heidelberg (2005)
4. Nishizaki, I, Sakawa, M: Fuzzy and Multiobjective Games for Conflict Resolution. Physica, Heidelberg (2001)
5. Campos, L: Fuzzy linear programming models to solve fuzzy matrix game. Fuzzy Sets Syst. **32**, 275–289 (1989)
6. Roy, SK, Mula, P, Mondal, SN: A new solution concept in credibilistic game. CiiT Int. J. Fuzzy Syst. **3**, 115–120 (2011)
7. Roy, SK: Game Theory Under MCDM and Fuzzy Set Theory. VDM (Verlag Dr. Muller), Germany (2010)
8. Mula, P, Roy, SK: Credibilistic bi-matrix game. J. Uncertain Syst. **6**(1), 71–80 (2012)
9. Liu, B: Theory and Practice of Uncertain Programming. Physica, Heidelberg (2002)
10. Mula, P, Roy, SK: Bifuzzy matrix game. CiiT. Int. J. Fuzzy Syst. **5**(1), 10–18 (2013)
11. Zhou, J, Liu, B: Analysis and algorithms of bifuzzy systems. Int. J. Uncertainty, Fuzziness Knowledge-Based Syst. **12**(3), 357–376 (2004)
12. Li, X, Liu, B: A sufficient and necessary condition for credibility measures. Int. Jour. Uncert. Fuzz. Know. Based Sys. **14**(5), 527–535 (2006)
13. Liu, B, Lui, YK: Expected value of fuzzy variable and fuzzy expected value models. IEEE Trans. Fuzzy Syst. **10**(4), 445–450 (2002)
14. Basar, T, Olsder, GJ: Dynamic Noncooperative Game Theory. Second Edition. Academic, San Diego (1995)

7

Polyrectangular theorem and independence of uncertain vectors

Baoding Liu

Correspondence:
liu@tsinghua.edu.cn
Department of Mathematical Sciences, Tsinghua University, Beijing 100084, China

Abstract

Uncertainty theory is a branch of mathematics for modeling human uncertainty. This paper gives a concept of polyrectangle in product uncertainty space and proves a polyrectangular theorem. This paper also proposes a concept of independence of uncertain vectors and discusses some mathematical properties of it. Finally, this paper presents a concept of normal uncertain vector with multivariate normal distribution.

Keywords: Uncertainty theory; Uncertain measure; Uncertain vector; Polyrectangular theorem

Background

Real decisions are usually made in the state of indeterminacy. For modeling indeterminacy, there exist two mathematical systems, one is probability theory [1], and the other is uncertainty theory [2]. Probability is interpreted as frequency, while uncertainty is interpreted as personal belief degree [3].

Uncertainty theory was founded by Liu [2] in 2007 and perfected by Liu [4] in 2009 with the fundamental concept of uncertain measure. After that, Gao [5], Zhang [6], and Peng and Iwamura [7] also discussed some mathematical properties of uncertain measure. In many applications of uncertainty theory, we have to deal with uncertain events which are not necessarily independent. In order to study the dependence of uncertain events, Liu [2] defined a conditional uncertain measure of an event after it has been learned that some other event has occurred. Since then, the tool of uncertain measure was well developed and became a rigorous footstone of uncertainty theory.

As another fundamental concept in uncertainty theory, the uncertain variable was presented by Liu [2] in 2007. In order to describe uncertain variable, Liu [2] also introduced the concept of uncertainty distribution. Later, Peng and Iwamura [8] proved a sufficient and necessary condition for uncertainty distribution. In addition, Liu [9] proposed the concept of inverse uncertainty distribution, and Liu [10] verified a sufficient and necessary condition for it. Furthermore, a measure inversion theorem was given by Liu [9] that may yield uncertain measures of some special events from the uncertainty distribution of the corresponding uncertain variable.

Following the independence of uncertain variables proposed by Liu [4], the operational law was given by Liu [9] for calculating the uncertainty distribution and inverse uncertainty distribution of strictly monotone function of independent uncertain variables.

In order to rank the uncertain variables, Liu [2] proposed the concept of expected value operator. In addition, the linearity of expected value operator was verified by Liu [9]. As an important contribution, Liu and Ha [11] derived a useful formula for calculating the expected values of monotone functions of uncertain variables. Based on the expected value operator, Liu [2] presented the concepts of variance and moments of uncertain variables. Furthermore, some variance formulas were given by Yao [12].

This paper will give a concept of polyrectangle in product uncertainty space and prove a polyrectangular theorem. This paper will also propose a concept of independence of uncertain vectors and discuss some mathematical properties of it. Finally, this paper will present a concept of normal uncertain vector with multivariate normal distribution.

Uncertainty theory

Let Γ be a nonempty set, and $\mathcal{L}$ a σ-algebra over Γ. Each element Λ in $\mathcal{L}$ is called an event. Liu [2] defined an *uncertain measure* by the following axioms:

Axiom 1. *Normality axiom:* $\mathcal{M}\{\Gamma\} = 1$ *for the universal set* Γ*;*

Axiom 2. *Duality axiom:* $\mathcal{M}\{\Lambda\} + \mathcal{M}\{\Lambda^c\} = 1$ *for any event* Λ*;*

Axiom 3. *Subadditivity axiom: For every countable sequence of events* $\Lambda_1, \Lambda_2, \cdots$*, we have*

$$\mathcal{M}\left\{\bigcup_{i=1}^{\infty} \Lambda_i\right\} \le \sum_{i=1}^{\infty} \mathcal{M}\{\Lambda_i\}. \tag{1}$$

The triplet $(\Gamma, \mathcal{L}, \mathcal{M})$ is called an *uncertainty space*. Furthermore, Liu [4] defined a product uncertain measure by the fourth axiom:

Axiom 4. *Product axiom: Let* $(\Gamma_k, \mathcal{L}_k, \mathcal{M}_k)$ *be uncertainty spaces for* $k = 1, 2, \cdots$*. The product uncertain measure* $\mathcal{M}$ *is an uncertain measure satisfying*

$$\mathcal{M}\left\{\prod_{k=1}^{\infty} \Lambda_k\right\} = \bigwedge_{k=1}^{\infty} \mathcal{M}_k\{\Lambda_k\}, \tag{2}$$

where Λ_k *are arbitrarily chosen events from* $\mathcal{L}_k$ *for* $k = 1, 2, \cdots$*, respectively.*

An *uncertain variable* is defined by Liu [2] as a function ξ from an uncertainty space $(\Gamma, \mathcal{L}, \mathcal{M})$ to the set of real numbers, such that $\{\xi \in B\}$ is an event for any Borel set B. In order to describe an uncertain variable in practice, the concept of *uncertainty distribution* is defined by Liu [2] as

$$\Phi(x) = \mathcal{M}\{\xi \le x\}, \quad \forall x \in \Re. \tag{3}$$

Peng and Iwamura [8] verified that a function $\Phi : \Re \to [0, 1]$ is an uncertainty distribution if and only if it is a monotone increasing function except $\Phi(x) \equiv 0$ and $\Phi(x) \equiv 1$. An uncertainty distribution $\Phi(x)$ is said to be regular if it is a continuous and strictly increasing function with respect to x at which $0 < \Phi(x) < 1$, and

$$\lim_{x \to -\infty} \Phi(x) = 0, \quad \lim_{x \to +\infty} \Phi(x) = 1. \tag{4}$$

Let ξ be an uncertain variable with regular uncertainty distribution $\Phi(x)$. Then, the inverse function $\Phi^{-1}(\alpha)$ is called the inverse uncertainty distribution of ξ [9]. It is also verified by Liu [10] that a function $\Phi^{-1}(\alpha):(0,1)\rightarrow\Re$ is an inverse uncertainty distribution if and only if it is a continuous and strictly increasing function with respect to α.

The *expected value* of an uncertain variable ξ is defined by Liu [2] as the following form

$$E[\xi]=\int_0^{+\infty}\mathcal{M}\{\xi\geq x\}dx-\int_{-\infty}^0\mathcal{M}\{\xi\leq x\}dx, \tag{5}$$

provided that at least one of the two integrals is finite. If ξ has an uncertainty distribution Φ, then the expected value may be calculated by

$$E[\xi]=\int_0^{+\infty}(1-\Phi(x))dx-\int_{-\infty}^0\Phi(x)dx. \tag{6}$$

Let ξ be an uncertain variable with finite expected value e. Then, the variance of ξ is defined by Liu [2] as

$$V[\xi]=E[(\xi-e)^2]. \tag{7}$$

If ξ has an uncertainty distribution Φ, then

$$V[\xi]=\int_0^{+\infty}(1-\Phi(e+\sqrt{x})+\Phi(e-\sqrt{x}))dx. \tag{8}$$

Independence is an extremely important concept in uncertainty theory. The uncertain variables $\xi_1,\xi_2,\cdots,\xi_n$ are said to be independent [4] if

$$\mathcal{M}\left\{\bigcap_{i=1}^n(\xi_i\in B_i)\right\}=\bigwedge_{i=1}^n\mathcal{M}\{\xi_i\in B_i\} \tag{9}$$

for any Borel sets $B_1,B_2,\cdots,B_n$ of real numbers. Equivalently, those uncertain variables are independent if and only if

$$\mathcal{M}\left\{\bigcup_{i=1}^n(\xi_i\in B_i)\right\}=\bigvee_{i=1}^n\mathcal{M}\{\xi_i\in B_i\}. \tag{10}$$

Let $\xi_1,\xi_2,\cdots,\xi_n$ be independent uncertain variables with uncertainty distributions $\Phi_1,\Phi_2,\cdots,\Phi_n$, respectively. Assume that the function $f(x_1,x_2,\cdots,x_n)$ is strictly increasing with respect to $x_1,x_2,\cdots,x_m$ and strictly decreasing with respect to $x_{m+1},x_{m+2},\cdots,x_n$. Liu [9] showed that $\xi=f(\xi_1,\xi_2,\cdots,\xi_n)$ has an inverse uncertainty distribution

$$\Psi^{-1}(\alpha)=f(\Phi_1^{-1}(\alpha),\cdots,\Phi_m^{-1}(\alpha),\Phi_{m+1}^{-1}(1-\alpha),\cdots,\Phi_n^{-1}(1-\alpha)). \tag{11}$$

In addition, Liu and Ha [11] proved that the uncertain variable ξ has an expected value

$$E[\xi]=\int_0^1 f(\Phi_1^{-1}(\alpha),\cdots,\Phi_m^{-1}(\alpha),\Phi_{m+1}^{-1}(1-\alpha),\cdots,\Phi_n^{-1}(1-\alpha))d\alpha. \tag{12}$$

Furthermore, Yao [12] proved that ξ has a variance

$$V[\xi]=\int_0^1(f(\Phi_1^{-1}(\alpha),\cdots,\Phi_m^{-1}(\alpha),\Phi_{m+1}^{-1}(1-\alpha),\cdots,\Phi_n^{-1}(1-\alpha))-E[\xi])^2d\alpha.$$

Independence of events

Consider an uncertainty space $(\Gamma, \mathcal{L}, \mathcal{M})$. The events $\Lambda_1, \Lambda_2, \cdots, \Lambda_n$ are said to be independent [9] if

$$\mathcal{M}\left\{\bigcap_{i=1}^{n} \Lambda_i^*\right\} = \bigwedge_{i=1}^{n} \mathcal{M}\{\Lambda_i^*\}, \tag{13}$$

where Λ_i^* are arbitrarily chosen from $\{\Lambda_i, \Lambda_i^c\}$, $i = 1, 2, \cdots, n$, respectively.

Note that (13) represents 2^n equations. For example, when $n = 2$, the four equations are

$$\begin{aligned}
&\mathcal{M}\{\Lambda_1 \cap \Lambda_2\} = \mathcal{M}\{\Lambda_1\} \wedge \mathcal{M}\{\Lambda_2\},\\
&\mathcal{M}\{\Lambda_1^c \cap \Lambda_2\} = \mathcal{M}\{\Lambda_1^c\} \wedge \mathcal{M}\{\Lambda_2\},\\
&\mathcal{M}\{\Lambda_1 \cap \Lambda_2^c\} = \mathcal{M}\{\Lambda_1\} \wedge \mathcal{M}\{\Lambda_2^c\},\\
&\mathcal{M}\{\Lambda_1^c \cap \Lambda_2^c\} = \mathcal{M}\{\Lambda_1^c\} \wedge \mathcal{M}\{\Lambda_2^c\}.
\end{aligned}$$

Theorem 1. *Let $(\Gamma_i, \mathcal{L}_i, \mathcal{M}_i)$ be uncertainty spaces and $\Lambda_i \in \mathcal{L}_i$ for $i = 1, 2, \cdots, n$. Then, the events*

$$\Gamma_1 \times \cdots \times \Gamma_{i-1} \times \Lambda_i \times \Gamma_{i+1} \times \cdots \times \Gamma_n, \quad i = 1, 2, \cdots, n \tag{14}$$

are always independent in the product uncertainty space. That is, the events

$$\Lambda_1, \Lambda_2, \cdots, \Lambda_n \tag{15}$$

are always independent if they are from different uncertainty spaces.

Proof. For simplicity, we only prove the case of $n = 2$. It follows from the product axiom that the product uncertain measure of the intersection is

$$\mathcal{M}\{(\Lambda_1 \times \Gamma_2) \cap (\Gamma_1 \times \Lambda_2)\} = \mathcal{M}\{\Lambda_1 \times \Lambda_2\} = \mathcal{M}_1\{\Lambda_1\} \wedge \mathcal{M}_2\{\Lambda_2\}.$$

Using $\mathcal{M}\{\Lambda_1 \times \Gamma_2\} = \mathcal{M}_1\{\Lambda_1\}$ and $\mathcal{M}\{\Gamma_1 \times \Lambda_2\} = \mathcal{M}_2\{\Lambda_2\}$, we obtain

$$\mathcal{M}\{(\Lambda_1 \times \Gamma_2) \cap (\Gamma_1 \times \Lambda_2)\} = \mathcal{M}\{\Lambda_1 \times \Gamma_2\} \wedge \mathcal{M}\{\Gamma_1 \times \Lambda_2\}.$$

Similarly, we may prove that

$$\begin{aligned}
\mathcal{M}\{(\Lambda_1 \times \Gamma_2)^c \cap (\Gamma_1 \times \Lambda_2)\} &= \mathcal{M}\{(\Lambda_1 \times \Gamma_2)^c\} \wedge \mathcal{M}\{\Gamma_1 \times \Lambda_2\},\\
\mathcal{M}\{(\Lambda_1 \times \Gamma_2) \cap (\Gamma_1 \times \Lambda_2)^c\} &= \mathcal{M}\{\Lambda_1 \times \Gamma_2\} \wedge \mathcal{M}\{(\Gamma_1 \times \Lambda_2)^c\},\\
\mathcal{M}\{(\Lambda_1 \times \Gamma_2)^c \cap (\Gamma_1 \times \Lambda_2)^c\} &= \mathcal{M}\{(\Lambda_1 \times \Gamma_2)^c\} \wedge \mathcal{M}\{(\Gamma_1 \times \Lambda_2)^c\}.
\end{aligned}$$

Thus, $\Lambda_1 \times \Gamma_2$ and $\Gamma_1 \times \Lambda_2$ are independent events. Furthermore, since Λ_1 and Λ_2 are understood as $\Lambda_1 \times \Gamma_2$ and $\Gamma_1 \times \Lambda_2$ in the product uncertainty space, respectively, the two events Λ_1 and Λ_2 are also independent. □

Polyrectangular theorem

Let $(\Gamma_1, \mathcal{L}_1, \mathcal{M}_1)$ and $(\Gamma_2, \mathcal{L}_2, \mathcal{M}_2)$ be two uncertainty spaces, $\Lambda_1 \in \mathcal{L}_1$ and $\Lambda_2 \in \mathcal{L}_2$. It follows from the product axiom that the rectangle $\Lambda_1 \times \Lambda_2$ has an uncertain measure

$$\mathcal{M}\{\Lambda_1 \times \Lambda_2\} = \mathcal{M}_1\{\Lambda_1\} \wedge \mathcal{M}_2\{\Lambda_2\}. \tag{16}$$

This section will extend this result to a more general case.

Definition 1. *Let* $(\Gamma_1, \mathcal{L}_1, \mathcal{M}_1)$ *and* $(\Gamma_2, \mathcal{L}_2, \mathcal{M}_2)$ *be two uncertainty spaces. A set on* $\Gamma_1 \times \Gamma_2$ *is called a polyrectangle if it has the form*

$$\Lambda = \bigcup_{i=1}^{m} (\Lambda_{1i} \times \Lambda_{2i}), \tag{17}$$

where $\Lambda_{1i} \in \mathcal{L}_1$, *and* $\Lambda_{2i} \in \mathcal{L}_2$ *for* $i = 1, 2, \cdots, m$, *and*

$$\Lambda_{11} \subset \Lambda_{12} \subset \cdots \subset \Lambda_{1m}, \tag{18}$$

$$\Lambda_{21} \supset \Lambda_{22} \supset \cdots \supset \Lambda_{2m}. \tag{19}$$

Theorem 2. *Polyrectangular Theorem: Let* $(\Gamma_1, \mathcal{L}_1, \mathcal{M}_1)$ *and* $(\Gamma_2, \mathcal{L}_2, \mathcal{M}_2)$ *be two uncertainty spaces. Then, the polyrectangle*

$$\Lambda = \bigcup_{i=1}^{m} (\Lambda_{1i} \times \Lambda_{2i}) \tag{20}$$

on the product uncertainty space $(\Gamma_1, \mathcal{L}_1, \mathcal{M}_1) \times (\Gamma_2, \mathcal{L}_2, \mathcal{M}_2)$ *has an uncertain measure*

$$\mathcal{M}\{\Lambda\} = \bigvee_{i=1}^{m} \mathcal{M}_1\{\Lambda_{1i}\} \wedge \mathcal{M}_2\{\Lambda_{2i}\}. \tag{21}$$

Proof. It is clear that the maximum rectangle in the polyrectangle Λ is one of $\Lambda_{1i} \times \Lambda_{2i}$, $i = 1, 2, \cdots, n$. Denote the maximum rectangle by $\Lambda_{1k} \times \Lambda_{2k}$.

Case I: If

$$\mathcal{M}\{\Lambda_{1k} \times \Lambda_{2k}\} = \mathcal{M}_1\{\Lambda_{1k}\};$$

then, the maximum rectangle in Λ^c is $\Lambda^c_{1k} \times \Lambda^c_{2,k+1}$, and

$$\mathcal{M}\{\Lambda^c_{1k} \times \Lambda^c_{2,k+1}\} = \mathcal{M}_1\{\Lambda^c_{1k}\} = 1 - \mathcal{M}_1\{\Lambda_{1k}\}.$$

Thus,

$$\mathcal{M}\{\Lambda_{1k} \times \Lambda_{2k}\} + \mathcal{M}\{\Lambda^c_{1k} \times \Lambda^c_{2,k+1}\} = 1.$$

Case II: If

$$\mathcal{M}\{\Lambda_{1k} \times \Lambda_{2k}\} = \mathcal{M}_2\{\Lambda_{2k}\},$$

then the maximum rectangle in Λ^c is $\Lambda^c_{1,k-1} \times \Lambda^c_{2k}$, and

$$\mathcal{M}\{\Lambda^c_{1,k-1} \times \Lambda^c_{2k}\} = \mathcal{M}_2\{\Lambda^c_{2k}\} = 1 - \mathcal{M}_2\{\Lambda_{2k}\}.$$

Thus,

$$\mathcal{M}\{\Lambda_{1k} \times \Lambda_{2k}\} + \mathcal{M}\{\Lambda^c_{1,k-1} \times \Lambda^c_{2k}\} = 1.$$

□

No matter what case happens, the sum of the uncertain measures of the maximum rectangles in Λ and Λ^c is always 1. It follows from the product axiom that (21) holds.

Independence of uncertain vectors

A k-dimensional uncertain vector is a function $\boldsymbol{\xi}$ from an uncertainty space $(\Gamma, \mathcal{L}, \mathcal{M})$ to the set of k-dimensional real vectors such that $\{\boldsymbol{\xi} \in B\}$ is an event for any k-dimensional

Borel set B [2]. It has been proved that $(\xi_1, \xi_2, \cdots, \xi_k)$ is an uncertain vector if and only if $\xi_1, \xi_2, \cdots, \xi_k$ are uncertain variables. The joint uncertainty distribution of an uncertain vector $(\xi_1, \xi_2, \cdots, \xi_k)$ is defined by Liu [2] as

$$\Phi(x_1, x_2, \cdots, x_k) = \mathcal{M}\{\xi_1 \le x_1, \xi_2 \le x_2, \cdots, \xi_k \le x_k\} \tag{22}$$

for any real numbers $x_1, x_2, \cdots, x_k$.

Theorem 3. *Let $\xi_1, \xi_2, \cdots, \xi_k$ be independent uncertain variables with uncertainty distributions $\Phi_1, \Phi_2, \cdots, \Phi_k$, respectively. Then, the uncertain vector $(\xi_1, \xi_2, \cdots, \xi_k)$ has a joint uncertainty distribution*

$$\Phi(x_1, x_2, \cdots, x_k) = \Phi_1(x_1) \wedge \Phi_2(x_2) \wedge \cdots \wedge \Phi_k(x_k) \tag{23}$$

for any real numbers $x_1, x_2, \cdots, x_k$.

Proof. Since $\xi_1, \xi_2, \cdots, \xi_k$ are independent uncertain variables, we have

$$\Phi(x_1, x_2, \cdots, x_k) = \mathcal{M}\left\{\bigcap_{i=1}^{k}(\xi_i \le x_i)\right\} = \bigwedge_{i=1}^{k} \mathcal{M}\{\xi_i \le x_i\} = \bigwedge_{i=1}^{k} \Phi_i(x_i)$$

for any real numbers $x_1, x_2, \cdots, x_k$. The theorem is proved. □

Definition 2. *The k-dimensional uncertain vectors $\boldsymbol{\xi}_1, \boldsymbol{\xi}_2, \cdots, \boldsymbol{\xi}_n$ are said to be independent if for any k-dimensional Borel sets $B_1, B_2, \cdots, B_n$, we have*

$$\mathcal{M}\left\{\bigcap_{i=1}^{n}(\boldsymbol{\xi}_i \in B_i)\right\} = \bigwedge_{i=1}^{n} \mathcal{M}\{\boldsymbol{\xi}_i \in B_i\}. \tag{24}$$

Theorem 4. *The k-dimensional uncertain vectors $\boldsymbol{\xi}_1, \boldsymbol{\xi}_2, \cdots, \boldsymbol{\xi}_n$ are independent if and only if*

$$\mathcal{M}\left\{\bigcup_{i=1}^{n}(\boldsymbol{\xi}_i \in B_i)\right\} = \bigvee_{i=1}^{n} \mathcal{M}\{\boldsymbol{\xi}_i \in B_i\} \tag{25}$$

for any k-dimensional Borel sets $B_1, B_2, \cdots, B_n$.

Proof. It follows from the duality of uncertain measure that $\boldsymbol{\xi}_1, \boldsymbol{\xi}_2, \cdots, \boldsymbol{\xi}_n$ are independent if and only if

$$\mathcal{M}\left\{\bigcup_{i=1}^{n}(\boldsymbol{\xi}_i \in B_i)\right\} = 1 - \mathcal{M}\left\{\bigcap_{i=1}^{n}(\boldsymbol{\xi}_i \in B_i^c)\right\}$$

$$= 1 - \bigwedge_{i=1}^{n} \mathcal{M}\{\boldsymbol{\xi}_i \in B_i^c\} = \bigvee_{i=1}^{n} \mathcal{M}\left\{\boldsymbol{\xi}_i \in B_i\right\}.$$

The theorem is thus proved. □

Theorem 5. *Let $\boldsymbol{\xi}_1, \boldsymbol{\xi}_2, \cdots, \boldsymbol{\xi}_n$ be independent uncertain vectors, and $f_1, f_2, \cdots, f_n$ vector-valued measurable functions. Then, $f_1(\boldsymbol{\xi}_1), f_2(\boldsymbol{\xi}_2), \cdots, f_n(\boldsymbol{\xi}_n)$ are independent uncertain vectors.*

Proof. For any Borel sets $B_1, B_2, \cdots, B_n$, it follows from the definition of independence that

$$\mathcal{M}\left\{\bigcap_{i=1}^{n}(f_i(\xi_i) \in B_i)\right\} = \mathcal{M}\left\{\bigcap_{i=1}^{n}(\xi_i \in f_i^{-1}(B_i))\right\}$$

$$= \bigwedge_{i=1}^{n} \mathcal{M}\{\xi_i \in f_i^{-1}(B_i)\} = \bigwedge_{i=1}^{n} \mathcal{M}\{f_i(\xi_i) \in B_i\}.$$

Thus, $f_1(\xi_1), f_2(\xi_2), \cdots, f_n(\xi_n)$ are independent uncertain variables. □

Multivariate normal distribution

An uncertain variable ξ is called normal if it has a normal uncertainty distribution

$$\Phi(x) = \left(1 + \exp\left(\frac{\pi(e - x)}{\sqrt{3}\sigma}\right)\right)^{-1}, \quad x \in \Re \tag{26}$$

denoted by $\mathcal{N}(e, \sigma)$, where e is the expected value, and $\sigma > 0$ is the standard variance.

Definition 3. *Let $\tau_1, \tau_2, \cdots, \tau_m$ be independent normal uncertain variables with expected value 0 and variance 1. Then, the uncertain vector*

$$\boldsymbol{\tau} = (\tau_1, \tau_2, \cdots, \tau_m) \tag{27}$$

is said to have a multivariate standard normal distribution.

It is easy to verify that a standard normal uncertain vector $(\tau_1, \tau_2, \cdots, \tau_m)$ has a joint uncertainty distribution

$$\Phi(x_1, x_2, \cdots, x_m) = \left(1 + \exp\left(-\frac{\pi(x_1 \wedge x_2 \wedge \cdots \wedge x_m)}{\sqrt{3}}\right)\right)^{-1} \tag{28}$$

for any real numbers $x_1, x_2, \cdots, x_m$. It is also easy to show that

$$\lim_{x_i \to -\infty} \Phi(x_1, x_2, \cdots, x_m) = 0, \text{ for each } i, \tag{29}$$

$$\lim_{(x_1, x_2, \cdots, x_m) \to +\infty} \Phi(x_1, x_2, \cdots, x_m) = 1. \tag{30}$$

Furthermore, the limit

$$\lim_{(x_1, \cdots, x_{i-1}, x_{i+1}, \cdots, x_m) \to +\infty} \Phi(x_1, x_2, \cdots, x_m) \tag{31}$$

is a standard normal distribution with respect to x_i.

Definition 4. *Let $(\tau_1, \tau_2, \cdots, \tau_m)$ be a standard normal uncertain vector, and let e_i, σ_{ij}, $i = 1, 2, \cdots, k, j = 1, 2, \cdots, m$ be real numbers. Define*

$$\xi_i = e_i + \sum_{j=1}^{m} \sigma_{ij}\tau_j \tag{32}$$

for $i = 1, 2, \cdots, k$. Then, the uncertain vector $(\xi_1, \xi_2, \cdots, \xi_k)$ is said to have a multivariate normal distribution.

That is, an uncertain vector $\boldsymbol{\xi}$ has a multivariate normal distribution if it can be represented in the form

$$\boldsymbol{\xi} = \boldsymbol{e} + \boldsymbol{\sigma}\boldsymbol{\tau} \tag{33}$$

for some real vector $\boldsymbol{e}$ and some real matrix $\boldsymbol{\sigma}$, where $\boldsymbol{\tau}$ is a standard normal uncertain vector. Note that $\boldsymbol{\xi}, \boldsymbol{e}$ and $\boldsymbol{\tau}$ are understood as column vectors. Please also note that for every index i, the component ξ_i is a normal uncertain variable with expected value e_i and standard variance

$$\sum_{j=1}^{m} |\sigma_{ij}|. \tag{34}$$

Theorem 6. *Assume $\boldsymbol{\xi}$ is a normal uncertain vector, c is a real vector, and D is a real matrix. Then,*

$$\eta = c + D\boldsymbol{\xi} \tag{35}$$

is another normal uncertain vector.

Proof. Since $\boldsymbol{\xi}$ is a normal uncertain vector, there exist a standard normal uncertain vector $\boldsymbol{\tau}$, a real vector $\boldsymbol{e}$, and a real matrix $\boldsymbol{\sigma}$, such that $\boldsymbol{\xi} = \boldsymbol{e} + \boldsymbol{\sigma}\boldsymbol{\tau}$. It follows that

$$\eta = c + D\boldsymbol{\xi} = c + D(\boldsymbol{e} + \boldsymbol{\sigma}\boldsymbol{\tau}) = (c + D\boldsymbol{e}) + (D\boldsymbol{\sigma})\boldsymbol{\tau}.$$

Hence, η is a normal uncertain vector. □

Conclusions

This paper gave a concept of polyrectangle in product uncertainty space and proved a polyrectangular theorem. This paper also proposed a concept of independence of uncertain vectors and discussed some mathematical properties of it. Finally, this paper presented a concept of normal uncertain vector.

Acknowledgements

This work was supported by the National Natural Science Foundation of China (grants no.91224008 and no.61273044).

References

1. Kolmogorov, AN: Grundbegriffe der Wahrscheinlichkeitsrechnung. Springer, Berlin (1933)
2. Liu, B: Uncertainty Theory, 2nd ed. Springer, Berlin (2007)
3. Liu, B: Why is there a need for uncertainty theory? J. Uncertain Syst. **6**(1), 3–10 (2012)
4. Liu, B: Some research problems in uncertainty theory. J. Uncertain Syst. **3**(1), 3–10 (2009)
5. Gao, X: Some properties of continuous uncertain measure. Int. J. Uncertain. Fuzz. **17**(3), 419–426 (2009)
6. Zhang, ZM: Some discussions on uncertain measure. Fuzzy Optimization Decis. Making. **10**(1), 31–43 (2011)
7. Peng, ZX, Iwamura, K: Some properties of product uncertain measure. J. Uncertain Syst. **6**(4), 263–269 (2012)
8. Peng, ZX, Iwamura, K: A sufficient and necessary condition of uncertainty distribution. J. Interdiscip. Math. **13**(3), 277–285 (2010)
9. Liu, B: Uncertainty Theory: A Branch of Mathematics for Modeling Human Uncertainty. Springer, Berlin (2010)
10. Liu, B: Toward uncertain finance theory. J. Uncertainty Anal. Appl. **1**(1) (2013)
11. Liu, YH, Ha, MH: Expected value of function of uncertain variables. J. Uncertain Syst. **4**(3), 181–186 (2010)
12. Yao, K: A formula to calculate the variance of uncertain variable (2013). http://orsc.edu.cn/online/130831.pdf

8

A type of uncertain differential equations with analytic solution

Kai Yao

Correspondence: yaokai@ucas.ac.cn
School of Management, University of Chinese Academy of Sciences, Beijing 100190, China

Abstract

Uncertain differential equation is a type of differential equation driven by the Liu process. So far, an analytic solution of linear uncertain differential equation has been obtained. This paper aims at proposing a method to solve a type of nonlinear uncertain differential equation.

Keywords: Uncertainty theory; Uncertain differential equation; Liu process

Introduction

Uncertainty theory, as a branch of axiomatic mathematics dealing with human's belief degree, was founded by Liu [1] in 2007 and refined by Liu [2] in 2010. During the past 6 years, many researchers have contributed in this area. For example, Peng and Iwamura [3] gave a sufficient condition for the uncertainty distribution of an uncertain variable. Liu and Ha [4] gave a formula to calculate the expected value of a function of multiple uncertain variables. Chen and Dai [5] showed that a normal uncertain variable has the maximum entropy given the expected value and variance. Especially, Liu [6] proposed uncertain programming as a type of mathematical programming involving uncertain variables.

In order to model the evolution of an uncertain phenomenon, Liu [7] proposed a concept of uncertain process. Meanwhile, Liu [7] gave an uncertain renewal process as an example. After that, Liu [2] proposed an uncertain renewal reward process, and Yao and Li [8] proposed an uncertain alternating renewal process. In addition, Zhang et al. [9] proposed an uncertain delayed renewal process. In 2009, Liu [10] mathematically defined a type of uncertain process, named canonical Liu process, which has independent and stationary uncertain normal increments and of which almost all the sample paths are Lipschitz continuous. In addition, Liu [11] proved the extreme value theorems for an independent increment uncertain process.

In 2009, Liu [10] founded an uncertain calculus to deal with the integral and differential of an uncertain process with respect to Liu process, which are called Liu integral and Liu differential afterwards. Then Liu and Yao [12] studied an uncertain integral with respect to multiple Liu processes. After that, Chen and Ralescu [13] proposed an uncertain integral with respect to the general Liu process. Inspired by the Liu integral, Yao [14] proposed an uncertain calculus with respect to an uncertain renewal process.

Uncertain differential equation was first proposed by Liu [7] as a type of differential equation driven by Liu process. Then Chen and Liu [15] gave a sufficient condition for an uncertain differential equation having a unique solution. After that, Gao [16] provided an existence and unique theorem under weaker conditions. In 2013, Yao et al. [17] gave a sufficient condition for it being stable. After that, Sheng [18] studied the stability in pth moment.

Nowadays, uncertain differential equation has been applied to many areas especially in finance. In 2009, Liu [10] assumed that the stock price follows a geometric Liu process in the short run and proposed a stock model in an uncertain environment. After that, Chen [19] derived its American option formulae. In 2011, Peng and Yao [20] proposed another model to describe the stock price displaying a mean-reverting property in the long run via uncertain differential equation. In 2013, Chen et al. [21] proposed a stock model with periodic dividends. Assuming that the interest rate follows an uncertain differential equation, Chen and Gao [22] presented an uncertain interest rate model and calculated its zero-coupon bond. Besides, Liu et al. [23] proposed an uncertain currency model via uncertain differential equation. For more recent developments in uncertain finance, please refer to Liu [24].

In 2010, Chen and Liu [15] gave an analytic solution for the linear uncertain differential equation. Then Liu [25] proposed a method to solve a special type of nonlinear uncertain differential equation. In addition, some numerical methods were designed to calculate the uncertainty distributions of the solution and the extreme values of the solution by Yao and Chen [26] and Yao [27]. In this paper, we will give an analytic method to solve another special type of uncertain differential equation. The rest of this paper is organized as follows. Some concepts and theorems about uncertainty theory, uncertain calculus, and uncertain differential equation will be introduced in the 'Uncertainty theory' section, 'Uncertain calculus' section, and 'Uncertain differential equation' section, respectively. Then a special type of uncertain differential equation is solved in the 'An analytic method' section. Finally, some remarks are made in the 'Conclusions' section.

Uncertainty theory

This section will briefly introduce some basic concepts in uncertainty theory, including uncertainty space, uncertain variable, independence, and the operational law.

Definition 1. (Liu [1]) Let $\mathcal{L}$ be a σ-algebra on a nonempty set Γ. A set function $\mathcal{M}$: $\mathcal{L} \rightarrow [0, 1]$ is called an uncertain measure if it satisfies the following axioms:

Axiom 1. (Normality Axiom) $\mathcal{M}\{\Gamma\} = 1$ for the universal set Γ.

Axiom 2. (Duality Axiom) $\mathcal{M}\{\Lambda\} + \mathcal{M}\{\Lambda^c\} = 1$ for any event Λ.

Axiom 3. (Subadditivity Axiom) For every countable sequence of events $\Lambda_1, \Lambda_2, \ldots$, we have

$$\mathcal{M}\left\{\bigcup_{i=1}^{\infty} \Lambda_i\right\} \leq \sum_{i=1}^{\infty} \mathcal{M}\{\Lambda_i\}.$$

Besides, the product uncertain measure on the product σ-algebre $\mathcal{L}$ is defined by Liu [10] as follows:

Axiom 4. (Product Axiom) Let $(\Gamma_k, \mathcal{L}_k, \mathcal{M}_k)$ be uncertainty spaces for $k = 1, 2, \ldots$ Then the product uncertain measure $\mathcal{M}$ on the product σ-algebra satisfies

$$\mathcal{M}\left\{\prod_{i=1}^{\infty} \Lambda_k\right\} = \bigwedge_{k=1}^{\infty} \mathcal{M}_k\{\Lambda_k\}$$

where Λ_k are arbitrarily chosen events from $\mathcal{L}_k$ for $k = 1, 2, \ldots$, respectively.

In order to represent the quantities with uncertainty, an uncertain variable was proposed as a real valued function on an uncertainty space.

Definition 2. (Liu [1]) An uncertain variable ξ is a measurable function from an uncertainty space $(\Gamma, \mathcal{L}, \mathcal{M})$ to the set of real numbers, i.e., for any Borel set B, the set $\{\gamma \mid \xi(\gamma) \in B\}$ is an event in $\mathcal{L}$.

The uncertainty distribution Φ of an uncertain variable is defined by $\Phi(x) = \mathcal{M}\{\xi \leq x\}$ for any real number x. The expected value of an uncertain variable ξ is

$$E[\xi] = \int_0^{+\infty} \mathcal{M}\{\xi \geq r\}dr - \int_{-\infty}^{0} \mathcal{M}\{\xi \leq r\}dr$$

provided that at least one of the two integrals is finite, and the variance of ξ is

$$V[\xi] = E\left[(\xi - E[\xi])^2\right].$$

An uncertain variable ξ is said to be normal if it has an uncertainty distribution

$$\Phi(x) = \left(1 + \exp\left(\frac{\pi(e-x)}{\sqrt{3}\sigma}\right)\right)^{-1}, \quad x \in \Re$$

where e and σ^2 are the expected value and variance of ξ, respectively.

Definition 3. (Liu [10]) The uncertain variables $\xi_1, \xi_2, \ldots, \xi_n$ are said to be independent if

$$\mathcal{M}\left\{\bigcap_{i=1}^{n} (\xi_i \in B_i)\right\} = \bigwedge_{i=1}^{n} \mathcal{M}\{\xi_i \in B_i\}$$

for any Borel sets $B_1, B_2, \ldots, B_n$.

Since the definition of independence for uncertain variables is quite different from that for random variables, the operational law of uncertain variables is also different.

Theorem 1. (Liu [2]) *Let $\xi_1, \xi_2, \ldots, \xi_n$ be independent uncertain variables with uncertainty distributions $\Phi_1, \Phi_2, \ldots, \Phi_n$, respectively. If $f(x_1, x_2, \ldots, x_n)$ is strictly increasing with respect to $x_1, x_2, \ldots, x_m$ and strictly decreasing with respect to $x_{m+1}, x_{m+2}, \ldots, x_n$, then $\xi = f(\xi_1, \xi_2, \ldots, \xi_n)$ is an uncertain variable with an inverse uncertainty distribution*

$$\Phi^{-1}(r) = f\left(\Phi_1^{-1}(r), \ldots, \Phi_m^{-1}(r), \Phi_{m+1}^{-1}(1-r), \ldots, \Phi_n^{-1}(1-r)\right).$$

Uncertain calculus

An uncertain process is essentially a sequence of uncertain variables indexed by time or space. As one of the most important types of uncertain processes, the canonical Liu process is defined as follows:

Definition 4. (Liu [10]) An uncertain process C_t is called a canonical Liu process if:

1. $C_0 = 0$ and almost all sample paths are Lipschitz continuous.
2. C_t has stationary and independent increments.
3. Every increment $C_{s+t} - C_s$ is a normal uncertain variable $\mathcal{N}(0, t)$ with expected value 0 and variance t^2 whose uncertainty distribution is

$$\Phi_t(x) = \left(1 + \exp\left(-\frac{\pi x}{\sqrt{3}t}\right)\right)^{-1}, \quad x \in \Re.$$

Based on the canonical Liu process, Liu [10] proposed an uncertain integral of an uncertain process with respect to the canonical Liu process and thus founded a theory of uncertain calculus.

Definition 5. (Liu [10]) Let X_t be an uncertain process and C_t be a canonical Liu process. For any partition of closed interval $[a, b]$ with $a = t_1 < t_2 < \cdots < t_{k+1} = b$, the mesh is written as

$$\Delta = \max_{1 \leq i \leq k} |t_{i+1} - t_i|.$$

Then the Liu integral of X_t is defined by

$$\int_a^b X_t \mathrm{d}C_t = \lim_{\Delta \to 0} \sum_{i=1}^{k} X_{t_i} \cdot (C_{t_{i+1}} - C_{t_i})$$

provided that the limit exists almost surely and is finite.

For example, the Liu integral of an integrable real function $f(t)$ is

$$\int_0^t f(s)\mathrm{d}C_s \sim \left(0, \int_0^t |f(s)|\mathrm{d}s\right)$$

at each time t.

Definition 6. (Liu [10]) Let X_t be an uncertain process and N_t be an uncertain renewal process. Then the Yao integral of X_t is defined by

$$\int_a^b X_t \mathrm{d}N_t = \sum_{a<t\leq b} X_{t-}(N_t - N_{t-})$$

provided that the sum exists almost surely and is finite.

For a continuously differentiable function $h(t, c, n)$, the uncertain process $Z_t = h(t, C_t, N_t)$ has an uncertain differential

$$\mathrm{d}Z_t = \frac{\partial h}{\partial t}(t, C_t, N_t)\mathrm{d}t + \frac{\partial h}{\partial c}(t, C_t, N_t)\mathrm{d}C_t + h(t, C_t, N_t) - h(t, C_t, N_{t-}).$$

For example, the uncertain differential of an uncertain process $X_t = \mu t + \sigma C_t + \gamma N_t$ is

$$\mathrm{d}X_t = \mu \mathrm{d}Z_t + \sigma \mathrm{d}C_t + \gamma(N_t - N_{t-}),$$

and the uncertain differential of an uncertain process $Y_t = tC_tN_t$ is

$$\mathrm{d}Y_t = C_tN_t\mathrm{d}t + tN_t\mathrm{d}C_t + tC_t(N_t - N_{t-}).$$

Uncertain differential equation

Definition 7. (Liu [7]) Suppose C_t is a canonical Liu process, and f and g are two given functions. Then

$$dX_t = f(t, X_t)dt + g(t, X_t)dC_t \tag{1}$$

is called an uncertain differential equation.

In 2010, Chen and Liu [15] solved the linear uncertain differential equation

$$dX_t = (u_{1t}X_t + u_{2t})dt + (v_{1t}X_t + v_{2t})dC_t$$

and obtained a solution

$$X_t = Y_t \left(X_0 + \int_0^t \frac{u_{2s}}{Y_s} ds + \int_0^t \frac{v_{2s}}{Y_s} dC_s \right)$$

where

$$Y_t = \exp\left(\int_0^t u_{1s} ds + \int_0^t v_{1s} dC_s \right).$$

After that, Liu [25] provided a method to solve the following types of uncertain differential equations:

$$dX_t = f(t, X_t)dt + \sigma_t X_t dC_t, \tag{2}$$

$$dX_t = \mu_t X_t dt + g(t, X_t)dC_t, \tag{3}$$

and showed that uncertain differential equation (2) has a solution $X_t = Y_t^{-1} Z_t$ where

$$Y_t = \exp\left(-\int_0^t \sigma_s dC_s \right)$$

and Z_t solves uncertain differential equation $dZ_t = Y_t f(t, Y_t^{-1} Z_t)dt$ with an initial value $Z_0 = X_0$, and uncertain differential equation (3) has a solution $X_t = Y_t^{-1} Z_t$ where

$$Y_t = \exp\left(-\int_0^t \alpha_s ds \right)$$

and Z_t solves uncertain differential equation $dZ_t = Y_t g(t, Y_t^{-1} Z_t)dC_t$ with an initial value $Z_0 = X_0$.

An analytic method

In this section, we will propose an analytic method to solve uncertain differential equations

$$dX_t = f(t, X_t)dt + \sigma_t dC_t$$

and

$$dX_t = \mu_t dt + g(t, X_t)dC_t.$$

Model I

Theorem 2. *Let f be a function of two variables, and σ_t be an integrable function on $\Re^+$. Then the uncertain differential equation*

$$dX_t = f(t, X_t)dt + \sigma_t dC_t$$

has a solution

$$X_t = Y_t + Z_t$$

where

$$Y_t = \int_0^t \sigma_s \mathrm{d}C_s$$

and Z_t is the solution of uncertain differential equation

$$\mathrm{d}Z_t = f(t, Y_t + Z_t)\mathrm{d}t, \quad Z_0 = X_0.$$

Proof. The uncertain process Y_t has an uncertain differential

$$\mathrm{d}Y_t = \mu_t \mathrm{d}t.$$

Then we have

$$\mathrm{d}(X_t - Y_t) = \mathrm{d}X_t - \mathrm{d}Y_t = f(t, X_t)\mathrm{d}t + \sigma_t \mathrm{d}C_t - \sigma_t \mathrm{d}C_t.$$

That is,

$$\mathrm{d}(X_t - Y_t) = f(t, X_t)\mathrm{d}t.$$

Defining $Z_t = X_t - Y_t$, we obtain $X_t = Y_t + Z_t$ and $\mathrm{d}Z_t = f(t, Y_t + Z_t)\mathrm{d}t$. Furthermore, since $Y_0 = 0$, the initial value Z_0 is just X_0. The theorem is verified. □

Remark 1. If σ_t degenerates to a constant σ, then $Y_t = \sigma C_t$. The uncertain differential equation

$$\mathrm{d}X_t = f(t, X_t)\mathrm{d}t + \sigma \mathrm{d}C_t$$

has a solution

$$X_t = Z_t + \sigma C_t$$

where Z_t solves the uncertain differential equation

$$\mathrm{d}Z_t = f(t, Z_t + \sigma C_t)\mathrm{d}t, \quad Z_0 = X_0.$$

Example 1. Let m_t, μ_t, and σ_t be some real functions on $\Re$. Consider the uncertain differential equation

$$\mathrm{d}X_t = (m_t - \mu_t X_t)\mathrm{d}t + \sigma_t \mathrm{d}C_t.$$

At first, we have

$$Y_t = \int_0^t \sigma_s \mathrm{d}C_s$$

and Z_t solves the uncertain differential equation

$$\mathrm{d}Z_t = \left(m_t - \mu_t \left(\int_0^t \sigma_s \mathrm{d}s + Z_t\right)\right) \mathrm{d}t,$$

so

$$\mathrm{d}Z_t + \mu_t Z_t \mathrm{d}t = \left(m_t - \mu_t \int_0^t \sigma_s \mathrm{d}C_s\right) \mathrm{d}t.$$

It follows from the fundamental theorem of uncertain calculus that

$$\mathrm{d}\left(\exp\left(\int_0^t \mu_s \mathrm{d}s\right) Z_t\right) = \exp\left(\int_0^t \mu_s \mathrm{d}s\right) \left(m_t - \mu_t \int_0^t \sigma_s \mathrm{d}C_s\right) \mathrm{d}t.$$

That is,

$$\exp\left(\int_0^t \mu_s \mathrm{d}s\right) Z_t - Z_0 = \int_0^t \exp\left(\int_0^s \mu_v \mathrm{d}v\right)\left(m_s - \mu_s \int_0^s \sigma_v \mathrm{d}C_v\right) \mathrm{d}s.$$

As a result,

$$\begin{aligned} Z_t &= \exp\left(-\int_0^t \mu_s \mathrm{d}s\right) Z_0 + \int_0^t \exp\left(-\int_s^t \mu_v \mathrm{d}v\right)\left(m_s - \mu_s \int_0^s \sigma_v \mathrm{d}C_v\right) \mathrm{d}s \\ &= \exp\left(-\int_0^t \mu_s \mathrm{d}s\right) X_0 + \int_0^t \exp\left(-\int_s^t \mu_v \mathrm{d}v\right)\left(m_s - \mu_s \int_0^s \sigma_v \mathrm{d}C_v\right) \mathrm{d}s. \end{aligned}$$

By Theorem 2, we have

$$X_t = Y_t + Z_t = \exp\left(-\int_0^t \mu_s \mathrm{d}s\right) X_0 + \int_0^t \sigma_s \mathrm{d}C_s + \int_0^t \exp\left(-\int_s^t \mu_v \mathrm{d}v\right)\left(m_s - \mu_s \int_0^s \sigma_v \mathrm{d}C_v\right) \mathrm{d}s.$$

Example 2. Let μ and σ be real numbers with $\mu \neq 0$. Consider the uncertain differential equation

$$\mathrm{d}X_t = \mu \exp(X_t)\mathrm{d}t + \sigma \mathrm{d}C_t.$$

At first, we have $Y_t = \sigma C_t$ and Z_t satisfies the uncertain differential equation

$$\mathrm{d}Z_t = \mu \exp(\sigma C_t + Z_t)\mathrm{d}t,$$

so

$$\exp(-Z_t)\mathrm{d}Z_t = \mu \exp(\sigma C_t)\mathrm{d}t.$$

It follows from the fundamental theorem of uncertain calculus that

$$\mathrm{d}\exp(-Z_t) = -\mu \exp(\sigma C_t)\mathrm{d}t.$$

That is,

$$\exp(-Z_t) - \exp(-Z_0) = -\mu \int_0^t \exp(\sigma C_s)\mathrm{d}s.$$

As a result,

$$Z_t = Z_0 - \ln\left(1 - \mu \int_0^t \exp\left(Z_0 + \sigma C_s\right) \mathrm{d}s\right) = X_0 - \ln\left(1 - \mu \int_0^t \exp\left(X_0 + \sigma C_s\right) \mathrm{d}s\right).$$

By Theorem 2, we have

$$X_t = Y_t + Z_t = X_0 + \sigma C_t - \ln\left(1 - \mu \int_0^t \exp\left(X_0 + \sigma C_s\right) \mathrm{d}s\right).$$

Model II

Theorem 3. *Let μ_t be an integrable function on $\Re^+$, and g be a function of two variables. Then the uncertain differential equation*

$$\mathrm{d}X_t = \mu_t \mathrm{d}t + g(t, X_t)\mathrm{d}C_t$$

has a solution

$$X_t = Y_t + Z_t$$

where

$$Y_t = \int_0^t \mu_s \mathrm{d}s$$

and Z_t is the solution of uncertain differential equation

$$dZ_t = g(t, Y_t + Z_t)dC_t, \quad Z_0 = X_0.$$

Proof. The uncertain process Y_t has an uncertain differential

$$dY_t = \mu_t dt.$$

Then we have

$$d(X_t - Y_t) = dX_t - dY_t = \mu_t dt + g(t, X_t)dC_t - \mu_t dC_t.$$

That is,

$$d(X_t - Y_t) = g(t, X_t)dC_t.$$

Defining $Z_t = X_t - Y_t$, we obtain $X_t = Y_t + Z_t$ and $dZ_t = g(t, Y_t + Z_t)dC_t$. Furthermore, since $Y_0 = 0$, the initial value Z_0 is just X_0. The theorem is verified. □

Remark 2. If μ_t degenerates to a constant μ, then $Y_t = \mu t$, and the uncertain differential equation

$$dX_t = \mu dt + g(t, X_t)dC_t$$

has a solution

$$X_t = Z_t + \mu t$$

where Z_t solves the uncertain differential equation

$$dZ_t = g(t, Z_t + \mu t)dt, \quad Z_0 = X_0.$$

Example 3. Let m_t, μ_t, and σ_t be some real functions on $\Re$. Consider the uncertain differential equation

$$dX_t = \mu_t dt + (m_t - \sigma_t X_t)dC_t.$$

At first, we have

$$Y_t = \int_0^t \mu_s ds$$

and Z_t solves the uncertain differential equation

$$dZ_t = \left(m_t - \sigma_t \left(\int_0^t \mu_s ds + Z_t\right)\right) dC_t,$$

so

$$dZ_t + \sigma_t Z_t dC_t = \left(m_t - \sigma_t \int_0^t \mu_s ds\right) dC_t.$$

It follows from the fundamental theorem of uncertain calculus that

$$d\left(\exp\left(\int_0^t \sigma_s dC_s\right) Z_t\right) = \exp\left(\int_0^t \sigma_s dC_s\right)\left(m_t - \sigma_t \int_0^t \mu_s ds\right) dC_t.$$

That is,

$$\exp\left(\int_0^t \sigma_s dC_s\right) Z_t - Z_0 = \int_0^t \exp\left(\int_0^s \sigma_v dC_v\right)\left(m_s - \sigma_s \int_0^s \mu_v dv\right) dC_s.$$

As a result,

$$Z_t = \exp\left(-\int_0^t \sigma_s \mathrm{d}C_s\right) Z_0 + \int_0^t \exp\left(-\int_s^t \sigma_v \mathrm{d}C_v\right)\left(m_s - \sigma_s \int_0^s \mu_v \mathrm{d}v\right) \mathrm{d}C_s$$
$$= \exp\left(-\int_0^t \sigma_s \mathrm{d}C_s\right) X_0 + \int_0^t \exp\left(-\int_s^t \sigma_v \mathrm{d}C_v\right)\left(m_s - \sigma_s \int_0^s \mu_v \mathrm{d}v\right) \mathrm{d}C_s.$$

By Theorem 3, we have

$$X_t = Y_t + Z_t = \exp\left(-\int_0^t \sigma_s \mathrm{d}C_s\right) X_0 + \int_0^t \mu_s \mathrm{d}s + \int_0^t \exp\left(-\int_s^t \sigma_v \mathrm{d}C_v\right)\left(m_s - \sigma_s \int_0^s \mu_v \mathrm{d}v\right) \mathrm{d}C_s.$$

Example 4. Let μ and σ be real numbers with $\sigma \neq 0$. Consider the uncertain differential equation

$$\mathrm{d}X_t = \mu \mathrm{d}t + \sigma \exp(X_t)\mathrm{d}C_t.$$

At first, we have $Y_t = \mu t$ and Z_t satisfies the uncertain differential equation

$$\mathrm{d}Z_t = \sigma \exp(\mu t + Z_t)\mathrm{d}C_t,$$

so

$$\exp(-Z_t)\mathrm{d}Z_t = \sigma \exp(\mu t)\mathrm{d}C_t.$$

It follows from the fundamental theorem of uncertain calculus that

$$\mathrm{d}\exp(-Z_t) = -\sigma \exp(\mu t)\mathrm{d}C_t.$$

That is,

$$\exp(-Z_t) - \exp(-X_0) = -\sigma \int_0^t \exp(\mu s)\mathrm{d}C_s.$$

As a result,

$$Z_t = Z_0 - \ln\left(1 - \sigma \int_0^t \exp(Z_0 + \mu s)\,\mathrm{d}C_s\right) = X_0 - \ln\left(1 - \sigma \int_0^t \exp(X_0 + \mu s)\,\mathrm{d}C_s\right).$$

By Theorem 3, we have

$$X_t = Y_t + Z_t = X_0 + \mu t - \ln\left(1 - \sigma \int_0^t \exp(X_0 + \mu s)\,\mathrm{d}C_s\right).$$

Conclusions

This paper proposed a method to solve a special type of uncertain differential equation and employed some examples to illustrate the method.

Acknowledgements
This work was supported by National Natural Science Foundation of China Grant Nos. 61273044 and 91224008.

References

1. Liu, B: Uncertainty Theory. 2nd ed. Springer, Berlin (2007)
2. Liu, B: Uncertainty Theory: A Branch of Mathematics for Modeling Human Uncertainty. Springer, Berlin (2010)
3. Peng, ZX, Iwamura, K: A sufficient and necessary condition of uncertainty distribution. J. Interdiscip. Math. **13**(3), 277–285 (2010)
4. Liu, YH, Ha, M: Expected value of function of uncertain variables. J. Uncertain Syst. **4**(3), 181–186 (2010)
5. Chen, X, Dai, W: Maximum entropy principle for uncertain variables. Int. J. Fuzzy Syst. **13**(3), 232–236 (2011)
6. Liu, B: Theory and Practice of Uncertain Programming. 2nd ed. Springer, Berlin (2009)
7. Liu, B: Fuzzy process, hybrid process and uncertain process. J. Uncertain Syst. **2**(1), 3–16 (2008)
8. Yao, K, Li, X: Uncertain alternating renewal process and its application. IEEE Trans. Fuzzy Syst. **20**(6), 1154–1160 (2012)
9. Zhang, X, Ning, Y, Meng, G: Delayed renewal process with uncertain interarrival times. Fuzzy Optimization Decis Making **12**(1), 79–87 (2013)

10. Liu, B: Some research problems in uncertainty theory. J. Uncertain Syst. **3**(1), 3–10 (2009)
11. Liu, B: Extreme value theorems of uncertain process with application to insurance risk model. Soft Comput. **17**(4), 549–556 (2013)
12. Liu, B, Yao, K: Uncertain integral with respect to multiple canonical processes. J. Uncertain Syst. **6**(4), 250–255 (2012)
13. Chen, X, Ralescu, DA: Liu process and uncertain calculus. J. Uncertainty Anal. Appl. **1**, Article 3 (2013)
14. Yao, K: Uncertain calculus with renewal process. Fuzzy Optimization Decis. Making **11**(3), 285–297 (2012)
15. Chen, X, Liu, B: Existence and uniqueness theorem for uncertain differential equations. Fuzzy Optimization Decis Making **9**(1), 69–81 (2010)
16. Gao, Y: Existence and uniqueness theorem on uncertain differential equations with local Lipschitz condition. J. Uncertain Syst. **6**(3), 223–232 (2012)
17. Yao, K, Gao, J, Gao, Y: Some stability theorems of uncertain differential equation. Fuzzy Optimization Decis Making **12**(1), 3–13 (2013)
18. Sheng, Y: Stability in the p-th moment for uncertain differential equation. J. Intell. Fuzzy Syst. (2013, in press)
19. Chen, X: American option pricing formula for uncertain financial market. Int. J. Oper. Res. **8**(2), 32–37 (2011)
20. Peng, J, Yao, K: A new option pricing model for stocks in uncertainty markets. Int. J. Oper. Res. **8**(2), 18–26 (2011)
21. Chen, X, Liu, YH, Ralescu, DA: Uncertain stock model with periodic dividends. Fuzzy Optimization Decis Making **12**(1), 111–123 (2013)
22. Chen, X, Gao, Y: Uncertain term structure model of interest rate. Soft Comput. **17**(4), 597–604 (2013)
23. Liu, YH, Chen, X, Ralescu, DA: Uncertain currency model and currency option pricing. Int. J. Intell. Syst. (2013, in press)
24. Liu, B: Toward uncertain finance theory. J. Uncertainty Anal. Appl. **1**, Article 1 (2013)
25. Liu, YH: An analytic method for solving uncertain differential equations. J. Uncertain Syst. **6**(4), 244–249 (2012)
26. Yao, K, Chen, X: A numerical method for solving uncertain differential equations. J. Intell. Fuzzy Syst. **25**(3), 825–832 (2013)
27. Yao, K: Extreme values and integral of solution of uncertain differential equation. J. Uncertainty Anal. Appl. **1**, Article 2 (2013)

9

Optimal redundancy allocation rule for improving system life

Debasis Bhattacharya[1] and Soma Roychowdhury[2*]

* Correspondence: srcdb@yahoo.com
[2]Indian Institute of Social Welfare and Business Management, Calcutta 700073, India
Full list of author information is available at the end of the article

Abstract

Enhancing system life span is an essential area of concern to reliability practitioners. Various techniques can be adopted to achieve this, of which redundancy allocation is a popular one. System life increases whenever a redundant component is added to any of the components of the system, but the amount of increase depends on the choice of component to which redundancy is allocated. We need to identify the best choice of component, to which when a redundant component is added, maximum increase in system life is achieved. In this paper, we propose a general rule for maximizing system life stochastically by properly allocating a redundant component. Knowledge about the order of component lives is enough to make the optimal decision here. This rule can be applied to any simple or complex systems, and any number of redundant components can be added to the system to have maximum possible improvement. Some useful coherent systems are studied here in the light of derived results.

MSC 2010: Primary 90B25; 62 N05; 60E15; Secondary 60 K10; 65 K10; 62G30

Keywords: Coherent system; Redundancy allocation; Stochastic ordering; System life

Introduction

Designing a system to perform its intended job at least up to a specified time, known as a target life or mission time, always has been a challenge to reliability practitioners. To enhance the random system life by allocating redundant components is a common practice. However, that does not ensure stochastic maximization of system life unless the redundancy is allocated properly. There has been no general rule for addressing this demanding issue of optimal allocation of redundancy that can be applied to any simple or complex coherent systems which maximizes the life span of a system stochastically. There may be a situation where no compromise with meeting the target life can be afforded. In such cases, the system needs to be so designed that it can operate satisfactorily at least for some specified length of time. Mention may be made of a missile flight, which would be required to function up to the time it takes in reaching its known destination or artificial satellites, and space explorers have to be safe for a predetermined mission time without any major failure. Another example could be military communication equipment, which must last all through its field operation of known length. Thus, designing a reliable system often requires that the system performs its intended job of specific duration without any major disruption, and malfunctioning of any of its constituent components may cause a great loss, which may not be restricted only to a loss of money or time due to malfunction.

It is important to note that the amount of increase in system life is different for a different choice of the components to which the redundant component is to be added. Thus, the issue boils down to attaching the redundant components in an effective manner so as to maximize the system life stochastically.

Generally, there are two commonly used types of redundancies - standby redundancy and active or parallel redundancy. In standby redundancy, a redundant component is attached in such a way that it starts functioning immediately after the failure of the component to which it is attached. Parallel redundancy is used when it is difficult or not possible to replace the failed components during operation of the system. The redundant components are connected in parallel with the components of the system, and they function simultaneously with the original components. In this paper, the active or parallel redundancy at component level is considered to increase the system life stochastically. It has been noted in Barlow and Proschan [1] that in the case of active redundancy, component-wise redundancy works much better than the system-wise redundancy.

The past work on reliability optimization can be classified as follows: papers solving the issue by focusing on the enhancement of component life [2-4] and by making provision of adding redundant components [5-8]. In some recent works, Sheikhalishahi et al. [9] presented a reliability redundancy allocation problem where they considered some particular system designs. Cao et al. [10] solved a redundancy allocation problem using a decomposition-based approach in series–parallel systems only. In this paper, we propose a general rule for making an optimal decision to choose a component of a given system to which the redundant component is to be added to stochastically maximize the system life. The method is applicable to any coherent systems in general.

The paper is organized as follows: the present section introduces the work and discusses earlier works in this area. The second section includes preliminaries necessary to develop a method for redundancy allocation, some related definitions, and the result. The third section formulates an optimum allocation rule, using the one which can identify the best choice of component of a system to which the redundant component is to be added to give maximum rise in system life. Applications of the rule to some important coherent systems are included in the fourth section. Finally, the fifth section concludes the work.

Preliminaries and notation

An n-component system is said to be coherent if every component is relevant, i.e., every component has some contribution towards the system performance and if the system is monotone, i.e., the performance of the system improves with the improvement of any component or a subset of components. For the formal definition of an n-component coherent system, one can refer to Barlow and Proschan [1].

A system life can be determined from its component lives. Here we decompose a coherent system in a number of subsystems in such a way that the system fails whenever any of the subsystems fail, and a subsystem fails when all of its components fail. Thus, the system life can be obtained from the subsystem lives.

Next, the concept of stochastic order and stochastic equality of two variables are given below which will be used in the sequel.

Let F and G be two probability distribution functions on real line R. Let two random variables X and Y be distributed according to F and G, respectively. F is stochastically larger than G if

$$F(x) \leq G(x) \text{ for all } x \in R.$$

In this case, we have $P(X \geq t) \geq P(Y \geq t)$ for all t. This means that X exceeds a fixed number with a higher probability than Y does. In other words, X is said to be stochastically larger than Y. In the notation, $X_i \geq_{st} X_j$.

Two random variables X and Y are said to be stochastically equal or simply equal in distribution if

$$P(X \leq t) = P(Y \leq t) \text{ for all } t.$$

It is denoted by $X =_{st} Y$.

In this paper, we develop a rule for selecting a component (optimal solution) for which system life is stochastically maximized, i.e., the chance of increase in system life being maximum is greatest when a redundant component is attached to it. The notion of stochastic ordering of component lives is used to formulate the rule for getting the solution. Let $X_1, X_2, \ldots, X_n$ be independently distributed random lives of the components of an n-component coherent system. By the definition of stochastic ordering of random variables, it can easily be seen that the order of the component lives implies and is implied by the order of component reliabilities, i.e., if reliability of a component is more than that of the other, then it is more likely to have longer life, or vice versa. If X_i be the life of the ith component having reliability $p_i(t)$ at time t, and X_j be the life of the jth component having reliability $p_j(t)$, then

$$\begin{aligned} & X_i \geq_{st} X_j \\ \text{if and only if} \quad & P(X_i > t) \geq P(X_j > t), \text{ for all } t \\ \text{i.e.,} \quad & p_i(t) \geq p_j(t), \text{ for all } t. \end{aligned} \tag{1}$$

Thus, in case the order of component reliabilities is known, the problem can be solved in a similar manner as will be discussed in this paper.

An implication of stochastic ordering between two random variables is now stated below, which will be used in interpreting the result discussed in Theorem 1 in the next section.

If $X_i \geq_{st} X_j$, then $P(X_i > X_j) \geq P(X_j > X_i)$, i.e., if X_i is stochastically larger than X_j, the chance of X_i being larger than X_j is more, since

$$\begin{aligned} P(X_i > X_j) &= \int_{x_j=-\infty}^{\infty} P(X_i > x_j)\, dF_{X_j}(x_j) \geq \int_{x_j=-\infty}^{\infty} P(X_j > x_j)\, dF_{X_j}(x_j), \quad \text{by (1)} \\ &= \int_{x_j=-\infty}^{\infty} (1 - F_{X_j}(x_j))\, dF_{X_j}(x_j) = 1 - \frac{1}{2} = \frac{1}{2}. \end{aligned} \tag{2}$$

In a group of competing system designs, if the system life for a certain design is stochastically larger than that of the other, the corresponding system reliability will also be more, by similar reasoning as in (1). If T_1 and T_2 are system reliabilities for two different designs, then

$$T_1 \geq_{st} T_2 \quad \text{if and only if} \quad P(T_1 > t) \geq P(T_2 > t), \text{for all } t. \tag{3}$$

In Theorem 1, we find the optimal solution that gives stochastically largest system life. By (3), the same solution will also maximize the system reliability.

Throughout the paper, by equality ('=') of random variables (random lives), we mean stochastic equality; by '≥' and '≤' between two random variables, we indicate 'stochastically larger' and 'stochastically smaller', respectively. Similarly, 'smaller (larger)' indicates 'stochastically smaller (larger)', and 'equal' indicates 'stochastically equal'.

Now consider X_1, X_2,..., X_n, independently distributed random lives of the components of an n-component system which is decomposed into k subsystems, C_1, C_2,..., C_k, of sizes n_1, n_2,..., n_k, respectively. Note that $\sum_{i=1}^{k} n_i \geq n$, when the subsystems are overlapping, i.e., one component may belong to more than one subsystem. Non-overlapping subsystems do not share any components. For a system with non-overlapping subsystems, $\sum_{i=1}^{k} n_i = n$.

Let $C_i = \{i_1, i_2, \ldots, i_{n_i}\}$, where i_j is the jth component of the ith subsystem C_i, $j = 1, 2, \ldots, n_i$, $i = 1, 2, \ldots, k$. Let Y_i be the life of the ith subsystem, C_i, i.e., the largest of the lives of all components belonging to C_i.

It is known that a coherent system is always monotone, which indicates that the reliability of a coherent system increases with the improvement of any component or a subset of components. When a redundant component is added to a system component in parallel, the component reliability increases. Thus, the system reliability increases (non-decreases) when a redundant component is added to any of the components of a coherent system. Using redundancy improves system life as well. Let us now prove the following result:

Result

The system life increases (non-decreases) when a redundant component is added to any of the components of the system.

Proof The system life can be written as

$$T = \min_{1 \leq i \leq k} \max_{j \in C_i} X_j = \min_{1 \leq i \leq k} Y_i \tag{4}$$

where $Y_i = \max_{j \in C_i} X_j$ is the life of the ith subsystem, which is the maximum of the component lives of the ith subsystem. The component having a maximum life among the lives of all components in the ith subsystem, by failing, will cause the subsystem to fail. The system life, as given in (4), will be the minimum of the lives of all subsystems.

Let U be the random life of the redundant component, which is independent of X_1, X_2,..., X_n .

Suppose $Y_{(1)}$ to be the smallest of the subsystem lives, i.e., $Y_{(1)} = \min_{1\le i\le k} Y_i$, and $Y_{(2)}$ to be the second smallest. When a redundant component having a random life U is added to a system component, one of the following cases may occur:

Case 1. The redundant component is added to the subsystem with the smallest life, where U will be one of the following:

(i) $U \le Y_{(1)}$
(ii) $Y_{(1)} < U < Y_{(2)}$
(iii) $Y_{(1)} < Y_{(2)} < U$

Case 2. The redundant component is added to some other subsystem (not to the one with the smallest life).

In case 1(i), again $Y_{(1)} = \min_{1\le i\le k} Y_i$. The system life will remain unchanged.

In 1(ii), $U = \min_{1\le i\le k} Y_i$. The system life will increase from $Y_{(1)}$ to U.

In 1(iii), $Y_{(2)} = \min_{1\le i\le k} Y_i$, and the system life will increase from $Y_{(1)}$ to $Y_{(2)}$.

In case 2, there will be no change in system life.

Hence, the result follows.

Our objective is to select the system component which gives maximum rise in the system life when a redundant component is added to it. The next section proves a theorem to get the optimal solution.

Optimum allocation rule

Suppose T_j to be the life of the jth subsystem C_j, if the redundant component is attached to a component belonging to C_j, $j = 1, 2,..., k$.

Now we see how the system life changes if a redundant component (s) is added to a component belonging to C_j.

If s is added to a component belonging to C_j, the life of the jth subsystem will become

$$T_j = Y_j + \max\left(U - Y_j, 0\right), j = 1, 2, ..., k. \tag{5}$$

Now suppose $Y_{(1)}, Y_{(2)},..., Y_{(k)}$ to be the ordered subsystem lives. Then, (4) becomes $T = Y_{(1)}$.

The following result helps select the subsystem to a component of which the redundant component is to be added in order to have maximum system life.

Theorem 1

Increase in system life is stochastically maximum when a redundant component is added to a component belonging to the subsystem with the smallest life.

Proof Here one of the following three cases may occur:

(i) $U \le Y_{(1)} \le Y_{(2)} \le ... \le Y_{(k)}$
(ii) $Y_{(1)} \le Y_{(2)} \le ... \le Y_{(k)} \le U$
(iii) $Y_{(1)} \le Y_{(2)} \le ... \le Y_{(i)} \le U \le Y_{(i+1)} \le ... \le Y_{(k)}, i = 1, 2, ..., (k-1)$

Now let us see how system life changes in the three scenarios.

(i) If $U \le Y_{(1)} \le Y_{(2)} \le \ldots \le Y_{(k)}$, then by (5), $T_j = Y_j$, for all $j = 1, 2, \ldots, k$, and system life, $T = Y_{(1)}$.
Thus, in this situation, the system life will remain the same no matter which subsystem component the redundant component is added to.

(ii) If $Y_{(1)} \le Y_{(2)} \le \ldots \le Y_{(k)} \le U$, by (5), $T_j = U$ for all $j = 1, 2, \ldots, k$, and therefore, if s is added to a component belonging to the subsystem having the smallest system life $Y_{(1)}$, the life of that subsystem will then be U, and hence, the system life will be the smallest of other $Y_{(i)}$'s, i.e., $T = Y_{(2)}$, which is greater than $Y_{(1)}$. However, if s is added to a component belonging to any other subsystem, the minimum of subsystem lives will then be $Y_{(1)}$, and hence, the system life will then be $T = Y_{(1)}$.
Thus, in this situation, the optimal choice of subsystem should be the subsystem having the smallest life. The redundant component should be added to a component belonging to the subsystem having the smallest life in order to increase the system life.

(iii) If $Y_{(1)} \le Y_{(2)} \le \ldots \le Y_{(i)} \le U \le Y_{(i+1)} \le \ldots \le Y_{(k)}$, by (5), $T_j = U$, for all j such that $Y_{(j)} \le U$, and $T_j = Y_{(j)}$, for all j such that $Y_{(j)} \ge U$. Therefore, if s is added to a component belonging to any subsystem having life $\ge U$, the system life will be $T = Y_{(1)}$. If s is added to a component belonging to a subsystem having life $\le U$, but $\ge Y_{(1)}$, then also the system life will be $T = Y_{(1)}$. However, if s is added to a component belonging to a subsystem having the smallest life, the system life will increase to $Y_{(2)}$.
Thus, in this case, also the optimum choice is the subsystem having the smallest life.

Combining all three cases, it is clear that the optimum choice of subsystem is the one that has the smallest life.

The statement 'increase in system life is stochastically maximum' in the theorem is equivalent to the 'chance of increase in system life being maximum is greatest', by the same reasoning as in (2).

Next, we determine which component of the selected subsystem should be chosen in order to get maximum system life. Suppose C_i to be the selected subsystem. Its life is

$$Y_i = \max_{j \in C_i} X_j = X_{(n_i)} \tag{6}$$

where $X_{(1)}, X_{(2)}, \ldots, X_{(n_i)}$ are the ordered component lives of subsystem C_i.

Adding s to a subsystem is the same as adding s to any of the n_i components, but since those components may also belong to some other subsystems, we need to take care of this fact while choosing the component to which s will be attached.

If $U \le X_{(n_i)}$, by (6), the subsystem life $Y_i = X_{(n_i)}$.

If $X_{(n_i)} \le U$, by (6), the subsystem life $Y_i = U$.

For a system with overlapping subsystems, the decision of choosing a component is based on other subsystems where the components belong to. We consider the lives of those subsystems that contain the components of the selected subsystem, and find the smallest of them. Following the same logic as in Theorem 1, the corresponding component is the component that maximizes the system life.

Suppose C_i to be the subsystem with the smallest life. Let $C_i = \{i_1, i_2, \ldots, i_{n_i}\}$, where i_1, $i_2, \ldots, i_{n_i}$ are n_i components belonging to C_i. Let the component i_j ($j = 1, 2, \ldots, n_i$) belong to other m_j (≥ 0) subsystems $C_1^{(i_j)}, C_2^{(i_j)}, \ldots, C_{m_j}^{(i_j)}$ (these are nothing but m_j subsystems from $C_1, C_2, \ldots, C_k$, except C_i), whose lives are, respectively, denoted by $Y_1^{(i_j)}, Y_2^{(i_j)}, \ldots, Y_{m_j}^{(i_j)}$.

Let $Y_{h_j}^{(i_j)}$ be the subsystem life, the smallest among the lives of all subsystems (except C_i) that contain the component i_j, i.e.,

$$Y_{h_j}^{(i_j)} = \min_{1 \leq l \leq m_j} Y_l^{(i_j)}, \quad j = 1, 2, \ldots., n_i$$

For each of components $i_1, i_2, \ldots, i_{n_i}$, we may get one such subsystem. Let their lives be $Y_{h_1}^{(i_1)}, Y_{h_2}^{(i_2)}, \ldots, Y_{h_{n_i}}^{(i_{n_i})}$, respectively. Following the logic used in proving Theorem 1, the component corresponding to the smallest life among $Y_{h_1}^{(i_j)}, Y_{h_2}^{(i_j)}, \ldots, Y_{h_{n_i}}^{(i_j)}$ will be chosen. Thus, we can write the following: Choose component i_j (to which the redundant component s will be attached to achieve maximum increase in system life) if

$$Y_{h_j}^{(i_j)} = \min_{1 \leq q \leq n_j} Y_{h_q}^{(i_q)} \tag{7}$$

Note that in case the subsystem with the second smallest life does not contain any of the components of the subsystem with the smallest life, it does not matter which component of the latter subsystem (one having the smallest life) the redundant component is added to, because in this case, the life of the subsystem with the second smallest life will not change, but the life of the other subsystems may increase when s is added. The system life will then be the minimum of $Y_{(2)}$, the second smallest subsystem life, and $\max(U, Y_{(1)})$, the larger between the redundant component life and the smallest subsystem life, i.e., system life $T = \min(Y_{(2)}, \max(U, Y_{(1)}))$.

In case of systems with non-overlapping subsystems, the redundant component can be added to any of the components belonging to the smallest life subsystem. The rise in system life will be the same, and that will be the maximum.

Note that the method discussed above can be used for adding any number of redundant components. The components should be added one at a time.

Application of the rule

Let us first consider some commonly used systems and see how the rule works to identify the component which gives maximum rise in system life when the redundant component is added to it.

In an n-component series system, there are n subsystems, each having a single component. Hence, by Theorem 1, the component (which is also a subsystem, in this case) with the smallest life is to be chosen for adding the redundant component in order to maximize the system life.

For an n-component parallel system, since there is only a single subsystem of size n and none of the components can belong to any other subsystem, the redundant component can be added to any of the components in parallel. The increase in system life will be the same, and that will be the maximum.

Consider a series–parallel system, where components 2 and 3 are in parallel and component 1 is in series to the parallel structure connecting components 2 and 3.

Here, there are two subsystems, {1} and {2, 3}, which are non-overlapping. If life of component 1 is less than that of the larger of components 2 and 3, then by Theorem 1, the redundant component should be added to component 1. Otherwise it should be added to any one of components 2 and 3 to get the maximum system life. Exactly similarly we can solve the problem for a hi-fi system, where components 1 and 2 are in parallel, components 4 and 5 are in parallel, and both of them (parallel structures) are in series with component 3, thus having a series–parallel structure with non-overlapping subsystems.

For a parallel–series system, where components 1 and 2 are in series and component 3 is in parallel to them (that series), the subsystems are {1, 3} and {2, 3}. Without loss of generality, suppose the life of component 1 to be smaller than that of component 2. Then, the life of subsystem {1, 3} will be less. This subsystem has two components, namely, 1 and 3, of which component 3 belongs to another subsystem. Hence, by (7), the redundant component should always be added to component 3, which gives maximum rise in system life.

An application to a bridge system

Let us consider a complex system, like a bridge system, as shown in Figure 1.

Let the order of the component lives be $X_2 < X_4 < X_1 < X_3 < X_5$. The subsystems of this system are {1, 2}, {1, 3, 5}, {2, 3, 4}, and {4, 5}, whose lives are, respectively, X_1, X_5, X_3, and X_5, of which X_1 is the smallest. Hence, by Theorem 1, subsystem {1, 2} is to be chosen. Its components, 1 and 2, also belong to subsystems {1, 3, 5} and {2, 3, 4}, respectively. Since $X_5 > X_3$, the life of subsystem {2, 3, 4} is smaller, and hence, by (7), component 2 is to be chosen to which the redundant component should be added so that the chance of getting maximum rise in system life is greatest.

For example, in particular, if $X_1 = 5$, $X_2 = 2$, $X_3 = 7$, $X_4 = 4$, and $X_5 = 9$, the system life, by (4), is 5. If we add the redundant component having life $U = 8$ to component 2, the system life becomes 8. If we added it to component 1, the system life would then be 7. If we added it to component 3 or 4 or 5, the system life would not have changed then, and that would be 5.

Now suppose the order of the component lives to be $X_2 < X_4 < X_1 < X_5 < X_3$. Then, the subsystem lives are, X_1, X_3, X_3, and X_5, respectively, for subsystems {1, 2}, {1, 3, 5},

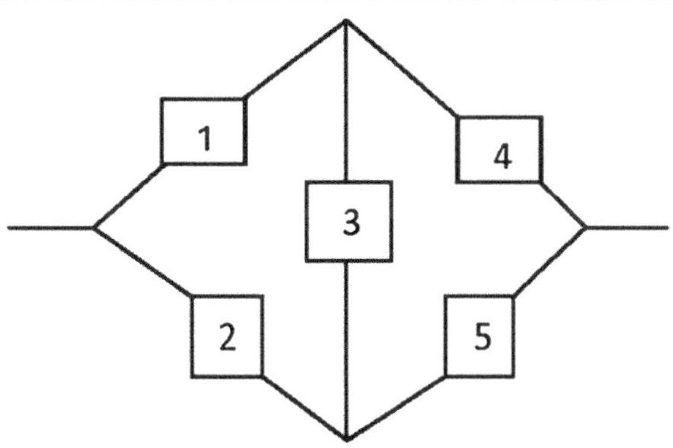

Figure 1 A bridge system.

{2, 3, 4}, and {4, 5}. Hence, the system life is X_1, being the smallest. By the theorem, we choose subsystem {1, 2}. Now since the lives of {1, 3, 5} and {2, 3, 4} are equal, the redundant component can be added to any one of components 1 and 2, and in each case, it will be the same and the increase in system life (and hence the system life) is stochastically maximum.

Conclusions and discussion

In this paper, we have developed a rule for selecting a component of a coherent system to which we can attach a redundant component to maximize the system life stochastically. If we have an idea about the order of the component lives, the rule proposed here can be used to make an optimal decision to choose the component. This rule can be applied to any simple or complex systems. However, proper care should be taken in achieving a trade-off between the benefit of enhancing the system life and the cost of achieving it subject to other constraints, such as constraints involving weight or volume.

Author details
[1]Visva-Bharati University, Santiniketan 731235, India. [2]Indian Institute of Social Welfare and Business Management, Calcutta 700073, India.

References

1. Barlow, RE, Proschan, F: Statistical Theory of Reliability and Life Testing: Probability Models. To Begin with Publisher, Silver Spring, MD (1981)
2. Mettas, A: Reliability allocation and optimization for complex systems, Proceedings, Annual Reliability and Maintainability Symposium, Los Angeles, California, USA (2000)
3. Huang, HZ: Fuzzy multi-objective optimization decision making of reliability of series system. Microelectron. Reliab. **37**(3), 447–449 (1997)
4. Gen, M, Cheng, R: Optimal design of system reliability using interval programming and genetic algorithms. Comput. Ind. Eng. **31**(1/2), 237–240 (1996)
5. Tillman, FA, Hwang, CL, Kuo, W: Optimization of Systems Reliability. Marcel Dekker, New York (1985)
6. Kuo, W, Prasad, R: An annotated overview of system-reliability optimization. IEEE. T. Reliab. **49**(2), 176–187 (2000)
7. Levitin, G: Redundancy optimization for multi-state system with fixed resource requirements and unreliable sources. IEEE. T. Reliab. **50**(1), 52–59 (2001)
8. Kulturel-Konak, S, Smith, AE, Coit, DW: Efficiently solving the redundancy allocation problem using tabu search. IEEE. Trans. **35**, 515–526 (2003)
9. Shekhalishahi, M, Ebrahimipour, V, Shiri, H, Zaman, H, Jeihoonian, M: A hybrid GA-PSO approach for reliability optimization in redundancy allocation problem. Int. J. Adv. Manuf. Technol. **68**, 317–338 (2013)
10. Cao, D, Murat, A, Chinnam, RB: Efficient exact optimization of multiobjective redundancy allocation problems in series-parallel systems. Reliab. Eng. Syst. Saf. **111**, 154–163 (2013)

Uncertainty relations for generalized metric adjusted skew information and generalized metric adjusted correlation measure

Kenjiro Yanagi[1*], Shigeru Furuichi[2] and Ken Kuriyama[3]

*Correspondence:
yanagi@yamaguchi-u.ac.jp
[1]Graduate School of Science and Engineering, Yamaguchi University, Une, Yamaguchi 755-8611, Japan
Full list of author information is available at the end of the article

Abstract

In this paper, we give a Heisenberg-type or a Schrödinger-type uncertainty relation for generalized metric adjusted skew information or generalized metric adjusted correlation measure. These results generalize the previous result of Furuichi and Yanagi (J. Math. Anal. Appl. 388:1147-1156, 2012).

AMS: Primary: 15A45, 47A63; secondary: 94A17

Keywords: Trace inequality; Metric adjusted skew information; Metric adjusted correlation measure

Introduction

We start from the Heisenberg uncertainty relation [1]:

$$V_\rho(A)V_\rho(B) \geq \frac{1}{4}|Tr\,[\rho[A,B]]\,|^2$$

for a quantum state (density operator) ρ and two observables (self-adjoint operators) A and B. The further stronger result was given by Schrödinger in [2,3]:

$$V_\rho(A)V_\rho(B) - |Re\left\{\mathrm{Cov}_\rho(A,B)\right\}|^2 \geq \frac{1}{4}|Tr\,[\rho[A,B]]\,|^2,$$

where the covariance is defined by $\mathrm{Cov}_\rho(A,B) \equiv Tr\,[\rho\,(A - Tr\,[\rho A]\,I)\,(B - Tr\,[\rho B]\,I)]$.

The Wigner-Yanase skew information represents a measure for non-commutativity between a quantum state ρ and an observable H. Luo introduced the quantity $U_\rho(H)$ representing a quantum uncertainty excluding the classical mixture [4]:

$$U_\rho(H) \equiv \sqrt{V_\rho(H)^2 - \left(V_\rho(H) - I_\rho(H)\right)^2},$$

with the Wigner-Yanase skew information [5]:

$$I_\rho(H) \equiv \frac{1}{2}Tr\left[(i[\rho^{1/2},H_0])^2\right] = Tr\,[\rho H_0^2] - Tr\,[\rho^{1/2}H_0\rho^{1/2}H_0]\,, \quad H_0 \equiv H - Tr\,[\rho H]\,I,$$

and then he successfully showed a new Heisenberg-type uncertainty relation on $U_\rho(H)$ in [4]:

$$U_\rho(A)U_\rho(B) \geq \frac{1}{4}|Tr\,[\rho[A,B]]|^2. \quad (1)$$

As stated in [4], the physical meaning of the quantity $U_\rho(H)$ can be interpreted as follows. For a mixed state ρ, the variance $V_\rho(H)$ has both classical mixture and quantum uncertainty. Also, the Wigner-Yanase skew information $I_\rho(H)$ represents a kind of quantum uncertainty [6,7]. Thus, the difference $V_\rho(H) - I_\rho(H)$ has a classical mixture so that we can regard that the quantity $U_\rho(H)$ has a quantum uncertainty excluding a classical mixture. Therefore, it is meaningful and suitable to study an uncertainty relation for a mixed state by the use of the quantity $U_\rho(H)$.

Recently, a one-parameter extension of the inequality (1) was given in [8]:

$$U_{\rho,\alpha}(A)U_{\rho,\alpha}(B) \geq \alpha(1-\alpha)|Tr\,[\rho[A,B]]|^2,$$

where

$$U_{\rho,\alpha}(H) \equiv \sqrt{V_\rho(H)^2 - \left(V_\rho(H) - I_{\rho,\alpha}(H)\right)^2},$$

with the Wigner-Yanase-Dyson skew information $I_{\rho,\alpha}(H)$ defined by

$$I_{\rho,\alpha}(H) \equiv \frac{1}{2}Tr\left[(i[\rho^\alpha, H_0])(i[\rho^{1-\alpha}, H_0])\right] = Tr\,[\rho H_0^2] - Tr\,[\rho^\alpha H_0 \rho^{1-\alpha} H_0]\,.$$

It is notable that the convexity of $I_{\rho,\alpha}(H)$ with respect to ρ was successfully proven by Lieb in [9]. The further generalization of the Heisenberg-type uncertainty relation on $U_\rho(H)$ has been given in [10] using the generalized Wigner-Yanase-Dyson skew information introduced in [11]. Recently, it is shown that these skew informations are connected to special choices of quantum Fisher information in [12]. The family of all quantum Fisher informations is parametrized by a certain class of operator monotone functions $\mathcal{F}_{op}$ which were justified in [13]. The Wigner-Yanase skew information and Wigner-Yanase-Dyson skew information are given by the following operator monotone functions:

$$f_{\mathrm{WY}}(x) = \left(\frac{\sqrt{x}+1}{2}\right)^2,$$

$$f_{\mathrm{WYD}}(x) = \alpha(1-\alpha)\frac{(x-1)^2}{(x^\alpha - 1)(x^{1-\alpha} - 1)}, \quad \alpha \in (0,1),$$

respectively. In particular, the operator monotonicity of the function f_{WYD} was proved in [14] (see also [15]). On the other hand, the uncertainty relation related to the Wigner-Yanase skew information was given by Luo [4], and the uncertainty relation related to the Wigner-Yanase-Dyson skew information was given by Yanagi [8]. In this paper, we generalize these uncertainty relations to the uncertainty relations related to quantum Fisher informations by using (generalized) metric adjusted skew information or correlation measure.

Operator monotone functions

Let $M_n(\mathbb{C})$ (respectively $M_{n,sa}(\mathbb{C})$) be the set of all $n \times n$ complex matrices (respectively all $n \times n$ self-adjoint matrices), endowed with the Hilbert-Schmidt scalar product $\langle A, B\rangle = Tr(A^*B)$. Let $M_{n,+}(\mathbb{C})$ be the set of strictly positive elements of $M_n(\mathbb{C})$ and $M_{n,+,1}(\mathbb{C})$ be the set of stricly positive density matrices, that is $M_{n,+,1}(\mathbb{C}) = \{\rho \in M_n(\mathbb{C}) | Tr\rho = 1, \rho > 0\}$. If it is not otherwise specified, from now on, we shall treat the case of faithful states, that is $\rho > 0$.

A function $f : (0, +\infty) \to \mathbb{R}$ is said to be operator monotone if, for any $n \in \mathbb{N}$ and $A, B \in M_{n,+}(\mathbb{C})$ such that $0 \leq A \leq B$, the inequalities $0 \leq f(A) \leq f(B)$ hold. An operator monotone function is said to be symmetric if $f(x) = xf(x^{-1})$ and normalized if $f(1) = 1$.

Definition 1. $\mathcal{F}_{op}$ is the class of functions $f : (0, +\infty) \to (0, +\infty)$ such that

1. $f(1) = 1$,
2. $tf(t^{-1}) = f(t)$,
3. f is operator monotone.

Example 1. Examples of elements of $\mathcal{F}_{op}$ are given by the following list:

$$f_{\mathrm{RLD}}(x) = \frac{2x}{x+1},\quad f_{\mathrm{WY}}(x) = \left(\frac{\sqrt{x}+1}{2}\right)^2,$$

$$f_{\mathrm{BKM}}(x) = \frac{x-1}{\log x},\quad f_{\mathrm{SLD}}(x) = \frac{x+1}{2},$$

$$f_{\mathrm{WYD}}(x) = \alpha(1-\alpha)\frac{(x-1)^2}{(x^\alpha - 1)(x^{1-\alpha}-1)},\quad \alpha \in (0,1).$$

Remark 1. Any $f \in \mathcal{F}_{op}$ satisfies

$$\frac{2x}{x+1} \le f(x) \le \frac{x+1}{2},\quad x > 0.$$

For $f \in \mathcal{F}_{op}$, define $f(0) = \lim_{x\to 0} f(x)$. We introduce the sets of regular and non-regular functions

$$\mathcal{F}_{op}^r = \left\{f \in \mathcal{F}_{op} | f(0) \neq 0\right\},\quad \mathcal{F}_{op}^n = \left\{f \in \mathcal{F}_{op} | f(0) = 0\right\}$$

and notice that trivially $\mathcal{F}_{op} = \mathcal{F}_{op}^r \cup \mathcal{F}_{op}^n$.

Definition 2. For $f \in \mathcal{F}_{op}^r$, we set

$$\tilde{f}(x) = \frac{1}{2}\left[(x+1) - (x-1)^2 \frac{f(0)}{f(x)}\right],\quad x > 0.$$

Theorem 1. ([12,16,17]) *The correspondence $f \to \tilde{f}$ is a bijection between $\mathcal{F}_{op}^r$ and $\mathcal{F}_{op}^n$.*

Metric adjusted skew information and correlation measure

In the Kubo-Ando theory of matrix means, one associates a mean to each operator monotone function $f \in \mathcal{F}_{op}$ by the formula

$$m_f(A,B) = A^{1/2} f(A^{-1/2} B A^{-1/2}) A^{1/2},$$

where $A, B \in M_{n,+}(\mathbb{C})$. Using the notion of matrix means, one may define the class of monotone metrics (also called quantum Fisher informations) by the following formula:

$$\langle A, B\rangle_{\rho,f} = Tr(A \cdot m_f(L_\rho, R_\rho)^{-1}(B)),$$

where $L_\rho(A) = \rho A, R_\rho(A) = A\rho$. In this case, one has to think of A, B as tangent vectors to the manifold $M_{n,+,1}(\mathbb{C})$ at the point ρ (see [12,13]).

Definition 3. For $A, B \in M_{n,sa}(\mathbb{C})$ and $\rho \in M_{n,+,1}(\mathbb{C})$, we define the following quantities:

$$\mathrm{Corr}_\rho^f(A,B) = Tr\left[\rho AB\right] - Tr\left[A \cdot m_{\tilde{f}}(L_\rho, R_\rho)B\right],$$

$$\mathrm{Corr}_\rho^{s(f)}(A,B) = \frac{f(0)}{2}\langle i[\rho,A]\,,i[\rho,B]\rangle_{\rho,f},$$
$$I_\rho^f(A) = \mathrm{Corr}_\rho^f(A,A),$$
$$C_\rho^f(A,B) = Tr\left[A\cdot m_f(L_\rho,R_\rho)B\right],$$
$$C_\rho^f(A) = C_\rho^f(A,A),$$
$$U_\rho^f(A) = \sqrt{V_\rho(A)^2 - (V_\rho(A) - I_\rho^f(A))^2},$$

The quantity $I_\rho^f(A)$ is known as metric adjusted skew information [18], and the metric adjusted correlation measure $\mathrm{Corr}_\rho^f(A,B)$ was also previously defined in [18].

Then we have the following proposition.

Proposition 1. ([16,19]) *For $A,B \in M_{n,sa}(\mathbb{C})$ and $\rho \in M_{n,+,1}(\mathbb{C})$, we have the following relations, where we put $A_0 = A - Tr[\rho A]I$ and $B_0 = B - Tr[\rho B]I$:*

1. $I_\rho^f(A) = I_\rho^f(A_0) = Tr[\rho A_0^2] - Tr[A_0\cdot m_{\tilde f}(L_\rho,R_\rho)A_0] = V_\rho(A) - C_\rho^{\tilde f}(A_0),$
2. $J_\rho^f(A) = Tr[\rho A_0^2] + Tr[A_0\cdot m_{\tilde f}(L_\rho,R_\rho)A_0] = V_\rho(A) + C_\rho^{\tilde f}(A_0),$
3. $0 \le I_\rho^f(A) \le U_\rho^f(A) \le V_\rho(A),$
4. $U_\rho^f(A) = \sqrt{I_\rho^f(A)\cdot J_\rho^f(A)},$
5. $Corr_\rho^f(A,B) = Corr_\rho^f(A_0,B_0) = Tr[\rho A_0B_0] - Tr[A_0\cdot m_{\tilde f}(L_\rho,R_\rho)B_0],$
6. $Corr_\rho^{s(f)}(A,B) = Corr_\rho^{s(f)}(A_0,B_0)$
$$= \tfrac{1}{2}Tr[\rho A_0B_0] + \tfrac{1}{2}Tr[\rho B_0A_0] - Tr[A_0\cdot m_{\tilde f}(L_\rho,R_\rho)B_0]$$
$$= \tfrac{1}{2}Tr[\rho A_0B_0] + \tfrac{1}{2}Tr[\rho B_0A_0] - C_\rho^{\tilde f}(A_0,B_0).$$

Now we modify the uncertainty relation given by [20].

Theorem 2. *For $f \in \mathcal{F}_{op}^r$, it holds*

$$I_\rho^f(A)\cdot I_\rho^f(B) \ge |Corr_\rho^{s(f)}(A,B)|^2,$$

where $A,B \in M_{n,sa}(\mathbb{C})$ and $\rho \in M_{n,+,1}(\mathbb{C})$.

Remark 2. Since Theorem 2 is easily given by using the Schwarz inequality, we omit the proof. In [20] we gave the uncertainty relation

$$U_\rho^f(A)\cdot U_\rho^f(B) \ge 4f(0)|\mathrm{Corr}_\rho^{s(f)}(A,B)|^2.$$

But since $4f(0) \le 1$ and $I_\rho^f(A) \le U_\rho^f(A)$, it is easily given by Theorem 2.

Theorem 3. ([20,21]) *For $f \in \mathcal{F}_{op}^r$, if*

$$\frac{x+1}{2} + \tilde f(x) \ge 2f(x), \tag{2}$$

then it holds

$$U_\rho^f(A)\cdot U_\rho^f(B) \ge f(0)|Tr(\rho[A,B])|^2, \tag{3}$$

$$U_\rho^f(A)\cdot U_\rho^f(B) \ge 4f(0)|Corr_\rho^f(A,B)|^2, \tag{4}$$

where $A, B \in M_{n,sa}(\mathbb{C})$ *and* $\rho \in M_{n,+,1}(\mathbb{C})$.

Remark 3. Though we cannot use the Schwarz inequality, we can get (4) in Theorem 3 by modifying the proof given by [20].

By putting

$$f_{\mathrm{WYD}}(x) = \alpha(1-\alpha)\frac{(x-1)^2}{(x^{\alpha}-1)(x^{1-\alpha}-1)}, \quad \alpha \in (0,1),$$

we obtain the following uncertainty relation.

Corollary 1. *For* $A, B \in M_{n,sa}(\mathbb{C})$ *and* $\rho \in M_{n,+,1}(\mathbb{C})$,

$$U_{\rho}^{f_{WYD}}(A) \cdot U_{\rho}^{f_{WYD}}(B) \geq \alpha(1-\alpha)|Tr\,(\rho[A,B])|^2,$$

$$U_{\rho}^{f_{WYD}}(A) \cdot U_{\rho}^{f_{WYD}}(B) \geq 4\alpha(1-\alpha)|Corr_{\rho}^{f_{WYD}}(A,B)|^2,$$

where

$$Corr_{\rho}^{f_{WYD}}(A,B) = Tr\,[\rho A_0 B_0] - \frac{1}{2}Tr\,[\rho^{\alpha}A_0\rho^{1-\alpha}B_0] - \frac{1}{2}Tr\,[\rho^{\alpha}B_0\rho^{1-\alpha}A_0].$$

Remark 4. Even if (2) does not necessarily hold, then

$$U_{\rho}^{f}(A) \cdot U_{\rho}^{f}(B) \geq f(0)^2|Tr\,[(\rho[A,B])|^2, \tag{5}$$

$$U_{\rho}^{f}(A) \cdot U_{\rho}^{f}(B) \geq 4f(0)^2|\mathrm{Corr}_{\rho}^{f}(A,B)|^2, \tag{6}$$

where $A, B \in M_{n,sa}(\mathbb{C})$ and $\rho \in M_{n,+,1}(\mathbb{C})$. Since $f(0) < 1$, it is easy to show that (5) and (6) are weaker than (3) and (4), respectively.

Generalized metric adjusted skew information and correlation measure

We give some generalizations of Heisenberg or Schrd̈inger uncertainty relations which include Theorem 3 as corollary.

Definition 4. ([22]) Let $g, f \in \mathcal{F}_{op}^{r}$ satisfy

$$g(x) \geq k\frac{(x-1)^2}{f(x)}$$

for some $k > 0$. We define

$$\Delta_g^f(x) = g(x) - k\frac{(x-1)^2}{f(x)} \in \mathcal{F}_{op}. \tag{7}$$

Definition 5. For $A, B \in M_{n,sa}(\mathbb{C})$ and $\rho \in M_{n,+,1}(\mathbb{C})$, we define the following quantities:

$$\mathrm{Corr}_{\rho}^{s(g,f)}(A,B) = k\langle i[\rho,A], i[\rho,B]\rangle_{\rho,f},$$

$$I_{\rho}^{(g,f)}(A) = \mathrm{Corr}_{\rho}^{s(g,f)}(A,A),$$

$$C_{\rho}^{f}(A,B) = Tr\,[A \cdot m_f(L_{\rho},R_{\rho})B],$$

$$C_{\rho}^{f}(A) = C_{\rho}^{f}(A,A),$$

$$U_\rho^{(g,f)}(A) = \sqrt{(C_\rho^g(A) + C_\rho^{\Delta_g^f}(A))(C_\rho^g(A) - C_\rho^{\Delta_g^f}(A))}.$$

The quantity $I_\rho^{(g,f)}(A)$ and $\mathrm{Corr}_\rho^{s(g,f)}(A,B)$ are said to be generalized metric adjusted skew information and generalized metric adjusted correlation measure, respectively.

Then we have the following proposition.

Proposition 2. *For $A, B \in M_{n,sa}(\mathbb{C})$ and $\rho \in M_{n,+,1}(\mathbb{C})$, we have the following relations, where we put $A_0 = A - Tr[\rho A]I$ and $B_0 = B - Tr[\rho B]I$:*

1. $I_\rho^{(g,f)}(A) = I_\rho^{(g,f)}(A_0) = C_\rho^g(A_0) - C_\rho^{\Delta_g^f}(A_0)$,
2. $J_\rho^{(g,f)}(A) = C_\rho^g(A_0) + C_\rho^{\Delta_g^f}(A_0)$,
3. $U_\rho^{(g,f)}(A) = \sqrt{I_\rho^{(g,f)}(A) \cdot J_\rho^{(g,f)}(A)}$,
4. $Corr_\rho^{s(g,f)}(A,B) = Corr_\rho^{s(g,f)}(A_0, B_0)$.

Theorem 4. *For $f \in \mathcal{F}_{op}^r$, it holds*

$$I_\rho^{(g,f)}(A) \cdot I_\rho^{(g,f)}(B) \geq |Corr_\rho^{s(g,f)}(A,B)|^2,$$

where $A, B \in M_{n,sa}(C)$ and $\rho \in M_{n,+,1}(\mathbb{C})$.

Proof of Theorem 4. We define for $X, Y \in M_n(\mathbb{C})$

$$\mathrm{Corr}_\rho^{s(g,f)}(X,Y) = k\langle i[\rho, X], i[\rho, Y]\rangle_{\rho,f}.$$

Since

$$\begin{aligned}\mathrm{Corr}_\rho^{s(g,f)}(X,Y) &= kTr((i[\rho,X])^* m_f(L_\rho, R_\rho)^{-1} i[\rho, Y]) \\ &= kTr((i(L_\rho - R_\rho)X)^* m_f(L_\rho, R_\rho)^{-1} i(L_\rho - R_\rho)Y) \\ &= Tr(X^* m_g(L_\rho, R_\rho)Y) - Tr(X^* m_{\Delta_g^f}(L_\rho, R_\rho)Y),\end{aligned}$$

it is easy to show that $\mathrm{Corr}_\rho^{s(g,f)}(X,Y)$ is an inner product in $M_n(\mathbb{C})$. Then we can get the result by using the Schwarz inequality.

□

Theorem 5. *For $f \in \mathcal{F}_{op}^r$, if*

$$g(x) + \Delta_g^f(x) \geq \ell f(x) \tag{8}$$

for some $\ell > 0$, then it holds

$$U_\rho^{(g,f)}(A) \cdot U_\rho^{(g,f)}(B) \geq k\ell |Tr(\rho[A,B])|^2, \tag{9}$$

where $A, B \in M_{n,sa}(\mathbb{C})$ and $\rho \in M_{n,+,1}(\mathbb{C})$.

In order to prove Theorem 5, we need the following lemmas.

Lemma 1. *If (7) and (8) are satisfied, then we have the following inequality:*

$$m_g(x,y)^2 - m_{\Delta_g^f}(x,y)^2 \geq k\ell (x-y)^2.$$

Proof of Lemma 1. By (7) and (8), we have

$$m_{\Delta_g^f}(x,y) = m_g(x,y) - k\frac{(x-y)^2}{m_f(x,y)}, \tag{10}$$

$$m_g(x,y) + m_{\Delta_g^f}(x,y) \geq \ell m_f(x,y). \tag{11}$$

Therefore, by (10) and (11),

$$\begin{aligned}
& m_g(x,y)^2 - m_{\Delta_g^f}(x,y)^2 \\
&= \left\{m_g(x,y) - m_{\Delta_g^f}(x,y)\right\}\left\{m_g(x,y) + m_{\Delta_g^f}(x,y)\right\} \\
&\geq k\frac{(x-y)^2}{m_f(x,y)}\ell m_f(x,y) \\
&= k\ell(x-y)^2.
\end{aligned}$$

We have the following expressions for the quantities $I_\rho^{(g,f)}(A)$, $J_\rho^{(g,f)}(A)$, $U_\rho^{(g,f)}(A)$, and $\mathrm{Corr}_\rho^{s(g,f)}(A,B)$ by using Proposition 2 and a mean $m_{\Delta_g^f}$.

Lemma 2. *Let $\{|\phi_1\rangle, |\phi_2\rangle, \ldots, |\phi_n\rangle\}$ be a basis of eigenvectors of ρ, corresponding to the eigenvalues $\{\lambda_1, \lambda_2, \ldots, \lambda_n\}$. We put $a_{jk} = \langle\phi_j|A_0|\phi_k\rangle$, $b_{jk} = \langle\phi_j|B_0|\phi_k\rangle$, where $A_0 \equiv A - Tr[\rho A]I$ and $B_0 \equiv B - Tr[\rho B]I$ for $A, B \in M_{n,sa}(\mathbb{C})$ and $\rho \in M_{n,+,1}(\mathbb{C})$. Then we have*

$$\begin{aligned}
I_\rho^{(g,f)}(A) &= \sum_{j,k} m_g(\lambda_j,\lambda_k)a_{jk}a_{kj} - \sum_{j,k} m_{\Delta_g^f}(\lambda_j,\lambda_k)a_{jk}a_{kj} \\
&= 2\sum_{j<k}\left\{(m_g(\lambda_j,\lambda_k) - m_{\Delta_g^f}(\lambda_j,\lambda_k)\right\}|a_{jk}|^2,
\end{aligned}$$

$$\begin{aligned}
J_\rho^{(g,f)}(A) &= \sum_{j,k} m_g(\lambda_j,\lambda_k)a_{jk}a_{kj} + \sum_{j,k} m_{\Delta_g^f}(\lambda_j,\lambda_k)a_{jk}a_{kj} \\
&\geq 2\sum_{j<k}\left\{m_g(\lambda_j,\lambda_k) + m_{\Delta_g^f}(\lambda_j,\lambda_k)\right\}|a_{jk}|^2,
\end{aligned}$$

$$U_\rho^{(g,f)}(A)^2 = \left(\sum_{j,k} m_g(\lambda_j,\lambda_k)|a_{jk}|^2\right)^2 - \left(\sum_{j,k} m_{\Delta_g^f}(\lambda_j,\lambda_k)|a_{jk}|^2\right)^2,$$

and

$$\begin{aligned}
& Corr_\rho^{s(g,f)}(A,B) \\
&= \sum_{j,k} m_g(\lambda_j,\lambda_k)a_{jk}b_{kj} - \sum_{j,k} m_{\Delta_g^f}(\lambda_j,\lambda_k)a_{jk}b_{kj} \\
&= \sum_{j<k}\left(m_g(\lambda_j,\lambda_k) - m_{\Delta_g^f}(\lambda_j,\lambda_k)\right)a_{jk}b_{kj} + \sum_{j<k}\left(m_g(\lambda_k,\lambda_j) - m_{\Delta_g^f}(\lambda_k,\lambda_j)\right)a_{kj}b_{jk}.
\end{aligned}$$

We are now in a position to prove Theorem 5.

Proof of Theorem 5. At first we prove (9). Since

$$Tr(\rho[A,B]) = \sum_{j,k}(\lambda_j - \lambda_k)a_{jk}b_{kj},$$

$$|Tr(\rho[A,B])| \le \sum_{j,k} |\lambda_j - \lambda_k||a_{jk}||b_{kj}|.$$

Then by Lemma 1, we have

$$\begin{aligned}
& k\ell |Tr(\rho[A,B])|^2 \\
\le & \left\{ \sum_{j,k} \sqrt{k\ell}|\lambda_j - \lambda_k||a_{jk}||b_{kj}| \right\}^2 \\
\le & \left\{ \sum_{j,k} \left(m_g(\lambda_j,\lambda_k)^2 - m_{\Delta_g^f}(\lambda_j,\lambda_k)^2 \right)^{1/2} |a_{jk}||b_{kj}| \right\}^2 \\
\le & \left\{ \sum_{j,k} \left(m_g(\lambda_j,\lambda_k) - m_{\Delta_g^f}(\lambda_j u,\lambda_k) \right) |a_{jk}|^2 \right\} \left\{ \sum_{j,k} \left(m_g(\lambda_j,\lambda_k) + m_{\Delta_g^f}(\lambda_j,\lambda_k) \right) |b_{kj}|^2 \right\} \\
= & I_\rho^{(g,f)}(A) J_\rho^{(g,f)}(B).
\end{aligned}$$

By a similar way, we also have

$$I_\rho^{(g,f)}(B) J_\rho^{(g,f)}(A) \ge k\ell |Tr(\rho[A,B])|^2.$$

Hence, we have the desired inequality (9). □

We give some examples satisfying the condition (8).

Example 2. Let

$$g(x) = \frac{x+1}{2},$$

$$f(x) = \alpha(1-\alpha)\frac{(x-1)^2}{(x^\alpha - 1)(x^{1-\alpha} - 1)}, \quad \alpha \in (0,1),$$

$$k = \frac{f(0)}{2} = \frac{\alpha(1-\alpha)}{2}, \quad \ell = 2.$$

Then

$$g(x) + \Delta_g^f(x) \ge 2f(x).$$

Proof of Example 2. In [10,21] we give

$$(x^{2\alpha} - 1)(x^{2(1-\alpha)} - 1) \ge 4\alpha(1-\alpha)(x-1)^2$$

for $x > 0$ and $0 \le \alpha \le 1$. Then we have

$$g(x) + \Delta_g^f(x) \ge 2f(x).$$

□

Example 3. Let

$$g(x) = \left(\frac{\sqrt{x}+1}{2} \right)^2,$$

$$f(x) = \alpha(1-\alpha)\frac{(x-1)^2}{(x^\alpha - 1)(x^{1-\alpha} - 1)}, \quad \alpha \in (0,1),$$

$$k = \frac{f(0)}{8} = \frac{\alpha(1-\alpha)}{8}, \quad \ell = \frac{3}{2}.$$

Then

$$g(x) + \Delta_g^f(x) \geq \frac{3}{2} f(x)$$

holds for $0 < \alpha < 1$.

Proof of Example 3. Since

$$\begin{aligned}
&\frac{1}{2}\left(\frac{1+\sqrt{x}}{2}\right)^2 - \frac{1}{8}(x^\alpha - 1)(x^{1-\alpha} - 1) \\
&= \frac{1}{8}(x + 2\sqrt{x} + 1 - x - 1 + x^\alpha + x^{1-\alpha}) \\
&= \frac{1}{8}(2\sqrt{x} + x^\alpha + x^{1-\alpha}) \\
&= \frac{1}{8}(x^{\alpha/2} + x^{(1-\alpha)/2})^2 \geq 0,
\end{aligned}$$

we have

$$2\left(\frac{1+\sqrt{x}}{2}\right)^2 \geq \frac{1}{8}(x^\alpha - 1)(x^{1-\alpha} - 1) + \frac{3}{2}\left(\frac{1+\sqrt{x}}{2}\right)^2.$$

Since

$$\alpha(1-\alpha)\frac{(x-1)^2}{(x^\alpha - 1)(x^{1-\alpha} - 1)} \leq \left(\frac{1+\sqrt{x}}{2}\right)^2,$$

we have

$$2\left(\frac{1+\sqrt{x}}{2}\right)^2 \geq \frac{1}{8}(x^\alpha - 1)(x^{1-\alpha} - 1) + \frac{3}{2}\alpha(1-\alpha)\frac{(x-1)^2}{(x^\alpha - 1)(x^{1-\alpha} - 1)}.$$

Then we have

$$g(x) + \Delta_g^f(x) \geq \frac{3}{2} f(x)$$

□

Example 4. Let

$$g(x) = \left(\frac{x^\gamma + 1}{2}\right)^{1/\gamma} \quad (\frac{3}{4} \leq \gamma \leq 1),$$

$$f(x) = \left(\frac{\sqrt{x}+1}{2}\right)^2,$$

$$k = \frac{f(0)}{4} = \frac{1}{16}, \quad \ell = 2.$$

Then $g(x) + \Delta_g^f(x) \geq 2f(x)$.

In order to prove Example 4, we need the following lemma.

Lemma 3. *For $x > 0$, we set the function of y as*

$$F(y) \equiv \left(\frac{1+x^y}{2}\right)^{1/y}.$$

Then $F(y)$ has following properties:

1. *$F(y)$ is monotone increasing for $y \in \mathbb{R}$.*

2. $F(y)$ is convex for $y < 0$.
3. $F(y)$ is concave for $y \geq 1/2$.

We give the proof of Lemma 3 in the Appendix.

Proof of Example 4. By Lemma 3,

$$2\left(\frac{1+x^{3/4}}{2}\right)^{4/3} \geq \frac{1+x}{2} + \left(\frac{1+\sqrt{x}}{2}\right)^2.$$

It follows from the monotonicity that

$$\left(\frac{1+x^y}{2}\right)^{1/y} \geq \left(\frac{1+x^{3/4}}{2}\right)^{4/3}$$

for $y \in [3/4, 1]$. Then

$$2\left(\frac{1+x^y}{2}\right)^{1/y} \geq \frac{1+x}{2} + \left(\frac{1+\sqrt{x}}{2}\right)^2$$

for $y \in [3/4, 1]$. Therefore, we have

$$2\left(\frac{1+x^y}{2}\right)^{1/y} - \left(\frac{\sqrt{x}-1}{2}\right)^2 \geq 2\left(\frac{\sqrt{x}+1}{2}\right)^2.$$

Hence, we have

$$g(x) + \Delta_g^f(x) \geq 2f(x).$$

□

Appendix

Proof of Lemma 3.

(i) Since $F(y) > 0$ for $x > 0$ and $t \in \mathbb{R}$, it is sufficient to prove $\frac{d}{dy}\log F(y) > 0$ for the proof of $F'(y) > 0$. We have

$$\frac{d}{dy}\log F(y) = \frac{1}{y^2}\left(\log 2 + \frac{x^y \log x^y}{1+x^y} - \log\left(1+x^y\right)\right).$$

Then we put

$$G(r) \equiv (r+1)\log 2 + r\log r - (r+1)\log(r+1), (r > 0),$$

where we put $x^y \equiv r > 0$. From elementary calculations, we have $G(r) \geq G(1) = 0$ which implies $\frac{d}{dy}\log F(y) > 0$.

(ii) We firstly set $f(y) \equiv \log F(y)$. Since $F(y) > 0$, we have only to prove $f''(y) > 0$ for the proof of $F''(y) > 0$. We set again $g(y) \equiv \frac{1+x^y}{2}, (x > 0, y < 0)$. Then we have $\frac{d^2}{dy^2}\log g(y) \equiv \frac{x^y(\log x)^2}{(1+x^y)^2} > 0$. In addition, by $f(y) = \frac{1}{y}\log g(y)$, we have

$$f'(y) = \frac{1}{y}\frac{g'(y)}{g(y)} - \frac{1}{y^2}\log g(y) > 0.$$

By $\frac{d^2}{dy^2}\log g(y) = \frac{g(y)g''(y)-\{g'(y)\}^2}{g(y)^2}$, we have

$$f''(y) = \frac{1}{y}\frac{g(y)g''(y)-\{g'(y)\}^2}{g(y)^2} - \frac{2}{y^2}\frac{g'(y)}{g(y)} + \frac{2}{y^3}\log g(y) = \frac{1}{y}\frac{d^2}{dy^2}\log g(y) - \frac{2}{y}f'(y).$$

We prove $f''(y) > 0$ for $y < 0$. We calculate

$$\begin{aligned} f''(y) &= \frac{1}{y}\frac{x^y(\log x)^2}{(1+x^y)^2} - \frac{2}{y}\frac{1}{y^2}\left(\log 2 + \frac{x^y \log x^y}{1+x^y} - \log\left(1+x^y\right)\right) \\ &= \frac{1}{y^3(1+x^y)^2}\left\{-2x^y\left(1+x^y\right)\log x^y + x^y\left(\log x^y\right)^2 + 2\left(1+x^y\right)^2\log\frac{1+x^y}{2}\right\}. \end{aligned}$$

Thus, if we put

$$h(y) \equiv -2x^y\left(1+x^y\right)\log x^y + x^y\left(\log x^y\right)^2 + 2\left(1+x^y\right)^2\log\frac{1+x^y}{2},$$

then we have only to prove $h(y) < 0$ for $y < 0$. Since we have $h(0) = 0$, we have only to prove $h'(y) > 0$ for $y < 0$. Here we have

$$h'(y) = -x^y\log x\left\{4x^y\log x^y - \left(\log x^y\right)^2 - 4\left(1+x^y\right)\log\frac{1+x^y}{2}\right\}.$$

If we set again

$$l(t) \equiv 4t\log t - \left(\log t\right)^2 - 4(t+1)\log\frac{t+1}{2},$$

where we put $x^y \equiv t > 0$, then we prove the following cases:

(a) If $x < 1$ (i.e., $t > 1$), then $l(t) > 0$.
(b) If $x > 1$ (i.e., $0 < t < 1$), then $l(t) < 0$.

For case (a), we calculate

$$l'(t) = \frac{1}{t}\left(4t\log 2 + (4t-2)\log t - 4t\log(t+1)\right)$$

and

$$l''(t) = \frac{2\left\{(t+1)\log t + t - 1\right\}}{t^2(t+1)} > 0, (t > 1).$$

Thus, we have $l'(t) \geq l'(1) = 0$, and then we have $l(t) \geq l(1) = 0$. For case (b), we easily find that

$$l''(t) = \frac{2\left\{(t+1)\log t + t - 1\right\}}{t^2(t+1)} < 0, (0 < t < 1).$$

Thus, we have $l'(t) \geq l'(1) = 0$, and then we have $l(t) \leq l(1) = 0$.

(iii) We calculate

$$\frac{d^2}{dy^2}F(y) = \frac{1}{y^4}\left(\frac{1+x^y}{2}\right)^{1/y} h(x,y),$$

where

$$\begin{aligned} h(x,y) &= (\log 2 - 2y)\log 2 + \frac{2\log 2}{1+x^y}\{x^y\log x^y - (1+x^y)\log(1+x^y)\} \\ &\quad + \frac{1}{(1+x^y)^2}\{x^y y^2(x^y+y)(\log x)^2\} \\ &\quad - \frac{1}{(1+x^y)^2}\{2x^y(1+x^y)(y+\log(1+x^y))\log x^y\} \\ &\quad + \{2y + \log(1+x^y)\}\log(1+x^y). \end{aligned}$$

We prove $h(x,y) \leq 0$ for $x > 0$ and $y \geq 1/2$. Then we have

$$\frac{dh(x,y)}{dx} = -\frac{x^{-1+y}y^2\log x}{(1+x^y)^3}\left\{(x^y(y-2) - y)\log x^y + 2(1+x^y)\log\left(\frac{1+x^y}{2}\right)\right\}.$$

Here we note that $\frac{dh(1,y)}{dx}=0$. We also put

$$g(x,y)=\left\{x^y(-2+y)-y\right\}\log x^y+2(1+x^y)\log\left(\frac{1+x^y}{2}\right).$$

If we have $g(x,y)\geq 0$ for $x>0$ and $y\geq 1/2$, then we have $\frac{dh(x,y)}{dx}\geq 0$ for $0<x\leq 1$ and $\frac{dh(x,y)}{dx}\leq 0$ for $x\geq 1$. Thus, we then obtain $h(x,y)\leq h(1,y)=0$ for $y\geq 1/2$, due to $\frac{dh(1,y)}{dx}=0$. Therefore, we have only to prove $g(x,y)\geq 0$ for $x>0$ and $y\geq 1/2$.

(a) For the case $0<x\leq 1$, we have

$$\frac{dg(x,y)}{dx}=\frac{y}{x}\left\{y(x^y-1)+(y-2)x^y\log x^y+2x^y\log\left(\frac{x^y+1}{2}\right)\right\}.$$

Since $g(1,y)=0$, if we prove $\frac{dg(x,y)}{dx}\leq 0$, then we can prove $g(x,y)\geq g(1,y)=0$ for $y\geq 1/2$ and $0<x\leq 1$. Since we have the relations

$$\frac{x-1}{\sqrt{x}}\leq\log x\leq\frac{2(x-1)}{x+1}\leq 0$$

for $0<x\leq 1$, we calculate

$$\begin{aligned}&y\left(x^y-1\right)+(y-2)x^y\log x^y+2x^y\log\left(\frac{x^y+1}{2}\right)\\&\leq y\left(x^y-1\right)+(y-2)x^y\frac{(x^y-1)}{x^{y/2}}+2x^y\frac{2\left(\frac{x^y+1}{2}-1\right)}{\frac{x^y+1}{2}+1}\\&=\frac{x^y-1}{x^y+3}\left\{3(y-2)x^{y/2}+(y-2)x^{3y/2}+3y+(y+4)x^y\right\}.\end{aligned}$$

Thus, we have only to prove

$$k(y)\equiv 3(y-2)x^{y/2}+(y-2)x^{3y/2}+3y+(y+4)x^y\geq 0$$

for $0<x\leq 1$ and $y\geq 1/2$. Since it is trivial $k(y)\geq 0$ for $y\geq 2$, we assume $1/2\leq y<2$ from here. To this end, we prove that $k_1(y)\equiv 3(y-2)x^{y/2}+(y-2)x^{3y/2}$ is monotone increasing for $1/2\leq y<2$ and $k_2(y)\equiv 3y+(y+4)x^y$ is also monotone increasing for $1/2\leq y<2$. We easily find that

$$\frac{dk_1(y)}{dy}=\frac{1}{2}x^{y/2}\left\{2(x^y+3)+3(x^y+1)(y-2)\log x\right\}>0,$$

for $0<x\leq 1$ and $1/2\leq y<2$.

We also have

$$\frac{dk_2(y)}{dy}=x^y+3+(y+4)x^y\log x.$$

Here we prove $\frac{dk_2(y)}{dy}\geq 0$ for $0<x\leq 1$ and $1/2\leq y<2$. We put again

$$k_3(x)\equiv x^y+3+(y+4)x^y\log x,$$

then we have

$$\frac{dk_3(x)}{dx}=x^{-1+y}\left\{2(y+2)+y(y+4)\log x\right\}.$$

Thus, we have

$$\frac{dk_3(x)}{dx}=0\Leftrightarrow x=e^{-\frac{2(y+2)}{y(y+4)}}\equiv\alpha_y.$$

Since $\frac{dk_3(x)}{dx} < 0$ for $0 < x < \alpha_y$ and $\frac{dk_3(x)}{dx} > 0$ for $\alpha_y < x \leq 1$, we have

$$k_3(x) \geq k_3(\alpha_y) = 3 - \frac{(y+4)e^{-\frac{2(y+2)}{y+4}}}{y} \equiv k_4(y).$$

Since we have $\frac{dk_4(y)}{dy} = \frac{8(y+2)e^{-\frac{2(y+2)}{y+4}}}{y^2(y+4)} > 0$, the function $k_4(y)$ is monotone increasing for y. Thus, we have

$$k_3(x) \geq k_3(\alpha_y) = 3 - \frac{(y+4)e^{-\frac{2(y+2)}{y+4}}}{y} \equiv k_4(y) \geq k_4(1/2) = 3 - \frac{9}{e^{10/9}} > 0$$

since $e^{10/9} \simeq 3.03773$. Therefore, $k_2(y)$ is also a monotone increasing function of y for $0 < x \leq 1$ and $1/2 \leq y < 2$. Thus, $k(y)$ is monotone increasing for $y \geq 1/2$, and then we have

$$k(y) \geq k(1/2) = -\frac{3}{2}\left(x^{1/4} - 1\right)^3 \geq 0.$$

(b) For the case $x \geq 1$, we firstly calculate

$$\frac{dg(x,y)}{dy} = \left(x^y - 1\right)\log x^y + \left\{y\left(x^y - 1\right) + (y-2)\,x^y \log x^y + 2x^y \log\left(\frac{1+x^y}{2}\right)\right\}\log x.$$

We put

$$p(x,y) \equiv (x^y - 1)y + x^y(y-2)\log x^y + 2x^y \log\left(\frac{1+x^y}{2}\right).$$

Then we calculate

$$\frac{dp(x,y)}{dx} = \frac{y}{x + x^{1-y}}\left\{(1+x^y)(y-2)\log x^y + 2\left(y(1+x^y) - 1 + (1+x^y)\log\left(\frac{1+x^y}{2}\right)\right)\right\}.$$

Then we put

$$q(x,y) = (y-2)\log x^y + 2\log\left(\frac{1+x^y}{2}\right) + 2y - \frac{2}{1+x^y}.$$

We have

$$\frac{dq(x,y)}{dy} = \frac{((1+x^y)^2 y - 2)\log x + (1+x^y)^2(\log x^y + 2)}{(1+x^y)^2} > 0$$

and then

$$q(x,y) \geq q(x,1/2) = 1 - \frac{2}{\sqrt{x}+1} + 2\log\left(\frac{1+\sqrt{x}}{2}\right) - \frac{3}{4}\log x.$$

Since we find

$$\frac{dq(x,1/2)}{dx} = \frac{(\sqrt{x}+3)(\sqrt{x}-1)}{4x(\sqrt{x}+1)^2} \geq 0$$

for $x \geq 1$, we have $q(x,y) \geq q(x,1/2) \geq q(1,1/2) = 0$. Therefore, we have $\frac{dp(x,y)}{dx} \geq 0$, which implies $p(x,y) \geq p(1,y) = 0$. Thus, we have $\frac{dg(x,y)}{dy} \geq 0$, and then we have $g(x,y) \geq g(x,1/2)$, where

$$g(x,1/2) = -\frac{1}{2}(3x^{1/2}+1)\log x^{1/2} + 2(x^{1/2}+1)\log\left(\frac{x^{1/2}+1}{2}\right).$$

To prove $g(x, 1/2) \geq 0$ for $x \geq 1$ and $y \geq 1/2$, we put $x^{1/2} \equiv z \geq 1$ and

$$r(z) \equiv -\frac{1}{2}(3z+1)\log z + 2(z+1)\log\left(\frac{z+1}{2}\right).$$

Since we have $r''(z) = \frac{(z-1)^2}{2z^2(z+1)} \geq 0$ and

$$r'(z) = \frac{1}{2z}\left\{z - 1 - 3z\log z + 4z\log\left(\frac{z+1}{2}\right)\right\},$$

we have $r'(1) = 0$ and then we have $r'(z) \geq 0$ for $z \geq 1$. Thus, we have $r(z) \geq 0$ for $z \geq 1$ by $r(1) = 0$. Finally, we have $g(x, y) \geq g(x, 1/2) \geq 0$, for $x \geq 1$ and$y \geq 1/2$.

□

Acknowledgements

The first author (KY) was partially supported by JSPS KAKENHI Grant Number 23540208. The second author (SF) was partially supported by JSPS KAKENHI Grant Number 24540146.

Author details

[1]Graduate School of Science and Engineering, Yamaguchi University, Une, Yamaguchi 755-8611, Japan. [2]College of Humanities and Science, Nihon University, Tokyo 156-8550, Japan. [3]Faculty of Education, Bukkyo University, Kyoto 603-8301, Japan.

References

1. Heisenberg, W: Uber den anschaulichen Inhalt der quantummechanischen Kinematik und Mechanik. Zeitschrift für Physik. **43**, 172–198 (1927)
2. Robertson, HP: The uncertainty principle. Phys. Rev. **34**, 163–164 (1929)
3. Schrödinger, E: About Heisenberg uncertainty relation. Proc. Prussian Acad. Sci. Phys. Math. **XIX**, 293 (1930)
4. Luo, S: Heisenberg uncertainty relation for mixed states. Phys. Rev. A. **72**, 042110 (2005)
5. Wigner, EP, Yanase, MM: Information content of distribution. Proc. Nat. Acad. Sci. **49**, 910–918 (1963)
6. Luo, S, Zhang, Q: Informational distance on quantum-state space. Phys. Rev. A. **69**, 032106 (2004)
7. Luo, S: Quantum versus classical uncertainty. Theor. Math. Phys. **143**, 681–688 (2005)
8. Yanagi, K: Uncertainty relation on Wigner-Yanase-Dyson skew information. J. Math. Anal. Appl. **365**, 12–18 (2010)
9. Lieb, EH: Convex trace functions and the Wigner-Yanase-Dyson conjecture. Adv. Math. **11**, 267–288 (1973)
10. Yanagi, K: Uncertainty relation on generalized Wigner-Yanase-Dyson skew information. Linear Algebra Appl. **433**, 1524–1532 (2010)
11. Cai, L, Luo, S: On convexity of generalized Wigner-Yanase-Dyson information. Lett. Math. Phys. **83**, 253–264 (2008)
12. Gibilisco, P, Hansen, F, Isola, T: On a correspondence between regular and non-regular operator monotone functions. Linear Algebra Appl. **430**, 2225–2232 (2009)
13. Petz, D: Monotone metrics on matrix spaces. Linear Algebra Appl. **244**, 81–96 (1996)
14. Petz, D, Hasegawa, H: On the Riemannian metric of α-entropies of density matrices. Lett. Math. Phys. **38**, 221–225 (1996)
15. Furuta, T: Elementary proof of Petz-Hasegawa theorem. Lett. Math. Phys. **101**, 355–359 (2012)
16. Gibilisco, P, Imparato, D, Isola, T: Uncertainty principle and quantum Fisher information, II. J. Math. Phys. **48**, 072109 (2007)
17. Kubo, F, Ando, T: Means of positive linear operators. Math. Ann. **246**, 205–224 (1980)
18. Hansen, F: Metric adjusted skew information. Proc. Nat. Acad. Sci. **105**, 9909–9916 (2008)
19. Gibilisco, P, Isola, T: On a refinement of Heisenberg uncertainty relation by means of quantum Fisher information. J. Math. Anal. Appl. **375**, 270–275 (2011)
20. Furuichi, S, Yanagi, K: Schrödinger uncertainty relation, Wigner-Yanase-Dyson skew information and metric adjusted correlation measure. J. Math. Anal. Appl. **388**, 1147–1156 (2012)
21. Yanagi, K: Metric adjusted skew information and uncertainty relation. J. Math. Anal. Appl. **380**, 888–892 (2011)
22. Gibilisco, P, Hiai, F, Petz, D: Quantum covariance, quantum Fisher information, and the uncertainty relations. IEEE Trans. Inf. Theory. **55**, 439–443 (2009)

11

A democratic preference aggregation model

Ismat Beg[1*] and Tabasam Rashid[2,3]

*Correspondence: begismat@yahoo.com
[1] Centre for Mathematics and Statistical Sciences, Lahore School of Economics, Lahore 53200,Pakistan
Full list of author information is available at the end of the article

Abstract

We propose a democratic preference aggregation function using trapezoidal fuzzy numbers. Examples are constructed to illustrate the proposed technique to optimize the problem by avoiding paradoxical outcomes without the fear of indecision.

2010 Mathematics Subject Classification: 91B52; 03B52

Keywords: Democracy; Preference aggregation; Condorcet paradox; Doctrinal paradox; Trapezoidal fuzzy number

Background

Democracy is based on the idea that groups make better decisions than individuals. However, majority voting is unable to ensure consistent social positions under all situations. How individual preferences can be aggregated into a collectively preferred alternative is recently studied by Arrow [1] and Sen [2]. In economics and formal epistemology, aggregation of individual preferences on logically related basic propositions into a collective preference is an emerging research area. Decision theory also studies the individual preference aggregation method. Preference aggregation is defined in social choice theory as forming a collective preference in a given set of alternatives. There are two kinds of proposition: the first kind is the premises and the other is the conclusion. Consider, for example, in a city traffic police department, there is a committee of five inspectors to decide to issue a driving license. The decision rule is such that a candidate C will be issued a driving license only if the candidate is physically fit and good in driving. We will say that issuing of a driving license to C is the conclusion (denoted by proposition D, resulting decision) while physically fit (denoted by proposition P) and good in driving (denoted by proposition Q) are the premises. The decision rule can be expressed in the formula $(P \wedge Q) \longleftrightarrow D$. Each inspector of the committee expresses his preference on P, Q, and D such that the rule $(P \wedge Q) \longleftrightarrow D$ is satisfied.

How shall we derive a group decision given the individuals' opinions on premises and conclusion? Assume that each individual expresses preference in the form of a binary decision (Yes/No) on the propositions. If we define the decision as the majority voting, then we may reach an inconsistent position, as shown in Table 1. In the literature on preference aggregation, such a problem is known as discursive dilemma and doctrinal paradox.

Although each inspector expresses a consistent preference, proposition-wise majority voting results in a majority for P and Q but a majority for $\neg D$, which is an inconsistent

Table 1 Discursive dilemma and doctrinal paradox

	P	Q	$(P \wedge Q) \longleftrightarrow D$	D
Inspectors 1 and 5	Yes	Yes	Yes	Yes
Inspectors 2 and 4	Yes	No	Yes	No
Inspector 3	No	Yes	Yes	No
Majority	Yes	Yes	Yes	?

collective result as it violates the rule $(P \wedge Q) \longleftrightarrow D$. The paradox rests with the fact that proposition-wise majority voting can lead a group of rational individuals to endorse an inconsistent collective preference. Such aggregation problems go beyond the specific example: it can be observed in all situations where individual binary decisions need to be combined into a preference of the group.

Condorcet [3] was the first to initiate a research on social choice theory that individual preferences can be aggregated into a collectively preferred alternative. Condorcet made a result that majority voting was a good method for a collectively preferred alternative, but they also found a problem in majority voting method. He suggested a method which is consisted of the pairwise comparison of each alternative. Majority voting decides the winner in each pair, and the collective result is obtained by the combination of all partial results.

Consider, for example, there are three alternatives x, y, and z and five voters V_1, V_2, V_3, V_4, and V_5. Let $\succ_i$ denote voter V_i's preference over $X = \{x, y, z\}$ and $\succ$ denote the collective preference result over X. $V_1 = \{x \succ_1 y, y \succ_1 z\}$, $V_2 = \{y \succ_2 z, z \succ_2 x\}$, $V_3 = \{z \succ_3 x, x \succ_3 y\}$, $V_4 = \{y \succ_2 z, z \succ_2 x\}$, and $V_5 = \{x \succ_1 y, y \succ_1 z\}$ are the five voters' preferences (see Table 2). Transitivity holds for each voter's preference, but unfortunately transitivity does not hold for collective preference 'V_S' (by majority method). This situation is called the Condorcet paradox.

When we combine individual decisions into a collective decision, we may have to face inconsistency in the collective level, like logical consistency and transitivity (in preference aggregation).

Realistic collective decision problems are handled with some extension beyond binary propositional logic into the realm of general multivalued logic (see [4-7]). Duddy and Piggins also used extension in propositional logic in degrees [8]. Beg and Butt [5] proved the (im)possibility theorem in fuzzy logic, similar to those of Arrow [1] and Sen [2]. There cannot exist any preference aggregation procedure that simultaneously satisfies certain consistency conditions [9].

Often, human preference is ambiguous and vague and cannot be estimated with an exact numeric value, so the binary values are not suitable to model real-world situations. Bellman and Zadeh [10] used the concept of fuzzy set theory to handle imprecision (or uncertainty) to solve the ambiguity and vagueness in human preference. Fuzzy logic

Table 2 Condorcet paradox

	$x \succ y$	$y \succ x$	$y \succ z$	$z \succ y$	$x \succ z$	$z \succ x$
V_1 and V_5	1	0	1	0	1	0
V_2 and V_4	0	1	1	0	0	1
V_3	1	0	0	1	0	1
V_S	1	0	1	0	0	1

can help in aggregation procedures to make the collective preference set democratic in nature. Trapezoidal fuzzy numbers are the best way to model preferences. Recently, a lot of work on trapezoidal fuzzy numbers has been done by Li et al. [11-13]. In this paper, trapezoidal fuzzy numbers are used for the truth value of the proposition. The rest of the paper is structured as follows: In the 'Methods' section, some basic relations and operations for trapezoidal fuzzy numbers are given and problems are reformulated in the setting of trapezoidal fuzzy numbers. The 'Results' section illustrates how the paradoxes are resolved in trapezoidal fuzzy numbers to find democratic fuzzy aggregation function. The 'Discussion' section presents the discussion in utility maximization. Conclusion of the whole work is given in the last section.

Methods

In some problems, propositions are vague and thus can have truth values between true and false. Beg and Butt [6] tried to use fuzzy logic for the solution of doctrinal and Condercet paradoxes. We can use trapezoidal fuzzy numbers to model the vagueness of human knowledge.

Definition 2.1. *A function 'A' given by*

$$A(x) = \begin{cases} 0 & if\ x < x_1 \text{or } x > x_4, \\ \frac{x-x_1}{x_2-x_1} & if\ x_1 \leq x < x_2, \\ 1 & if\ x_2 \leq x < x_3, \\ \frac{x-x_4}{x_3-x_4} & if\ x_3 \leq x \leq x_4, \end{cases}$$

where $x_1 < x_2 < x_3 < x_4$, *is called a trapezoidal fuzzy number. Symbolically, A is denoted by* (x_1, x_2, x_3, x_4) *(see p. 330 in* [14]).

Let $D[0, 1]$ denote the set of all trapezoidal fuzzy numbers such that $0 \leq x_1$ and $x_4 \leq 1$. We assume that individual preferences take values on $D[0, 1]$.

Definition 2.2. ([15]) *We define an operator* Δ *on trapezoidal fuzzy numbers, which is given by*

$$A\Delta B = (0 \vee (x_1 + y_1 - 1), 0 \vee (x_2 + y_2 - 1), 0 \vee (x_3 + y_3 - 1), 0 \vee (x_4 + y_4 - 1)),$$

where $A = (x_1, x_2, x_3, x_4)$, $B = (y_1, y_2, y_3, y_4)$, *and* $A, B \in D[0, 1]$. *Also,* $A = B$ if $x_1 = y_1$, $x_2 = y_2$, $x_3 = y_3$, *and* $x_4 = y_4$.

Definition 2.3. *The constant 'k' multiple of trapezoidal fuzzy number A is defined as* $kA = k(x_1, x_2, x_3, x_4) = (k \times x_1, k \times x_2, k \times x_3, k \times x_4)$ *where* $A = (x_1, x_2, x_3, x_4)$, $A \in D[0, 1]$, *and* $k \in [0, 1]$.

Definition 2.4. ([15]) *The relation* $\preceq$ *in* $D[0, 1]$ *is introduced as follows: Let* $A = (x_1, x_2, x_3, x_4)$, $B = (y_1, y_2, y_3, y_4)$, *and* $A, B \in D[0, 1]$:

If $x_4 < y_4$, then $A \prec B$;
If $x_4 = y_4$ and

(i) $x_3 < y_3$, then $A \prec B$;

(ii) $x_3 = y_3$ and

(a) $x_2 < y_2$, then $A \prec B$;
(b) $x_2 = y_2$ and $x_1 < y_1$, then $A \prec B$.

Obviously, $D[0,1]$ is an ordered set.

Definition 2.5. ([15]) *In* $D[0,1]$,

$$\min(A,B) = (\min(x_1,y_1), \min(x_2,y_2), \min(x_3,y_3), \min(x_4,y_4))$$

and

$$\max(A,B) = (\max(x_1,y_1), \max(x_2,y_2), \max(x_3,y_3), \max(x_4,y_4))$$

are called min *and* max *operators for trapezoidal fuzzy numbers, respectively.*

Definition 2.6. ([15]) *A fuzzy implication* $\Longrightarrow$ *for trapezoidal fuzzy numbers is a map*

$$\Longrightarrow: D[0,1] \times D[0,1] \to D[0,1]$$

satisfying

(i) $[(0,0,0,0) \Longrightarrow (0,0,0,0)] = (1,1,1,1)$,
(ii) $[(0,0,0,0) \Longrightarrow (1,1,1,1)] = (1,1,1,1)$,
(iii) $[(1,1,1,1) \Longrightarrow (0,0,0,0)] = (0,0,0,0)$,
(iv) $[(1,1,1,1) \Longrightarrow (0,0,0,0)] = (1,1,1,1)$.

An example of a fuzzy implication for trapezoidal fuzzy numbers A and B is

$$A \Longrightarrow B = \begin{cases} (1-\max(|x_1-y_1|,|x_2-y_2|,|x_3-y_3|,|x_4-y_4|), \\ 1-\max(|x_1-y_1|,|x_2-y_2|,|x_3-y_3|,|x_4-y_4|), \\ 1-\min(|x_1-y_1|,|x_2-y_2|,|x_3-y_3|,|x_4-y_4|), & ;\text{if } A \succ B \\ 1-\min(|x_1-y_1|,|x_2-y_2|,|x_3-y_3|,|x_4-y_4|)) \\ (1,1,1,1) & ;\text{if } A \preceq B. \end{cases}$$

Definition 2.7. *The weighted average operator 'avg' of trapezoidal fuzzy numbers is defined as follows: Let* $A_1 = (x_{11}, x_{12}, x_{13}, x_{14})$, $A_2 = (x_{21}, x_{22}, x_{23}, x_{24})$, ..., $A_n = (x_{n1}, x_{n2}, x_{n3}, x_{n4})$, *and* $A_1, A_2, \ldots, A_n \in D[0,1]$:

$$\begin{aligned} &\mathrm{avg}(A_1, A_2, \ldots, A_n) \\ &= \left(\frac{w_1 \times x_{11} + w_2 \times x_{21} + \cdots + w_n \times x_{n1}}{w_1 + w_2 + \cdots + w_n}, \frac{w_1 \times x_{12} + w_2 \times x_{22} + \cdots + w_n \times x_{n2}}{w_1 + w_2 + \cdots + w_n}, \right. \\ &\left. \frac{w_1 \times x_{13} + w_2 \times x_{23} + \cdots + w_n \times x_{n3}}{w_1 + w_2 + \cdots + w_n}, \frac{w_1 \times x_{14} + w_2 \times x_{24} + \cdots + w_n \times x_{n4}}{w_1 + w_2 + \cdots + w_n}\right), \end{aligned}$$

where w_i *is the weight assigned to truth value* A_i.

Note that if $w_i = 1$ for all i, then $\mathrm{avg}(A_1, A_2, \ldots, A_n)$ is denoted by $\mathrm{average}(A_1, A_2, \ldots, A_n)$.

Example 2.8. Here we reformulate our problem in a trapezoidal fuzzy number, which is illustrated and summarized in Table 1. Table 3 is a clear illustration of Table 1 for trapezoidal fuzzy numbers and also the average of inspectors' decision for each proposition. In this case, we get at least one solution of doctrinal paradox for trapezoidal fuzzy numbers.

One important property of majority selection: min((0.45, 0.48, 0.54, 0.59), (0.15, 0.19, 0.27, 0.31), (0.75, 0.79, 0.85, 0.89)) $\preceq (\theta_1(x_1), \theta_1(x_2), \theta_1(x_3), \theta_1(x_4)) \preceq$ max((0.45, 0.48, 0.54, 0.59), (0.15, 0.19, 0.27, 0.31), (0.75, 0.79, 0.85, 0.89)), min((0.61, 0.65, 0.71, 0.75), (0.61, 0.65, 0.71, 0.75), (0.05, 0.09, 0.16, 0.22)) $\preceq (\theta_2(x_1), \theta_2(x_2), \theta_2(x_3), \theta_2(x_4)) \preceq$ max((0.61, 0.65, 0.71, 0.75), (0.61, 0.65, 0.71, 0.75), (0.05, 0.09, 0.16, 0.22)), and min((0.45, 0.48, 0.54, 0.59), (0.15, 0.19, 0.27, 0.31), (0.05, 0.09, 0.16, 0.22)) $\preceq (\theta_3(x_1), \theta_3(x_2), \theta_3(x_3), \theta_3(x_4)) \preceq$ max((0.45, 0.48, 0.54, 0.59), (0.15, 0.19, 0.27, 0.31), (0.05, 0.09, 0.16, 0.22)) is in Table 3.

Simply, $(0.15, 0.19, 0.27, 0.31) \preceq (\theta_1(x_1), \theta_1(x_2), \theta_1(x_3), \theta_1(x_4)) \preceq (0.75, 0.79, 0.85, 0.89)$, $(0.05, 0.09, 0.16, 0.22) \preceq (\theta_2(x_1), \theta_2(x_2), \theta_2(x_3), \theta_2(x_4)) \preceq (0.61, 0.65, 0.71, 0.75)$ and $(0.05, 0.09, 0.16, 0.22) \preceq (\theta_3(x_1), \theta_3(x_2), \theta_3(x_3), \theta_3(x_4)) \preceq (0.45, 0.48, 0.54, 0.59)$ such that $0.15 \leq \theta_1(x_1) \leq 0.75$, $0.19 \leq \theta_1(x_2) \leq 0.79$, $0.27 \leq \theta_1(x_3) \leq 0.85$, $0.31 \leq \theta_1(x_4) \leq 0.89$, $0.05 \leq \theta_2(x_1) \leq 0.61$, $0.09 \leq \theta_2(x_2) \leq 0.65$, $0.16 \leq \theta_2(x_3) \leq 0.71$, $0.22 \leq \theta_2(x_4) \leq 0.75$, $0.05 \leq \theta_3(x_1) \leq 0.45$, $0.09 \leq \theta_3(x_2) \leq 0.48$, $0.16 \leq \theta_3(x_3) \leq 0.54$, and $0.22 \leq \theta_3(x_4) \leq 0.59$, but the values of $\theta_i(x_j)$ will satisfy the condition for trapezoidal fuzzy numbers of $D[0,1]$.

At the same time, let $\pi(P)$ denote the degree of truth of the proposition P, and the fuzzy integrity constraint is $\{\pi(P)\Delta\pi(Q) \Longrightarrow \pi(R)\}$. Suppose that the committee members are rational and they never violate the fuzzy integrity constraints (IC) (see List [16]). Now by using the above given Δ operator and fuzzy implication $\Longrightarrow$, the IC can be translated as $\pi(R) \geq \max(0, \pi(P) + \pi(Q) - 1)$: Here $\pi(R) = f_i(\pi(P), \pi(Q))$ is a particular rule of inference for individual i (see [17]).

Results

We assume that decision makers are rational and they have freedom to fully express their opinions on a proposition with which they do not agree or disagree. So any number from $D[0,1]$ that best representstheir opinions can be opted. A finite set of n individuals and a finite set X of propositions over which individuals have to make their preferences are called an *agenda*. A preference set A_i for an individual i is an n-tuple containing the degree of truth for each proposition. Let $A_i = (h_{i1}, h_{i2}, \ldots, h_{i|X|})$, where $|X|$ denotes the

Table 3 Revisit of Table 1 for trapezoidal fuzzy numbers

	P	Q	R	$(P\Delta Q) \Longrightarrow R$
Inspectors 1 and 5	(0.45, 0.48, 0.54, 0.59)	(0.61, 0.65, 0.71, 0.75)	(0.45, 0.48, 0.54, 0.59)	(1, 1, 1, 1)
Inspectors 2 and 4	(0.15, 0.19, 0.27, 0.31)	(0.61, 0.65, 0.71, 0.75)	(0.15, 0.19, 0.27, 0.31)	(1, 1, 1, 1)
Inspector 3	(0.75, 0.79, 0.85, 0.89)	(0.05, 0.09, 0.16, 0.22)	(0.05, 0.09, 0.16, 0.22)	(1, 1, 1, 1)
Average	(0.45, 0.49, 0.55, 0.60)	(0.42, 0.46, 0.53, 0.57)	(0.22, 0.25, 0.32, 0.37)	(1, 1, 1, 1)
Majority	$(\theta_1(x_1), \theta_1(x_2), \theta_1(x_3), \theta_1(x_4))$	$(\theta_2(x_1), \theta_2(x_2), \theta_2(x_3), \theta_2(x_4))$	$(\theta_3(x_1), \theta_3(x_2), \theta_3(x_3), \theta_3(x_4))$	(1, 1, 1, 1)

cardinality of X and $h_{ij} \in D[0,1]$ for $j \in (1, 2, \ldots, |X|)$. A profile is an n-tuple $(A_1, A_2, \ldots, A_n)$ of individual preference sets. Function f is an aggregation that assigns to each profile $(A_1, A_2, \ldots, A_n)$ a collective preference set $(h_1^*, h_2^*, \ldots, h_{|X|}^*) = f(A_1, A_2, \ldots, A_n)$. Here $h_j^* \in D[0,1]$ for $j \in (1, 2, \ldots, |X|)$.

The truth value of some proposition $\varphi \in X$ for the preference set A_i is $A_i(\varphi)$. The truth value of some proposition $\varphi \in X$ for the collective preference set $f(A_1, A_2, \ldots, A_n)$ is $f(A_1, A_2, \ldots, A_n)(\varphi) = f(A_1(\varphi), A_2(\varphi), \ldots, A_n(\varphi))$.

If $f(A_1, A_2, \ldots, A_n) = A_i$ for some $i \in (1, 2, \ldots, n)$ and every $(A_1, A_2, \ldots, A_n)$, then f is 'dictatorship.'

Function f is 'manipulable' if and only if there exist some voter i, proposition φ, and profile $(A_1, A_2, \ldots, A_n)$ such that $A_i(\varphi) \neq f(A_1, A_2, \ldots, A_i, \ldots, A_n)(\varphi)$, but $A_i(\varphi) = f(A_1, A_2, \ldots, A_i^*, \ldots, A_n)(\varphi)$ for some alternate preference set A_i^*.

Function f is 'independent' if and only if for all propositions $\varphi \in X$ there is a function $g_\varphi : D[0,1]^n \to D[0,1]$ such that for all $(A_1, A_2, \ldots, A_n)$, we have $f(A_1, A_2, \ldots, A_n)(\varphi) = g_\varphi(A_1(\varphi), A_2(\varphi), \ldots, A_n(\varphi))$.

Preference aggregation formally investigates how to aggregate a finite number of preference bases into a collective one. This formal framework consists of a propositional language $\aleph$ which is built up from a finite set P of propositional letters standing for atomic propositions. An interpretation is a function from P to $D[0,1]$. Function $\pi(.)$ represents the truth function that maps elements in set P to $D[0,1]$. Let the preference base K_i for the agent i be the following set $(\pi_i(p_1), \pi_i(p_2), \ldots, \pi_i(p_{|P|}))$, where $|P|$ denotes the cardinality of P. A preference set is the set $E = \{K_1, K_2, \ldots, K_n\}$. Given a set of IC in a fuzzy setting, ψ maps E and IC into a new (collective) preference base $\psi_{\text{IC}}(E)$, and this process is called fuzzy aggregation for trapezoidal fuzzy numbers.

Let W denote the set of all interpretations and the distance between interpretations is a real-valued function

$$d : W \times W \to \Re$$

such that for all $w, w', w'' \in W$:

1. $d(w, w') \geq 0$.
2. $d(w, w') = 0$ if and only if $w = w'$.
3. $d(w, w') = d(w', w)$.
4. $d(w, w'') \leq d(w, w') + d(w', w'')$.

One possible choice for distance function is the Hamming distance:

$$\begin{aligned} d^*(w, w') &= \frac{1}{4} \sum_{\forall x \in P} |w(x) - w'(x)| \\ &= \frac{1}{4} \sum_{\forall x \in P} (|w(x_1) - w'(x_1)| + |w(x_2) - w'(x_2)| \\ &\quad + |w(x_3) - w'(x_3)| + |w(x_4) - w'(x_4)|). \end{aligned}$$

Another possible choice for distance function is the square distance:

$$\begin{aligned} d^{**}(w, w') &= \sum_{\forall x \in P} (w(x_1) - w'(x_1))^2 + (w(x_2) - w'(x_2))^2 \\ &\quad + (w(x_3) - w'(x_3))^2 + (w(x_4) - w'(x_4))^2). \end{aligned}$$

Now let us define a preference aggregation operator in this framework. For any interpretation $w \in W$ and any profile of preference basis $K \in K^n$, the distance between an interpretation and a profile can now be defined as

$$D^d(w, K) = \sum_{\forall i} d(w, K_i).$$

Our objective is to choose w which minimizes this distance and does not violate any IC in the fuzzy setting. To minimize the distance is the same as to minimize a measure of disagreement in the society by bringing the collective preference set close to the individual preference sets as possible. Individual disagreement brings about individual disutility. Thus, we seek to minimize the societal disutility which is assumed to be the sum of individual disutilities. We have many distance and dissimilarity measures in the literature like distance 'd_h' [18], distances 'd_{LR}' and 'd_f' [19,20], distance 'D_{MLR}' [21], etc. In this paper, we have chosen the Hamming distance d^* and square distance d^{**}. Choosing any distance or dissimilarity measure is solely at our discretion provided that it satisfies certain normative principles.

Let w be any arbitrary interpretation. In this case, $w(P) = (\theta_1, \theta_2, \ldots, \theta_{|P|})$, where we have $\theta_i \in D[0, 1]$ for all $1 \leq i \leq |P|$ and $|P|$ denotes the cardinality of P.

We can think a collective preference set $\bar{F}$ which is the weighted average of individual preference sets. We believe that the closer our collective preference set F is to $\bar{F}$, the more legitimate and democratic is our final preference. Now we assert that an optimal fuzzy aggregation function is democratic if the solution is as close as possible to the average of individual preferences. Such a view is held only for the purpose of illustration in our paper, and there could be different possible ways for democratization.

A fuzzy aggregation function is democratic as well as optimal if we make the solution of Table 3 as close as possible to the avg(inspector 1, inspector 2, inspector 3, inspector 4, inspector 5). This could be easily achieved by adding the penalty to the objective function. Here we use the distance function $d(w,\text{avg})$ as penalty. So $d(w,\text{avg})$ is the degree of democracy.

Now d^* is a generalization of the Hamming distance, and we can use it in trapezoidal fuzzy numbers to avoid the doctrinal paradox. We can formulate the fuzzy aggregation as an optimization problem which can be stated as

Minimize

$$D^{d^*}(w, K) = \sum_{\forall i} d^*(w, K_i)$$

and

$$d^*(w, \text{avg})$$

subject to the fuzzy IC.

Here $w(P) = (\theta_1, \theta_2, \ldots, \theta_{|P|})$, $\min\{K_1(j), K_2(j), \ldots, K_n(j)\} \leq \theta_j \leq \max\{K_1(j), K_2(j), \ldots, K_n(j)\}$, and $\theta_j \in D[0, 1]$ for $j \in \{1, 2, 3, \ldots, |P|\}$, where $K_i(j)$ denotes the jth element of the preference base K_i.

The above optimization problem helps us to avoid doctrinal paradox and find an optimal fuzzy aggregation function. We say that an aggregation function is optimal if the collective preference set is close to the individual preference set and weighted average of individual preference sets as possible. Finding a collective social choice function in Table 3 now becomes an optimization problem which can have multiple optimal solutions. The problem in Table 3 can be democratized in d^* and d^{**}. The optimal fuzzy aggregation function gives the solution for Table 3 as $(\theta_1, \theta_2, \theta_3) = ((0.45, 0.48, 0.54, 0.59), (0.42, 0.621, 0.621, 0.621), (0.151, 0.249, 0.32, 0.367))$ with minimum $D^{d^*} + d^*(w,\text{average}) = 1.8527$. The optimal fuzzy aggregation function gives the solution for Table 3 as $(\theta_1, \theta_2, \theta_3) = ((0.45, 0.487, 0.553, 0.598), (0.422, 0.462, 0.462, 0.572), (0.217, 0.252, 0.405, 0.405))$ with minimum $D^{d^{**}} + d^{**}(w, \text{average}) = 1.9164$. The fact that there is at least one solution to the problem shows that doctrinal paradox cannot occur in this case.

We like our aggregation procedure to be strategy proof. Impossibility theorems proved by Dietrich and List [22] were similar to the Gibbard-Satterthwaite theorem on strategy-proof aggregation rules. Given these theorems, we do not claim that our distance-based aggregation method is strategy proof. In fact, Dietrich [9] has proved that independence and monotonicity are properties of an aggregator that result in strategy proofness. Since we do not claim that our distance-based aggregator is independent and monotone simultaneously, the strategy proofness of our aggregator is clear. The nature of our objective function in the optimization problem is such that if an individual was to submit an insincere preference (in an attempt to manipulate the collective preference), any deviation of the collective preference set from this insincere preference has a penalty in the objective function. In this situation, there appears to be a partial corrective mechanism whereby our aggregation method is not easily prone to manipulation.

Example 3.1. Here we reformulate our problem in trapezoidal fuzzy number form, which is illustrated and summarized in Table 2. Table 4 is a clear illustration of Table 2 for trapezoidal fuzzy numbers. Also, the average of voters' decision for each proposition is calculated. Now we get at least one solution of the Condorcet paradox for trapezoidal fuzzy numbers. Suppose that the individuals have a preference structure as shown in Table 4.

Now consider Table 4. Assume that there is a small economy with three individuals and three goods x, y, and z. Individual binary relations over $X = \{x, y, z\}$, namely V_1, V_2, V_3, V_4, and V_5, are linear orders. Any optimal fuzzy social preference aggregation function maps the individual preference set into a social preference set that must be a linear order. Accordingly, this becomes an optimization problem which minimizes the sum of the distances of social preference from the individual preferences and weighted

Table 4 Revisit of Table 2 for trapezoidal fuzzy numbers

	$x \succ y$	$y \succ x$	$y \succ z$	$z \succ y$	$x \succ z$	$z \succ x$
V_1 and V_5	(0.45,0.48, 0.54,0.59)	(0.15,0.19, 0.27,0.31)	(0.45,0.48, 0.54,0.59)	(0.05,0.09, 0.16,0.22)	(0.39,0.42, 0.44,0.48)	(0.15,0.19, 0.27,0.31)
V_2 and V_4	(0.75,0.79, 0.85,0.89)	(0.61,0.65, 0.71,0.75)	(0.15,0.19, 0.27,0.31)	(0.05,0.09, 0.16,0.22)	(0.45,0.48, 0.54,0.59)	(0.25,0.32, 0.35,0.39)
V_3	(0.75,0.79, 0.85,0.89)	(0.05,0.09, 0.16,0.22)	(0.25,0.32, 0.35,0.39)	(0.15,0.19, 0.27,0.31)	(0.61,0.65, 0.71,0.75)	(0.15,0.19, 0.27,0.31)
Average	(0.65,0.69, 0.75,0.79)	(0.27,0.31, 0.38,0.43)	(0.28,0.33, 0.39,0.43)	(0.08,0.12, 0.20,0.25)	(0.48,0.52, 0.56,0.61)	(0.18,0.23, 0.30,0.34)
V_S	θ_1	θ_2	θ_3	θ_4	θ_5	θ_6

average of individual preferences using d^* (subject to fuzzy IC of linear order). Here preference aggregation is molded by representing preference ordering as truth values over trapezoidal fuzzy numbers.

The problem in Table 4 can be democratized in d^* and d^{**}. The optimal fuzzy preference function gives the solution for Table 4 as $(\theta_1, \theta_2, \theta_3, \theta_4, \theta_5, \theta_6) = ((0.69, 0.69, 0.75, 0.79), (0.15, 0.19, 0.27, 0.31), (0.25, 0.32, 0.39, 0.42), (0.05, 0.09, 0.16, 0.22), (0.45, 0.48, 0.56, 0.59), (0.17, 0.22, 0.29, 0.31))$ with minimum distance $D^{d^*} + d^*(w,\text{average}) = 1.925$. The optimal fuzzy preference function gives the solution for Table 4 as $(\theta_1, \theta_2, \theta_3, \theta_4, \theta_5, \theta_6) = ((0.65, 0.69, 0.75, 0.79), (0.27, 0.31, 0.38, 0.43), (0.28, 0.33, 0.39, 0.43), (0.08, 0.12, 0.2, 0.25), (0.48, 0.52, 0.56, 0.61), (0.18, 0.23, 0.3, 0.34))$ with minimum distance $D^{d^{**}} + d^{**}(w, \text{average}) = 1.2864$. Note that the uniqueness of the optimal solution is not guaranteed.

Discussion

We have assumed that agents have only 'epistemic' preferences, i.e., they only care about the 'distance' between the collective preference set which is collectively calculated and the individual preference set they personally decide (see [23]). This distance was originally measured by d^* where P is a finite set of propositional letters standing for atomic propositions. We re-scale this distance to form a new distance measure:

$$d(p,q) = \frac{d^*(p,q)}{4 \times |P|}.$$

Now this re-scaling makes $d(p,q) \in [0,1]$. Let p_i be the profile of individual i. Individual i receives fuzzy 'utility' if the collective preference set is collectively accepted, given by the formula $u_i(q) = \eta(d_i(p_i, q)) \in [0,1]$, where $\eta : [0,1] \to [0,1]$ is the strong fuzzy negation (see [14]) that satisfies:

1. $\eta(0) = 1, \eta(1) = 0.$
2. η is nonincreasing.
3. $\eta(\eta(x)) = x.$

For the sake of simplicity, assume that $\eta(x) = 1 - x$. Now $u_i(q) = 1 - d_i(p_i, q)$ implies that $\sum_{\forall i} u_i(q) + \sum_{\forall i} d_i(p_i, q) = n$. Choosing a collective preference set q which minimizes $\sum_{\forall i} d_i(p_i, q)$ is therefore equivalent to choosing q which maximizes $\sum_{\forall i} u_i(q)$, i.e., the sum of individual utilities. As a result, optimization problems in Tables 3 and 4 can be viewed as social utility maximization problems.

Conclusion

Policy makers could be allowed to opt for trapezoidal fuzzy numbers to express their preferences. It not only gives freedom of expression to the decision makers but also provides us with a wider range of fuzzy connectives that can be used according to the nature of the problem. In this paper, a democratic fuzzy aggregation approach for trapezoidal fuzzy numbers is used to find an interpretation having the least distance with the profile of individual preference sets and avg set. The authors believe that the preference aggregation modeling for trapezoidal fuzzy numbers will have tremendously useful applications. Our aggregation procedure is an improvement of Pigozzi's [24] aggregation method (in the context of binary logic) because cases of dictatorship are highly unlikely in our method. Our aggregation procedure did not give a unique solution. A tie-breaking method can be used in the case of our optimal solution as it is not unique. Assume that we have a preference set $\bar{C} = (\text{average}(\pi_i(P_1)), \text{average}(\pi_i(P_2)), \ldots, \text{average}(\pi_i(P_{|P|})))$. Set $\bar{C}$ might violate the fuzzy integrity constraints. Solutions could be narrowed down by picking solutions which are at a minimal distance from $\bar{C}$ with the help of the tie-breaking method. Another useful method is the aggregation function from the optimal solution, which has minimal disagreement with other aggregation functions, using a dissimilarity or distance measure. However, the important question is, how well behaved is our aggregation operator. We want the collective preference set to be responsive to the individual preference set, and we also want the collective preference set to obey rationality constraints. However, it does not ensure collective rationality. Distance-based operators are used to obtain collective rationality results in a situation of tie or indecision. Fuzzy aggregation methods are useful to construct democratic fuzzy social preference aggregation functions as already given in the previous discussions.

Competing interests

Both authors declare that they have no competing interests.

Authors' contributions

IB and TR contributed equally and significantly in this research. Both authors read and approved the final manuscript.

Acknowledgements

The authors sincerely thank the learned referees and Professor Xiang Li for the careful reading and thoughtful comments. The present version of the paper owes much to their precise and kind remarks.

Author details

[1]Centre for Mathematics and Statistical Sciences, Lahore School of Economics, Lahore 53200,Pakistan . [2]Department of Science and Humanities, National University of Computer and Emerging Sciences, Lahore 54000, Pakistan. [3]University of Management and Technology, Lahore 54000, Pakistan.

References

1. Arrow, KJ: Social Choice and Individual Values. second edition. Wiley, New York (1963)
2. Sen, A: Collective Choice and Social Welfare. Holden Day, San Francisco (1970)
3. Condorcet, MJAN: de Marquis: Essai sur l'Application de l'Analyse à la Probabilité des Décisions Rendues à la Pluralità des Voix. l'Imprimerie Royale, Paris (1785)
4. Dietrich, F: A generalized model of judgment aggregation. Soc. Choice Welf. **28**(4), 529–565 (2007)
5. Beg, I, Butt, N: (Im)possibility theorems in fuzzy framework. Critical Review: A Publication of Society for Mathematics of Uncertainty. **IV**(2010), 1–11 (2010)
6. Beg, I, Butt, N: Belief merging and judgement aggregation in fuzzy setting. Adv. Fuzzy Syst, 5 (2012). doi:10.1155/2012/635043
7. Beg, I, Khalid, A: Belief aggregation in fuzzy framework. J. Fuzzy Math. **20**(4), 911–924 (2012)
8. Duddy, C, Piggins, A: Many-valued judgment aggregation, characterizing the possibility/impossibility boundary for an important class of agendas, Vol. 0154. School of Economics, National University of Ireland, Galway (2009)
9. Dietrich, F: Judgment aggregation: (im)possibility theorems. J. Econ. Theory. **126**(1), 286–298 (2006)
10. Bellman, RE, Zadeh, LA: Decision making in a fuzzy environment. Manage. Sci. **17**(4), 141–164 (1970)
11. Li, X: Credibilistic Programming. Springer, Heidelberg (2013)
12. Li, X, Liu, B: Foundation of credibilistic logic. Fuzzy Optimization Decis. Mak. **8**(1), 91–102 (2009)
13. Li, X, Wang, HS: Credibilistic satisfiability and credibilistic entailment. J. Multiple-Valued Logic Soft Comput. **16**(1–2), 105–134 (2010)
14. Nguyen, HT, Walker, E: A First Course in Fuzzy Logic. third edition. Chapman & Hall/CRC, Boca Raton (2006)
15. Beg, I, Rashid, T: A generalized model of judgment and preference aggregation. Fuzzy Econ. Rev. (2013), in press
16. List, C: Group knowledge and group rationality: a judgment aggregation perspective. Episteme. **2**(1), 25–38 (2005)
17. Claussen, CA, Roisland, O: Collective economic decisions and the discursive paradox. Norges Bank Working Paper. Norges Bank, Oslo (2005)
18. Hathaway, RJ, Bezdek, JC, Pedrycz, W: A parametric model for fusing heterogeneous fuzzy data. IEEE Trans. Fuzzy Syst. **4**(3), 270–281 (1996)
19. Yang, MS, Hwang, PY, Chen, DH: Fuzzy clustering algorithms for mixed feature variables. Fuzzy Sets Syst. **141**, 301–317 (2004)
20. Yang, MS, Ko, CH: On a class of fuzzy c-numbers clustering procedures for fuzzy data. Fuzzy Sets Syst. **84**, 49–60 (1996)

21. Hung, WL, Yang, MS, Lee, ES: Cell formation using fuzzy relational clustering algorithm. Math. Comput. Model. **53**, 1776–1787 (2011)
22. Dietrich, F, List, C: Strategy-proof judgment aggregation. Econ. Philos. **23**(3), 269–300 (2007)
23. van, Hees, M: The limits of epistemic democracy. Soc. Choice Welf. **28**(4), 649–666 (2007)
24. Pigozzi, G: Belief merging and the discursive dilemma: an argument-based account to paradoxes of judgment aggregation. Synthese. **152**(2), 285–298 (2006)

Fuzzy fixed point theorems for fuzzy mappings via β-admissible with applications

Supak Phiangsungnoen[1], Wutiphol Sintunavarat[2] and Poom Kumam[1*]

*Correspondence:
poom.kum@kmutt.ac.th
[1] Department of Mathematics, Faculty of Science, King Mongkut's University of Technology Thonburi (KMUTT), 126 Pracha Uthit Rd., Bang Mod, Thung Khru, Bangkok 10140, Thailand
Full list of author information is available at the end of the article

Abstract

The object of this paper is to utilize the notion of β-admissible in the sense of Mohammadi et al. (Fixed Point Theory Appl. 2013:24, 2013) to prove a fuzzy fixed point theorem and present some corollaries. We also give illustrative examples which demonstrate the validity of hypotheses of our results and applications to fuzzy fixed points for fuzzy mappings in partially ordered metric spaces are studied.

Keywords: Fuzzy set; Fuzzy mapping; Fuzzy fixed point; β-admissible mappings

Introduction

In real world problems, fuzzy concepts were used to modeled mathematically for the purpose of automation, like regulation, production control and many others. The concept of fuzzy set was introduced by Zadeh [1]. Several researches were conducted on the generalizations of the concept of a fuzzy set. Heilpern [2] introduced the concept of fuzzy contraction mappings which maps from an arbitrary set to a certain subfamily of fuzzy sets in a metric linear space X. He also proved the existence of a fuzzy fixed point theorem which is a generalization of Nadler's [3] fixed point theorem for multivalued mappings.

In 2001, Estruch and Vidal [4] proved a fuzzy fixed point theorem for fuzzy contraction mappings (in the Heilperns sense) over a complete metric space. Afterward, many mathematicians [5-9] generalized the result given in [4]. Recently, Abbas and Turkoglu [10] established the existence of a fuzzy fixed point theorem for fuzzy mapping of a generalized contractive mapping on a complete metric space, which generalized and extended the results in [4] and many known results in [2,6].

On the other hand, the concept of an β-admissible mapping was introduced by Samet et al. [11]. They also give the existence of fixed point theorems via this concept. Afterward, Asl et al. [12] extended the concept of β-admissible for single-valued mappings to multivalued mappings. Recently, Mohammadi et al. [13] introduced the concept of β-admissible for multivalued mappings which is different from the notion of β_*-admissible which has been provided in [12]. Recently, Azam et al. [14], obtained common fixed point theorems for Chatterjea fuzzy mappings on closed balls in a complete metric space. The investigation is based on the fact that fuzzy fixed point results can be obtained simply from the fixed point theory of mappings on closed balls.

In this paper, we prove a fuzzy fixed point for a generalized contractive fuzzy mapping on a complete metric space by using the concept of β-admissible mapping in sense of

Mohammadi et al. [13]. We also give illustrative example which demonstrate the validity of hypotheses of our results. As an application, we deduce fixed point theorems in partially ordered metric spaces.

Preliminaries

In this section, we recall some basic definitions and preliminaries that will be needed in this paper.

Definition 1 ([1]). Let X be a space of points with generic elements of X denoted by x and $I = [0,1]$. A fuzzy subset of X is characterized by a membership function such that each element in X is associated with a real number in the interval I.

Definition 2 ([2]). Let (X,d) be a metric space and A be a fuzzy set in X. If X is endowed with topology, for $\alpha \in [0,1]$, the α-level set of A, denoted by $[A]_\alpha$, is defined as

$$[A]_\alpha = \{x : A(x) \geq \alpha\},$$

where $\alpha \in (0,1]$ and for $\alpha = 0$ we have

$$[A]_0 = \overline{\{x : A(x) > 0\}},$$

where $\overline{B}$ denotes the closure of the non-fuzzy set B.

Definition 3 ([15]). Let X be a nonempty set. For $x \in X$, we write $\{x\}$ the characteristic function of the ordinary subset $\{x\}$ of X. For $\alpha \in (0,1]$, the fuzzy point x_α of X is the fuzzy set of X given by

$$x_\alpha(y) = \begin{cases} \alpha & ; \quad x = y \\ 0 & ; \quad x \neq y \end{cases} .$$

In the sequel, $I = [0,1]$, (X,d) and I^X denote the metric space and the collection of all fuzzy subsets of X. For $A, B \in I^X$, a fuzzy set A is said to be more accurate than a fuzzy set B (denoted by $A \subset B$) if and only if $Ax \leq Bx$ for each x in X, where $A(x)$ and $B(x)$ denote the membership function of A and B, respectively. For $x \in X$, $S \subset X$, $A, B \in I^X$ and $\alpha \in [0,1]$, we define

$$d(x,S) = \inf\{d(x,a); a \in S\},$$

$$p_\alpha(x,A) = \inf\{d(x,a); a \in [A]_\alpha\},$$

$$p_\alpha(A,B) = \inf\{d(a,b); a \in [A]_\alpha, b \in [B]_\alpha\},$$

$$p(A,B) = \sup_\alpha p_\alpha(A,B),$$

$$D_\alpha(A,B) = H([A]_\alpha, [B]_\alpha) = \max\left\{ \sup_{a \in [A]_\alpha} d(a,[B]_\alpha), \sup_{b \in [B]_\alpha} d(b,[A]_\alpha) \right\},$$

$$d_\infty(A,B) = \sup_\alpha D_\alpha(A,B),$$

and

$$W_\alpha(X) = \{C \in I^X : [C]_\alpha \text{ is nonempty and compact}\}.$$

It is easy to see that H is the Hausdorff metric on $W_\alpha(X)$ induced by the metric d.

Definition 4 ([4]). Let (X, d) be a metric space. $T : X \to I^X$ and $\alpha \in [0, 1]$. A fuzzy point x_α in X is called fuzzy fixed point of T if $x_\alpha \subset Tx$. If $\{x\} \subset Tx$, then x is called fixed point of T.

Definition 5 ([12]). Let X be a nonempty set, $T : X \to 2^X$, where 2^X is a collection of nonempty subsets of X and $\beta : X \times X \to [0, \infty)$. We say that T is β_*-admissible if

$$\text{for } x, y \in X, \quad \beta(x, y) \geq 1 \Longrightarrow \beta_*(Tx, Ty) \geq 1,$$

where

$$\beta_*(Tx, Ty) := \inf\{\beta(a, b) : a \in Tx \text{ and } b \in Ty\}.$$

Definition 6 ([13]). Let X be a nonempty set, $T : X \to 2^X$, where 2^X is a collection of nonempty subsets of X and $\beta : X \times X \to [0, \infty)$. We say that T is β-admissible whenever for each $x \in X$ and $y \in Tx$ with $\beta(x, y) \geq 1$, we have $\beta(y, z) \geq 1$ for all $z \in Ty$.

Remark 1. If T is β^*-admissible mapping, then T is also β-admissible mapping.

Example 1. Let $X = \mathbb{R}$ be endowed with the usual metric d. Define mappings $T : X \to 2^X$ and $\beta : X \times X \to [0, \infty)$ by

$$T(x) = \begin{cases} [x, x+1], & \text{if } x \geq 0; \\ (-\infty, 6x], & \text{if } x < 0 \end{cases}$$

and

$$\beta(x, y) = \begin{cases} 1, & \text{if } x, y \geq 0; \\ 0, & \text{if otherwise.} \end{cases}$$

Then T is β-admissible.

Fuzzy fixed point theorem

In what follows, we give some well known lemma which is useful for the proof.

Lemma 1 ([2]). Let (X, d) be a metric space, $x \in X$ and $A \in I^X$. For $\alpha \in [0, 1]$, if $p_\alpha(x, A) = 0$ and $[A]_\alpha$ is closed subset of X, then $x_\alpha \subset A$.

Now, we introduce the concept of β-admissible in sense of Mohammadi et al. [13] for fuzzy mapping.

Definition 7. Let (X, d) be a metric space, $\beta : X \times X \to [0, \infty)$, $\alpha \in [0, 1]$ and $T : X \to W_\alpha(X)$. A mapping T is said to be β-admissible if for each $x \in X$ and $y \in [Tx]_\alpha$, with $\beta(x, y) \geq 1$, we have $\beta(y, z) \geq 1$ for all $z \in [Ty]_\alpha$.

Next, we prove the existence of fuzzy fixed point theorem for some generalized type of contraction fuzzy mapping. For our result, we let Ψ be the family of non-decreasing functions $\psi : [0, \infty) \to [0, \infty)$ such that $\sum_{n=1}^{\infty} \psi^n(t) < \infty$ for each $t > 0$. It is easy to see that for $\psi \in \Psi$, $\psi(t) < t$ for all $t > 0$ and $\psi(0) = 0$.

Theorem 1. Let (X,d) be a complete metric space, $\alpha \in [0,1]$ and T be fuzzy mapping from X to $W_\alpha(X)$. Suppose that there exist $\psi \in \Psi$ and $\beta : X \times X \to [0,\infty)$ such that

$$\beta(x,y)D_\alpha(Tx,Ty) \leq \psi(M(x,y)) + L\min\left\{p_\alpha(x,Tx), p_\alpha(y,Ty), p_\alpha(x,Ty), p_\alpha(y,Tx)\right\} \tag{1}$$

for all $x, y \in X$, where $L \geq 0$ and

$$M(x,y) = \max\left\{d(x,y), p_\alpha(x,Tx), p_\alpha(y,Ty), \frac{p_\alpha(x,Ty)+p_\alpha(y,Tx)}{2}\right\}.$$

If the following condition holds,

(i) T is β-admissible,
(ii) there exist $x_0 \in X$ and $x_1 \in [Tx_0]_\alpha$ such that $\beta(x_0,x_1) \geq 1$,
(iii) if $\{x_n\}$ is sequence in X such that $\beta(x_n,x_{n+1}) \geq 1$ and $x_n \to u$ as $n \to \infty$, then $\beta(x_n,u) \geq 1$,
(iv) ψ is continuous,

then there exists $x \in X$ such that x_α is a fuzzy fixed point of T.

Proof. For $x_0 \in X$ and $x_1 \in [Tx_0]_\alpha$ in condition (ii), we get $\beta(x_0,x_1) \geq 1$. Since $[Tx_1]_\alpha$ is nonempty compact subset of X, there exists $x_2 \in [Tx_1]_\alpha$, such that

$$d(x_1,x_2) = p_\alpha(x_1,Tx_1) \leq D_\alpha(Tx_0,Tx_1). \tag{2}$$

From (2) and the fact that $\beta(x_0,x_1) \geq 1$, we have

$$\begin{aligned} d(x_1,x_2) &\leq D_\alpha(Tx_0,Tx_1)\\ &\leq \beta(x_0,x_1)D_\alpha(Tx_0,Tx_1)\\ &\leq \psi(M(x_0,x_1)) + L\min\left\{p_\alpha(x_0,Tx_0), p_\alpha(x_1,Tx_1), p_\alpha(x_0,Tx_1), p_\alpha(x_1,Tx_0)\right\}\\ &\leq \psi(M(x_0,x_1)) + L\min\left\{p_\alpha(x_0,x_1), p_\alpha(x_1,x_2), p_\alpha(x_0,x_2), 0\right\}\\ &= \psi(M(x_0,x_1)). \end{aligned}$$

By the same argument, for $x_2 \in X$, we have $[Tx_2]_\alpha$, which is a nonempty compact subset of X and then there exists $x_3 \in [Tx_2]_\alpha$ such that

$$d(x_2,x_3) = p_\alpha(x_2,Tx_2) \leq D_\alpha(Tx_1,Tx_2). \tag{3}$$

For $x_0 \in X$ and $x_1 \in [Tx_0]_\alpha$ with $\beta(x_0,x_1) \geq 1$, by definition of β-admissible, we get

$$\beta(x_1,x_2) \geq 1. \tag{4}$$

From (1), (3) and (4), we have

$$\begin{aligned} d(x_2,x_3) &\leq D_\alpha(Tx_1,Tx_2)\\ &\leq \beta(x_1,x_2)D_\alpha(Tx_1,Tx_2)\\ &\leq \psi(M(x_1,x_2)) + L\min\left\{p_\alpha(x_1,Tx_1), p_\alpha(x_2,Tx_2), p_\alpha(x_1,Tx_2), p_\alpha(x_2,Tx_1)\right\}\\ &\leq \psi(M(x_0,x_1)) + L\min\left\{p_\alpha(x_1,x_2), p_\alpha(x_2,x_3), p_\alpha(x_1,x_3), 0\right\}\\ &= \psi(M(x_1,x_2)). \end{aligned}$$

By induction, we can construct a sequence $\{x_n\}$ in X such that, for each $n \in \mathbb{N}$, $x_n \in [Tx_{n-1}]_\alpha$ with $\beta(x_{n-1},x_n) \geq 1$ and

$$d(x_n,x_{n+1}) \leq \psi(M(x_{n-1},x_n)),$$

where

$$\begin{aligned} M(x_{n-1},x_n) &= \max\left\{d(x_{n-1},x_n), p_\alpha(x_{n-1},Tx_{n-1}), p_\alpha(x_n,Tx_n), \frac{p_\alpha(x_{n-1},Tx_n)+p_\alpha(x_n,Tx_{n-1})}{2}\right\} \\ &\leq \max\left\{d(x_{n-1},x_n), d(x_n,x_{n+1}), \frac{d(x_{n-1},x_{n+1})}{2}\right\} \\ &= \max\left\{d(x_{n-1},x_n), d(x_n,x_{n+1})\right\}. \end{aligned}$$

Hence,

$$d(x_n,x_{n+1}) \leq \psi(\max\{d(x_{n-1},x_n), d(x_n,x_{n+1})\}) \tag{5}$$

for all $n \in \mathbb{N}$. If there exists $n^* \in \mathbb{N}$ which $p_\alpha(x_{n^*}, Tx_{n^*}) = 0$, then from Lemma 1, we have $(x_{n^*})_\alpha \subset Tx_{n^*}$, that is $(x_{n^*})_\alpha$ is a fuzzy fixed point of T. Therefore, we suppose that for each $n \in \mathbb{N}$, $p_\alpha(x_n, Tx_n) > 0$ and thus $d(x_{n-1}, x_n) > 0$ for all $n \in \mathbb{N}$. So, if $d(x_n, x_{n+1}) > d(x_{n-1}, x_n)$ for some $n \in \mathbb{N}$, then from (5) and $\psi(t) < t$ for $t \in (0, \infty)$, we have

$$d(x_n,x_{n+1}) \leq \psi\left(d(x_n,x_{n+1})\right) < d(x_n,x_{n+1})$$

which is a contradiction. Therefore, we have

$$\begin{aligned} d(x_n,x_{n+1}) &\leq \psi\left(d(x_{n-1},x_n)\right) \\ &\leq \psi\left(\psi\left(d(x_{n-2},x_{n-1})\right)\right) \\ &\ \ \vdots \\ &\leq \psi^n(d(x_0,x_1)). \end{aligned} \tag{6}$$

Next, we will show that $\{x_n\}$ is a Cauchy sequence in X. Since continuous function ψ is belong to Ψ, there exist $\varepsilon > 0$ and positive integer $h = h(\varepsilon)$ such that

$$\sum_{n\geq h} \psi^n(d(x_0,x_1)) < \varepsilon.$$

Let $m > n > h$. Using the triangular inequality, previous relation and (6), we have

$$d(x_n,x_m) \leq \sum_{k=n}^{m-1} d(x_k,x_{k+1}) \leq \sum_{k=n}^{m-1} \psi^k(d(x_0,x_1)) \leq \sum_{n\geq h} \psi^n(d(x_0,x_1)) < \varepsilon.$$

This implies that $\{x_n\}$ is a Cauchy sequence in X. By completeness of X, there exists $x \in X$ such that $x_n \to x$ as $n \to \infty$.

Finally, we show that $p_\alpha(x, Tx) = 0$. Assume on the contrary that $p_\alpha(x, Tx) > 0$. By condition (iii), we have $\beta(x_n, x) \geq 1$ for all $n \in \mathbb{N}$. Now we have

$$\begin{aligned} p_\alpha(x,Tx) &\leq d(x_n,x_{n+1}) + p_\alpha(x_{n+1},Tx) \\ &\leq d(x_n,x_{n+1}) + D_\alpha(Tx_n,Tx) \\ &\leq d(x_n,x_{n+1}) + \beta(x_n,x)D_\alpha(Tx_n,Tx) \\ &\leq d(x_n,x_{n+1}) + \psi\left(\max\left\{d(x_n,x), p_\alpha(x_n,Tx_n), p_\alpha(x,Tx), \frac{p_\alpha(x_n,Tx)+p_\alpha(x,Tx_n)}{2}\right\}\right) \\ &\quad + L\min\left\{p_\alpha(x_n,Tx_n), p_\alpha(x,Tx), p_\alpha(x_n,Tx), p_\alpha(x,Tx_n)\right\} \\ &\leq d(x_n,x_{n+1}) + \psi\left(\max\left\{d(x_n,x), d(x_n,x_{n+1}), p_\alpha(x,Tx), \frac{p_\alpha(x_n,Tx)+d(x,x_{n+1})}{2}\right\}\right) \\ &\quad + L\min\left\{p_\alpha(x_n,x_{n+1}), p_\alpha(x,Tx), p_\alpha(x_n,Tx), p_\alpha(x,Tx_n)\right\}. \end{aligned}$$

Letting $n \to \infty$, it follows that

$$p_\alpha(x, Tx) \leq \psi\left(p_\alpha(x, Tx)\right) < p_\alpha(x, Tx),$$

which is a contradiction. Therefore, we have $p_\alpha(x, Tx) = 0$. Hence, by Lemma 1, $x_\alpha \subset Tx$. This complete the proof. □

Next, we give some examples to support the validity of our result.

Example 2. Let $X = [0, 1]$ and $d : X \times X \to [0, \infty)$ as $d(x, y) = |x - y|$ for all $x, y \in X$. Then (X, d) is a complete metric space. Let us define $T : X \to I^X$ by

$$(Tx)(t) = \begin{cases} 0 & , 0 \leq t < \frac{1}{6}, \\ \frac{3}{4} & , \frac{1}{6} \leq t \leq \frac{x+1}{6}, \\ \frac{1}{2} & , \frac{x+1}{6} < t \leq 1. \end{cases}$$

Let $\alpha = \dfrac{3}{4}$. We observe that

$$[Tx]_\alpha = [Tx]_{\frac{3}{4}} = \left[\frac{1}{6}, \frac{x+1}{6}\right]$$

for all $x \in X$. Therefore, T is fuzzy mapping from X to $W_\alpha(X)$.

Define $\beta : X \times X \to [0, \infty)$ by $\beta(x, y) = 2$ for all $x, y \in X$. Then it is easy to check that T is an β-admissible. For each $x, y \in X$, we get

$$\begin{aligned} \beta(x, y) D_\alpha(Tx, Ty) &= \beta(x, y) H([Tx]_\alpha, [Ty]_\alpha) \\ &= \frac{1}{3}|x - y| \\ &= \frac{1}{3} d(x, y) \\ &< \frac{1}{2} d(x, y) \\ &\leq \psi(M(x, y)) + L \min\left\{p_\alpha(x, Tx), p_\alpha(y, Ty), p_\alpha(x, Ty), p_\alpha(y, Tx)\right\}, \end{aligned}$$

where $\psi(t) = \dfrac{t}{2}$ for all $t > 0$ and $L \geq 0$. It is easy to see that conditions (ii) and (iii) in Theorem 1 hold. Therefore, all conditions of Theorem 1 hold. Thus, T has an α-fuzzy fixed point $x \in X$, that is, a point $x = \frac{1}{6}$.

By using Remark 1, we get the following result.

Theorem 2. Let (X, d) be a complete metric space, $\alpha \in [0, 1]$ and T be fuzzy mapping from X to $W_\alpha(X)$. Suppose that there exist $\psi \in \Psi$ and $\beta : X \times X \to [0, \infty)$ such that

$$\beta(x, y) D_\alpha(Tx, Ty) \leq \psi(M(x, y)) + L \min\left\{p_\alpha(x, Tx), p_\alpha(y, Ty), p_\alpha(x, Ty), p_\alpha(y, Tx)\right\} \tag{7}$$

for all $x, y \in X$, where $L \geq 0$ and

$$M(x, y) = \max\left\{d(x, y), p_\alpha(x, Tx), p_\alpha(y, Ty), \frac{p_\alpha(x, Ty) + p_\alpha(y, Tx)}{2}\right\}.$$

If the following condition holds,

(i) T is β^*-admissible,
(ii) there exist $x_0 \in X$ and $x_1 \in [Tx_0]_\alpha$ such that $\beta(x_0, x_1) \geq 1$,
(iii) if $\{x_n\}$ is sequence in X such that $\beta(x_n, x_{n+1}) \geq 1$ and $x_n \to u$ as $n \to \infty$, then $\beta(x_n, u) \geq 1$,
(iv) ψ is continuous,

then there exists $x \in X$ such that x_α is a fuzzy fixed point of T.

In Theorems 1 and 2, we take $\psi(t) = \theta t$, where $\theta \in (0, 1)$ then we have the following corollary which is a fuzzy extension of fixed point theorem given by Berinde [16].

Corollary 1. Let (X, d) be a complete metric space, $\alpha \in [0, 1]$ and T be fuzzy mapping from X to $W_\alpha(X)$. Suppose that there exists $\beta : X \times X \to [0, \infty)$ such that

$$\beta(x, y)D_\alpha(Tx, Ty) \leq \theta M(x, y) + L \min\left\{p_\alpha(x, Tx), p_\alpha(y, Ty), p_\alpha(x, Ty), p_\alpha(y, Tx)\right\} \quad (8)$$

for all $x, y \in X$, where, $\theta \in (0, 1)$, $L \geq 0$ and

$$M(x, y) = \max\left\{d(x, y), p_\alpha(x, Tx), p_\alpha(y, Ty), \frac{p_\alpha(x, Ty) + p_\alpha(y, Tx)}{2}\right\}.$$

If the following condition holds,

(i) T is β-admissible (or β^*-admissible),
(ii) there exist $x_0 \in X$ and $x_1 \in [Tx_0]_\alpha$ such that $\beta(x_0, x_1) \geq 1$,
(iii) if $\{x_n\}$ is sequence in X such that $\beta(x_n, x_{n+1}) \geq 1$ and $x_n \to u$ as $n \to \infty$, then $\beta(x_n, u) \geq 1$,

then there exists $x \in X$ such that x_α is a fuzzy fixed point of T.

If we take $L = 0$ in Theorems 1 and 2 and Corollary 1, then we have the following corollaries:

Corollary 2. Let (X, d) be a complete metric space, $\alpha \in [0, 1]$ and T be fuzzy mapping from X to $W_\alpha(X)$. Suppose that there exist $\psi \in \Psi$ and $\beta : X \times X \to [0, \infty)$ such that

$$\beta(x, y)D_\alpha(Tx, Ty) \leq \psi(M(x, y)) \quad (9)$$

for all $x, y \in X$ and

$$M(x, y) = \max\left\{d(x, y), p_\alpha(x, Tx), p_\alpha(y, Ty), \frac{p_\alpha(x, Ty) + p_\alpha(y, Tx)}{2}\right\}.$$

If the following condition holds,

(i) T is β-admissible (or β^*-admissible),
(ii) there exist $x_0 \in X$ and $x_1 \in [Tx_0]_\alpha$ such that $\beta(x_0, x_1) \geq 1$,
(iii) if $\{x_n\}$ is sequence in X such that $\beta(x_n, x_{n+1}) \geq 1$ and $x_n \to u$ as $n \to \infty$, then $\beta(x_n, u) \geq 1$,
(iv) ψ is continuous,

then there exists $x \in X$ such that x_α is a fuzzy fixed point of T.

Corollary 3. Let (X,d) be a complete metric space, $\alpha \in [0,1]$ and T be fuzzy mapping from X to $W_\alpha(X)$. Suppose that there exists $\beta : X \times X \to [0,\infty)$ such that

$$\beta(x,y)D_\alpha(Tx,Ty) \leq \theta \max\left\{d(x,y), p_\alpha(x,Tx), p_\alpha(y,Ty), \frac{p_\alpha(x,Ty)+p_\alpha(y,Tx)}{2}\right\} \quad (10)$$

for all $x, y \in X$, where $\theta \in (0,1)$. If the following condition holds,

(i) T is β-admissible (or β^*-admissible),
(ii) there exist $x_0 \in X$ and $x_1 \in [Tx_0]_\alpha$ such that $\beta(x_0,x_1) \geq 1$,
(iii) if $\{x_n\}$ is sequence in X such that $\beta(x_n,x_{n+1}) \geq 1$ and $x_n \to u$ as $n \to \infty$, then $\beta(x_n,u) \geq 1$,

then there exists $x \in X$ such that x_α is a fuzzy fixed point of T.

If we set $\beta(x,y) = 1$ for all $x, y \in X$ in Theorem 1 or Theorem 2, we get the following result:

Corollary 4. Let (X,d) be a complete metric space, $\alpha \in [0,1]$ and T be fuzzy mapping from X to $W_\alpha(X)$. Suppose that there exist $\psi \in \Psi$ such that

$$D_\alpha(Tx,Ty) \leq \psi(M(x,y)) + L\min\{p_\alpha(x,Tx), p_\alpha(y,Ty), p_\alpha(x,Ty), p_\alpha(y,Tx)\} \quad (11)$$

for all $x, y \in X$, where $L \geq 0$ and

$$M(x,y) = \max\left\{d(x,y), p_\alpha(x,Tx), p_\alpha(y,Ty), \frac{p_\alpha(x,Ty)+p_\alpha(y,Tx)}{2}\right\}.$$

Then there exists $x \in X$ such that x_α is a fuzzy fixed point of T.

If $\psi(t) = \theta t$, where $\theta \in (0,1)$ and $\beta(x,y) = 1$ for all $x, y \in X$ in Theorem 1 or Theorem 2, then we have the following corollary.

Corollary 5. Let (X,d) be a complete metric linear space, $\alpha \in [0,1]$ and T be fuzzy mapping from X to $W_\alpha(X)$ such that

$$D_\alpha(Tx,Ty) \leq \theta \max\left\{d(x,y), p_\alpha(x,Tx), p_\alpha(y,Ty), \frac{p_\alpha(x,Ty)+p_\alpha(y,Tx)}{2}\right\} \quad (12)$$

for all $x, y \in X$, where $\theta \in (0,1)$. Then there exists $x \in X$ such that x_α is a fuzzy fixed point of T.

Applications

In this section, we give the existence of fuzzy fixed point theorems in a partially ordered metric space.

Before presenting our results, we give the following definitions.

Definition 8. Let X be a nonempty set. Then $(X,d,\preceq)$ is said to be an ordered metric spaces if (X,d) is a metric space and $(X,\preceq)$ is a partially ordered set.

Definition 9. Let $(X,\preceq)$ be a partially ordered set. Then $x, y \in X$ are said to be comparable if $x \preceq y$ or $y \preceq x$ holds.

For partially ordered set $(X, \preceq)$, we define

$$\overline{\wedge} := \{(x, y) \in X \times X : x \preceq y \text{ or } y \preceq x\}.$$

Definition 10. A partially ordered set $(X, \preceq)$ is said to satisfy the order sequential limit property if $(x_n, x) \in \overline{\wedge}$ for all $n \in \mathbb{N}$, whenever a sequence $x_n \to x$ as $n \to \infty$ and $(x_n, x_{n+1}) \in \overline{\wedge}$ for all $n \in \mathbb{N}$.

Example 3. Let $X = \mathbb{R}$ and $\preceq = \leq$ be a partial ordered on X. It is easy to check that $(X, d, \preceq)$ satisfy the order sequential limit property.

Definition 11. Let $(X, \preceq)$ be a partially ordered set and $\alpha \in [0, 1]$. A fuzzy mapping $T : X \to W_\alpha(X)$ is said to be comparative if for each $x \in X$ and $y \in [Tx]_\alpha$, with $(x, y) \in \overline{\wedge}$, we have $(y, z) \in \overline{\wedge}$ for all $z \in [Ty]_\alpha$.

Here we give the existence of fuzzy fixed point theorem for fuzzy mapping in ordered metric spaces.

Theorem 3. Let $(X, d, \preceq)$ be a complete partially ordered metric space, $\alpha \in [0, 1]$ and T be fuzzy mapping from X to $W_\alpha(X)$. Suppose that there exists $\psi \in \Psi$ such that

$$D_\alpha(Tx, Ty) \leq \psi(M(x, y)) + L \min\{p_\alpha(x, Tx), p_\alpha(y, Ty), p_\alpha(x, Ty), p_\alpha(y, Tx)\} \quad (13)$$

for all $(x, y) \in \overline{\wedge}$, where $L \geq 0$ and

$$M(x, y) = \max\left\{d(x, y), p_\alpha(x, Tx), p_\alpha(y, Ty), \frac{p_\alpha(x, Ty) + p_\alpha(y, Tx)}{2}\right\}.$$

If the following condition holds,

(I) T is comparative fuzzy mapping,
(II) there exist $x_0 \in X$ and $x_1 \in [Tx_0]_\alpha$ such that $(x_0, x_1) \in \overline{\wedge}$,
(III) X satisfies the order sequential limit property,
(IV) ψ is continuous,

then there exists $x \in X$ such that x_α is a fuzzy fixed point of T.

Proof. Consider the mapping $\beta : X \times X \to [0, \infty)$ defined by

$$\beta(x, y) = \begin{cases} 1 \text{ if } (x, y) \in \overline{\wedge}, \\ 0 \text{ if } (x, y) \notin \overline{\wedge}. \end{cases} \quad (14)$$

From condition (II), we get $\beta(x_0, x_1) \geq 1$ and then the condition (ii) in Theorem 1 holds. It follows from T is comparative fuzzy mapping that T is β-admissible mapping. By (18), we have, for all $x, y \in X$,

$$\beta(x, y)D_\alpha(Tx, Ty) \leq \psi(M(x, y)) + L \min\{p_\alpha(x, Tx), p_\alpha(y, Ty), p_\alpha(x, Ty), p_\alpha(y, Tx)\}. \quad (15)$$

Since X satisfies the order sequential limit property, the condition (iii) in Theorem 1 is also holds. Now all the hypotheses of Theorem 1 are satisfied and thus the existence of the fuzzy fixed point of fuzzy mapping T follows from Theorem 1. □

By using the same technique in the proof of Theorem 3 with Corollaries 1, 2 and 3, we get the following results.

Corollary 6. Let $(X, d, \preceq)$ be a complete partially ordered metric space, $\alpha \in [0,1]$ and T be fuzzy mapping from X to $W_\alpha(X)$ such that

$$D_\alpha(Tx, Ty) \leq \theta M(x,y) + L \min\{p_\alpha(x,Tx), p_\alpha(y,Ty), p_\alpha(x,Ty), p_\alpha(y,Tx)\} \tag{16}$$

for all $(x,y) \in \overline{\wedge}$, where $L \geq 0$ and

$$M(x,y) = \max\left\{d(x,y), p_\alpha(x,Tx), p_\alpha(y,Ty), \frac{p_\alpha(x,Ty) + p_\alpha(y,Tx)}{2}\right\}.$$

If the following condition holds,

(I) T is comparative fuzzy mapping,
(II) there exist $x_0 \in X$ and $x_1 \in [Tx_0]_\alpha$ such that $(x_0, x_1) \in \overline{\wedge}$,
(III) X satisfies the order sequential limit property,

then there exists $x \in X$ such that x_α is a fuzzy fixed point of T.

Corollary 7. Let $(X, d, \preceq)$ be a complete partially ordered metric space, $\alpha \in [0,1]$ and T be fuzzy mapping from X to $W_\alpha(X)$. Suppose that there exists $\psi \in \Psi$ such that

$$D_\alpha(Tx, Ty) \leq \psi(M(x,y)) \tag{17}$$

for all $(x,y) \in \overline{\wedge}$, where $L \geq 0$ and

$$M(x,y) = \max\left\{d(x,y), p_\alpha(x,Tx), p_\alpha(y,Ty), \frac{p_\alpha(x,Ty) + p_\alpha(y,Tx)}{2}\right\}.$$

If the following condition holds,

(I) T is comparative fuzzy mapping,
(II) there exist $x_0 \in X$ and $x_1 \in [Tx_0]_\alpha$ such that $(x_0, x_1) \in \overline{\wedge}$,
(III) X satisfies the order sequential limit property,
(IV) ψ is continuous,

then there exists $x \in X$ such that x_α is a fuzzy fixed point of T.

Corollary 8. Let $(X, d, \preceq)$ be a complete partially ordered metric space, $\alpha \in [0,1]$ and T be fuzzy mapping from X to $W_\alpha(X)$ such that

$$D_\alpha(Tx, Ty) \leq \theta M(x,y) \tag{18}$$

for all $(x,y) \in \overline{\wedge}$, where

$$M(x,y) = \max\left\{d(x,y), p_\alpha(x,Tx), p_\alpha(y,Ty), \frac{p_\alpha(x,Ty) + p_\alpha(y,Tx)}{2}\right\}.$$

If the following condition holds,

(I) T is comparative fuzzy mapping,
(II) there exist $x_0 \in X$ and $x_1 \in [Tx_0]_\alpha$ such that $(x_0, x_1) \in \overline{\wedge}$,
(III) X satisfies the order sequential limit property,

then there exists $x \in X$ such that x_α is a fuzzy fixed point of T.

Acknowledgements
The second authors gratefully acknowledge the financial support provided by Thammasat University under the TU Research Scholar, Contract No. 2/10/2557.

Author details
[1]Department of Mathematics, Faculty of Science, King Mongkut's University of Technology Thonburi (KMUTT), 126 Pracha Uthit Rd., Bang Mod, Thung Khru, Bangkok 10140, Thailand. [2]Department of Mathematics and Statistics, Faculty of Science and Technology, Thammasat University Rangsit Center, Pathumthani 12121, Thailand.

References

1. Zadeh, LA: Fuzzy sets. Inf. Control. **8**, 103–112 (1965)
2. Heilpern, S: Fuzzy mappings and fixed point theorems. J. Math. Anal. Appl. **83**, 566–569 (1981)
3. Nadler, SB Jr: Multivalued contraction mapping. Pac. J. Math. **30**(2), 475–488 (1969)
4. Estruch, VD, Vidal, A: A note on fixed fuzzy points for fuzzy mappings. Rend Istit. Univ. Trieste. **32**, 39–45 (2001)
5. Azam, A, Beg, I: Common fixed points of fuzzy maps. Math. Comput. Model. **49**, 1331–1336 (2009)
6. Turkoglu, D, Rhoades, BE: A fixed fuzzy point for fuzzy mapping in complete metric spaces. Math. Commun. **10**, 115–121 (2005)
7. Sedghi, S, Shobe, N, Altun, I: A fixed fuzzy point for fuzzy mappings in complete metric spaces. Math. Commun. **13**, 289–294 (2008)
8. Phiangsungnoen, S, Sintunavarat, W, Kumam, P: Common α-fuzzy fixed point theorems for fuzzy mappings via $\beta_{\mathcal{F}}$-admissible pair. J. Intell. Fuzzy Syst (2014). doi.org/10.3233/IFS-141218
9. Phiangsungnoen, S, Sintunavarat, W, Kumam, P: Fuzzy fixed point theorems in Hausdorff fuzzy metric spaces. J. Inequalities Appl. **2014**, 201 (2014)
10. Abbas, M, Turkoglu, D: Fixed point theorem for a generalized contractive fuzzy mapping. J. Intell. Fuzzy Syst. **26**(1), 33–36 (2014)
11. Samet, B, Vetro, C, Vetro, P: Fixed point theorems for α-ψ-contractive type mappings. Nonlinear Anal. **75**, 2154–2165 (2012)
12. Asl, JH, Rezapour, S, Shahzad, N: On fixed points of α-ψ-contractive multifunctions. Fixed Point Theory Appl. **2012**, 212 (2012)
13. Mohammadi, B, Rezapour, S, Shahzad, N: Some results on fixed points of α-ψ-Ciric generalized multifunctions. Fixed Point Theory Appl. **2013**, 24 (2013)
14. Azam, A, Hussain, S, Arshad, M: Common fixed points of Chatterjea type fuzzy mappings on closed balls. Neural Comput. Appl. **21**(1 Supplement), 313–317 (2012)
15. Pu, P-M, Liu, Y-M: Fuzzy topology. I. Neighborhood structure of a fuzzy point and Moore-smith convergence. J. Math. Anal. Appl. **76**, 571–599 (1980)
16. Berinde, V: General constructive fixed point theorems for *Ćirić*- type almost contractions in metric spaces. Carpathian J. Math. **24**(2), 10–19 (2008)

13

Uncertainty theory based multiple objective mean-entropy-skewness stock portfolio selection model with transaction costs

Rupak Bhattacharyya[1*], Amitava Chatterjee[2] and Samarjit Kar[2]

* Correspondence: mathsrup@gmail.com
[1]Department of Applied Science & Humanities, Global Institute of Management & Technology, Krishnagar, West Bengal 741102, India
Full list of author information is available at the end of the article

Abstract

Purpose: The aim of this paper is to develop a mean-entropy-skewness stock portfolio selection model with transaction costs in an uncertain environment.

Methods: Since entropy is free from reliance on symmetric probability distributions and can be computed from nonmetric data, it is more general than others as a competent measure of risk. In this work, returns of securities are assumed to be uncertain variables, which cannot be estimated by randomness or fuzziness. The model in the uncertain environment is formulated as a nonlinear programming model based on uncertainty theory. Also, some other criteria like short-and long-term returns, dividends, number of assets in the portfolio, and the maximum and minimum allowable capital invested in stocks of any company are considered. Since there is no efficient solution methodology to solve the proposed model, assuming the returns as some special uncertain variables, the original portfolio selection model is transformed into an equivalent deterministic model, which can be solved by any state-of-the-art solution methodology.

Results: The feasibility and effectiveness of the proposed model is verified by a numerical example extracted from Bombay Stock Exchange, India. Returns are considered in the form of trapezoidal uncertain variables. A genetic algorithm is used for simulation.

Conclusions: The efficiency of the portfolio is evaluated by looking for risk contraction on one hand and expected return and skewness augmentation on the other hand. An empirical application has served to illustrate the computational tractability of the approach and the effectiveness of the proposed algorithm.

Keywords: Uncertainty modeling; Mean-entropy-skewness portfolio selection model; Uncertain variables; Trapezoidal uncertain variable; Genetic algorithm

Introduction

The Markowitz [1] formulation of modern portfolio theory has been the most impact-making development in mathematical finance management to date. Since returns are uncertain in nature, the allocation of capital in different risky assets to minimize the risk and to maximize the return is the main concern of it.

In most of the significant works on portfolio selection, the first-order moment of return distribution about the origin, *i.e.*, the mean, quantifies the return, and the second-order moment about the mean, i.e., the variance, quantifies the risk.

Consideration of variance as risk is erroneous as it equally suggests penalties for up and down deviations from the mean. To face this problem, Markowitz [2] recommended semi-variance, a downside risk measure. Another alternative definition of risk is the probability of an adverse outcome [3]. The popular risk measure value at risk [4,5] is in fact an alternative expression of the definition by Roy [3]. Different authors like Philippatos and Wilson [6], Philippatos and Gressis [7], Nawrocki and Harding [8], Simonelli [9], Huang [10], Qin *et al.* [11], and Bhattacharyya *et al.* [12] used entropy as an alternative measure of risk to replace the variance proposed by Markowitz [1]. Uncertainty causes loss and so investors dislike uncertainty. Since entropy is a measure of uncertainty, it is used to measure risk. Entropy is more general than others as an efficient measure of risk because entropy is free from reliance on symmetric distributions and can be computed from nonmetric data.

One of the important theoretical difficulties of these studies is that they assume that asset returns are normally distributed or the utility function is quadratic or that the higher moments are irrelevant to the investors' decision. However, some experimental studies show that portfolio returns are generally not normally distributed. As a result, a natural extension of the mean-variance model is to add the skewness as a factor for consideration in portfolio management. The importance of higher order moments in portfolio selection was suggested by Samuelson [13]. However, considerations of skewness in portfolio selection problem were started by 1990 and were done by Lai [14], Konno and Suzuki [15], Chunhachinda *et al.* [16], Liu *et al.* [17], Prakash *et al.* [18], Briec *et al.* [19], Yu *et al.* [20], Li *et al.* [21], Bhattacharyya *et al.* [22], Bhattacharyya and Kar [23,24], Bhattacharyya [25], and others. Consideration of a mean-entropy-skewness model in portfolio selection problem is introduced by Bhattacharyya *et al.* [12]. They have constructed three portfolio selection models in fuzzy environment using the credibility theory approach.

In most of the abovementioned research works on portfolio selection, the common assumptions are that the investor has enough historical data and that the situation of asset markets in the future can be reflected with certainty by asset data in the past. However, it cannot always be made with certainty. Basically, the usual feature of a financial environment is uncertainty. Mostly, it is realized as risk uncertainty and is modeled by stochastic approaches. However, the term uncertainty has a second aspect-vagueness (imprecision or ambiguity), which can be modeled by fuzzy methodology. In this respect, to tackle the uncertainty in the financial market, stochastic-fuzzy and fuzzy-stochastic methodologies are extensively used in portfolio modeling. Authors like Konno and Suzuki [15], Leon *et al.* [26], Vercher et al. [27], Bhattacharyya *et al.* [12,22], Dey and Bhattacharyya [28], etc. used fuzzy numbers to replace uncertain returns of securities, and they define portfolio selection as a mathematical programming problem in order to select the best alternative. Huang [29] measures portfolio risk by credibility measure and proposed two credibility theory-based mean-variance models. Huang [30] also proposed a mean-semi-variance model for describing the asymmetry of fuzzy returns. She extends the risk definition of variance and chance to a random fuzzy environment and formulates optimization models where security returns are fuzzy random variables.

So, in attempts dealing with portfolio selection problems, randomness and fuzziness are considered as the two basic types of uncertainty contained in security returns. It has become a common practice that when security returns cannot be reflected by historical data, fuzzy variables can be used to show experts' knowledge and estimation of security returns. However, illogicality will come into view if fuzzy variables are used to describe the subjective estimation of security returns. For example, a stock return is considered as a triangular fuzzy variable $\xi = (-0.2, 0.3, 0.7)$. Using possibility theory (or credibility theory), the return is exactly 0.3 with belief degree 1 in possibility measure (or 0.5 in credibility measure). However, this conclusion is unacceptable because the belief degree of exactly 0.3 is almost 0. In addition, the return being exactly 0.3 and not exactly 0.3 has the same belief degree in either possibility measure or credibility measure, which implies that the two events will happen equally likely. This conclusion is quite astonishing and hard to accept.

Again, philosophically, though randomness and fuzziness are two basic types to represent uncertain phenomena, in real life, there are some situations where uncertainty behaves neither randomly nor fuzzily. For example, the occurrence chance of a security price falling in the interval of [100, 110] is 30%, and the occurrence chance of the security price in the interval of [110, 120] is 20%. Then what is the occurrence chance of the security price in the interval of [100, 120]? A survey shows that some people believe that the occurrence chance should be somewhere that is not less than 30% but not greater than 50%. In this case, the security price is neither random nor fuzzy. Recently, Liu [31] proposed an uncertain measure and developed an uncertainty theory, which can be used to handle subjective imprecise quantity. Much research works have been done on the development of uncertainty theory and related theoretical works. Though some considerable amounts of publications have been done in the field of uncertainty theory, not much work has been done in the portfolio selection problem. Huang [32] proposed a mean-risk model for uncertain portfolio selection. Yan [33] found out the deterministic forms of mean-variance portfolio section models corresponding to different special uncertain variables like rectangular uncertain variable, triangular uncertain variable, trapezoidal uncertain variable, and normal uncertain variable. In this study, security returns are considered as uncertain variables, which are characterized by identification functions, and instead of possibility/credibility measure, uncertain measure is used to handle the uncertain events.

Not all the relevant information for an investment decision can be confined in terms of explicit return, risk, and skewness. By capturing additional and alternative decision criteria, a portfolio that is dominated with respect to expected return, skewness, and risk may frame for the shortfall in these three important factors by a very good act on one or several other criteria. As a result, portfolio selection models that consider more criteria than the standard expected return and variance objectives of the Markowitz model have become well liked. Ehrgott *et al.* [34] proposed a model having five criteria, *viz.*, short-and long-term return, dividend, ranking, and risk, and used a multicriteria decision making approach to solve the portfolio selection problem. Fang *et al.* [35] proposed a portfolio rebalancing model with transaction costs based on fuzzy decision theory considering three criteria: return, risk, and liquidity.

The main focus of this paper is to propose a mean-entropy-skewness portfolio selection framework with transaction cost having returns in the form of uncertain variables. In addition, it incorporates some useful constraints in the model to make the model more realistic. In addition, this paper provides a real application by using data from

Bombay Stock Exchange (BSE), where we consider returns as trapezoidal uncertain variables.

The rest of the paper is organized as follows. We review the necessary knowledge about uncertainty variables and develop some essential results in the 'Uncertainty theory: related topics' section. In the 'Mean-entropy-skewness model formulation' section, a tri-objective mean-entropy-skewness portfolio selection model is formulated with constraints on short-term and long-term returns, dividends, number of assets in the portfolio, and the maximum and minimum allowable capital invested in stocks of any company. The model is then converted into a single-objective constrained optimization problem with weights over mean, skewness, and entropy. To solve the proposed optimization problem, we provide a genetic algorithm in the 'Genetic algorithm' section. In the 'Case study: Bombay Stock Exchange' section, a case study from Bombay Stock Exchange is done to illustrate the method. The same section also contains a comparative study with other relevant models. Finally, in the last section, some concluding remarks are specified.

Uncertainty theory: related topics

In this paper, the concept of uncertainty theory has been introduced in the field of stock portfolio selection. This section contains only those definitions and theorems on uncertainty theory which are directly used for the formation of this article. The concepts of uncertain measure, uncertain variable, uncertain space, first and second identification functions, rectangular uncertain variable, triangular uncertain variable, exponential uncertain variable, bell-shaped uncertain variable, linear uncertain variable, zigzag uncertain variable, normal uncertain variable, lognormal uncertain variable, and others would be useful to understand the backbone of the article and can be obtained from Liu [31].

Definition 1. A trapezoidal uncertain variable is defined to be the uncertain variable which is fully determined by the four-tuple *(a, b, c, d)* of crisp numbers with $a < b < c < d$, and whose first identification function is

$$\lambda(x) = \begin{cases} \dfrac{x-a}{2(b-a)} & if \;\; a \le x \le b \\ 0.5 & if \;\; b \le x \le c \\ \dfrac{d-x}{2(d-c)} & if \;\; c \le x \le d. \end{cases}$$

Definition 2. The uncertainty distribution $\Phi : \mathbb{R} \to [0, 1]$ of an uncertain variable ξ is defined by

$$\Phi(x) = M\{\xi \le x\}.$$

Definition 3. An uncertain variable ξ is said to have an empirical uncertainty distribution if

$$\phi(x) = \begin{matrix} 0 & if \, x < x_i \\ \alpha_i + \dfrac{(\alpha_{i+1}-\alpha_i)(x-x_i)}{(x_{i+1}-x_i)} & if \, x_i \le x \le x_{i+1}, \, 1 \le i \le n \\ 1 & if \, x > x_n \end{matrix}$$

and is denoted by $\varepsilon(x_1, \alpha_1, x_2, \alpha_2, \ldots, x_n, \alpha_n)$, where $x_1 < x_2 < \ldots < x_n$ and $0 \le \alpha_1 \le \alpha_2 \le \ldots \le \alpha_n \le 1$.

Example 1. The trapezoidal uncertain variable $\breve{\xi} = (a, b, c, d)$ follows the empirical uncertain distribution given by

$$\phi(x) = \begin{cases} 0 & \text{if } x < a \\ \dfrac{x-a}{2(b-a)} & \text{if } a \leq x \leq b \\ 0.5 & \text{if } b \leq x \leq c\,. \\ 0.5 + \dfrac{0.5(x-c)}{d-c} & \text{if } c \leq x \leq d \\ 1 & \text{if } x > d \end{cases}$$

Definition 4. The uncertain variables $\breve{\xi}_1, \breve{\xi}_2, \ldots, \breve{\xi}_n$ are said to be independent if

$$M\left\{ \bigcap_{i=1}^{n} \breve{\xi}_i \right\} = \min_{1 \leq i \leq n} M\{\breve{\xi}_i \in B_i\}$$

for Borel sets $B_1, B_2, \ldots, B_n$ of real numbers. Here M denotes the uncertain measure.

Definition 5. Let $\breve{\xi}$ be an uncertain variable. Then the expected value of $\breve{\xi}$ is given by

$$E(\breve{\xi}) = \int_0^{+\infty} M\{\breve{\xi} \geq r\}dr - \int_{-\infty}^{0} M\{\breve{\xi} \leq r\}dr$$

provided that at least one of the two integrals is finite.

Definition 6. Let $\breve{\xi}$ be an uncertain variable. Then the entropy of $\breve{\xi}$ is given by

$$H[\breve{\xi}] = \int_{-\infty}^{\infty} \{-\Phi(x).\ln(\Phi(x)) - (1-\Phi(x))\ln(1-\Phi(x))\}dx.$$

Definition 7. Let $\breve{\xi}$ be an uncertain variable with finite expected value e. Then the variance and skewness of $\breve{\xi}$ are respectively given by

$$V[\breve{\xi}] = E\left\{ (\breve{\xi} - e)^2 \right\},$$

$$S[\breve{\xi}] = E\left\{ (\breve{\xi} - e)^3 \right\}.$$

Example 2. If $\breve{\xi} = (a, b, c, d)$ is a trapezoidal uncertain variable then

$$E[\breve{\xi}] = \frac{a+b+c+d}{4},$$

$$H[\breve{\xi}] = \frac{b-a+d-c}{2} + (c-b)\ln 2,$$

$$S[\breve{\xi}] = \frac{\{(d-a)^2 - (c-b)^2\}\{(d-c)-(d-c)\}}{32}.$$

Theorem 1. *Let $\breve{\xi}_1, \breve{\xi}$ be two independent uncertain variables with finite expected values. Then for any real numbers a and b, we have* $E[a\breve{\xi}_1 + b\breve{\xi}] = aE[\breve{\xi}_1]bE[\breve{\xi}]$.

Theorem 2. *Let $\breve{r}_i = (a_i, b_i, c_i, d_i), (i = 1, 2, ..., n)$ be n independent uncertain trapezoidal variables and let x_i (i = 1, 2,..., n) be n real variables. Then*

$$E[\breve{r}_1 x_1 + \breve{r}_2 x_2 + ... + \breve{r}_n x_n] = \frac{1}{4}\sum_{i=1}^{n}(a_i + b_i + c_i + d_i)x_i,$$

$$H[\breve{r}_1 x_1 + \breve{r}_2 x_2 + ... + \breve{r}_n x_n] = \frac{1}{2}\sum_{i=1}^{n}\left[(b_i - a_i) + (d_i - c_i) + 2(c_i - b_i)\ln 2\right]x_i,$$

$$S[\breve{r}_1 x_1 + \breve{r}_2 x_2 + ... + \breve{r}_n x_n] = \frac{1}{32}\sum_{i=1}^{n}\left[\{(d_i - a_i)^2 - (c_i - b_i)^2\}\{(d_i - c_i) - (b_i - a_i)\}\right]x_i^3.$$

Proof. As $\breve{r}_i = (a_i, b_i, c_i, d_i), (i = 1, 2, ..., n)$ are n independent uncertain trapezoidal variables and x_i (*i* = *1, 2,..., n*) are n real variables, we have

$$\breve{r}_1 x_1 + \breve{r}_2 x_2 + ... + \breve{r}_n x_n$$

$$= \sum_{i=1}^{n} \breve{r}_i x_i = \sum_{i=1}^{n} (a_i, b_i, c_i, d_i)x_i$$

$$= \sum_{i=1}^{n} (a_i x_i, b_i x_i, c_i x_i, d_i x_i) = \left(\sum_{i=1}^{n} a_i x_i, \sum_{i=1}^{n} b_i x_i, \sum_{i=1}^{n} c_i x_i, \sum_{i=1}^{n} d_i x_i\right).$$

Hence, $\breve{r}_1 x_1 + \breve{r}_2 x_2 + ... + \breve{r}_n x_n$ is a trapezoidal uncertain variable. Combining the above result with the results obtained in Example 2, we are with the theorem.

Mean-entropy-skewness model formulation

In this section, we will first describe the assumptions and notations used in the construction of the paper. Then the objective functions of the models will be constructed in the next subsection. In the third subsection, we will discuss the constraints used in our portfolio selection model. The fourth subsection will include three different mathematical models for different situations.

Assumptions and notations

Let us consider a financial market with n risky assets offering uncertain returns. An investor allocates his wealth among these risky assets.

For the ith risky asset (*i* = 1, 2,..., *n*), let us use the following notations:

x_i = portion of the total capital invested in ith security

$\breve{p}_i$ = uncertain variable representing the closing price of the ith security at present

$\breve{p}'_i$ = uncertain variable representing the estimated closing price of the ith security in the next year

d_i = the estimated dividends in the next year

$\breve{r}_i = \frac{\breve{p}'_i + d_i - \breve{p}_i}{\breve{p}_i}$ = uncertain variable representing the return of the ith security

$R_i^{(12)}$ = the average 12 month performance

$R_i^{(36)}$ = the average 36 month performance

k_i = the constant transaction cost per change in a proportion, $k_i \geq 0$

$$y_i = \begin{cases} 1 & \text{if the } i\text{th asset is contained in the portfolio} \\ 0 & \text{if the } i\text{th asset is not contained in the portfolio.} \end{cases}$$

Formulation of objective functions

It is impossible to predict future returns of stocks in any budding security market. The arithmetic mean of historical data is in general considered as the expected return of securities, which yield us a crisp value. However, for this technique, two main problems need to be solved. Firstly, if historical data for a long period are considered, the influence of earlier historical data is the same as that of recent data, whereas recent data of a security is more important than the earlier historical data. Secondly, if the historical data of a security are not adequate, due to the lack of information, the estimations of the statistical parameters are not adequate. For these reasons, the expected return of a security is considered here as an uncertain variable instead of the crisp arithmetic mean of historical data. Similarly, in an uncertain environment, the risk (entropy) and skewness cannot be predicted exactly. Therefore, the entropy and skewness are also considered here as uncertain variables.

Let us consider the transaction cost c_i to be a V-shaped function of the difference between a given portfolio $x^0 = \left(x_1^0, x_2^0, \ldots, x_n^0\right)$ and a new portfolio $x = (x_1, x_2, \ldots, x_n)$ and is incorporated explicitly into the portfolio return. Thus, the transaction cost of ith risky asset can be expressed as

$$c_i = k_i\left|x_i - x_i^0\right|, i = 1, 2, 3, \ldots, n.$$

Hence the total transaction cost is

$$\sum_{i=1}^{n} c_i = \sum_{i=1}^{n} k_i\left|x_i - x_i^0\right|.$$

The expected return of portfolio $x = (x_1, x_2, \ldots, x_n)$ with transaction cost is thus given by

$$\text{Re}(x) = E[(\breve{r}_1 - c_1)x_1 + (\breve{r}_2 - c_2)x_2 + \ldots + (\breve{r}_n - c_n)x_n].$$

The entropy of portfolio $x = (x_1, x_2, \ldots, x_n)$ is given by

$$\text{En}(x) = H\left[\breve{r}_1x_1 + \breve{r}_2x_2 + \ldots + \breve{r}_nx_n\right].$$

The skewness of portfolio $x = (x_1, x_2, \ldots, x_n)$ is given by

$$\text{Sk}(x) = S\left[\breve{r}_1x_1 + \breve{r}_2x_2 + \ldots + \breve{r}_nx_n\right].$$

We consider the portfolio selection problem as a tri-objective optimization problem. As discussed earlier, the objectives we consider are

$$\begin{cases} \text{Max} & \text{Re}(x) \\ \text{Min} & \text{En}(x) \\ \text{Max} & \text{Sk}(x). \end{cases}$$

Construction of the constraints

For the portfolio $x = (x_1, x_2,..., x_n)$, the expected short-term return is expressed as

$$R_{\mathrm{st}}(x) = \sum_{i=1}^{n} R_i^{(12)} x_i - \sum_{i=1}^{n} k_i |x_i - x_i^0|.$$

For the portfolio $x = (x_1, x_2,..., x_n)$, the expected long-term return is expressed as

$$R_{\mathrm{lt}}(x) = \sum_{i=1}^{n} R_i^{(36)} x_i - \sum_{i=1}^{n} k_i |x_i - x_i^0|.$$

Since investors plan their asset allocation on short-term, long-term, or both cases, they should prefer a portfolio having at least a minimum short-term, long-term, or both types of return. For that reason, we consider the following two constraints:

$$R_{\mathrm{st}}(x) \geq \varsigma,$$
$$R_{\mathrm{lt}}(x) \geq \tau,$$

where ς and τ will be allocated by the investor.

Dividend is the payment made by a company to its shareholders. It is the portion of corporate profits paid out to the investors. For the portfolio $x = (x_1, x_2,..., x_n)$, the annual dividend is expressed as

$$D(x) = \sum_{i=1}^{n} d_i x_i.$$

Clearly, investors would like to have a portfolio that yields them a high dividend. Keeping in mind this fact, we propose the following constraint:

$$D(x) \geq d,$$

where d will be allocated by the investor.

The well-known capital budget constraint on the assets is presented by

$$\sum_{i=1}^{n} x_i = 1.$$

The maximum and minimum fractions of the capital budget being allocated to each of the assets in the portfolio depend upon factors like price relative to the asset in comparison with the average of the price of all the assets in the chosen portfolio, minimal lot size that can be traded in the market, the past performance of the price of the asset, information available about the issuer of the asset, trends in the business of which it is a division, etc. That is, an investor will have to look upon a host of the basics affecting the commerce. Different investors having different views may allocate the same overall capital budget differently.

Let the maximum fraction of the capital that can be invested in a single asset i be M_i. Then

$$x_i \leq M_i y_i \forall i = 1, 2, ..., n.$$

Let the minimum fraction of the capital that can be invested in a single asset i be m_i. Then

$x_i \geq m_i y_i \ \forall i = 1, 2, \ldots, n.$

The investor would like to pick up the assets among all the assets in a given set that in his subjective estimate are likely to yield the greatest performance. Thus it is not necessary that all the assets in the given set may configure in the portfolio. Investors can thus consider the number of assets they can effectively handle in a portfolio.

Let the number of assets held in a portfolio be k. Then

$$\sum_{i=1}^{n} y_i = k.$$

As no short selling is considered, we have

$x_i \geq 0 \ \forall i = 1, 2, \ldots, n.$

If X is the set of feasible portfolios, then we have,

$$X = \{x = \{x_1, \ x_2, \ldots, \ x_n\} \text{ such that } R_{st}(x) \geq \varsigma, \ R_{lt}(x) \geq \tau, \ D(x) \geq d, \ x_i \leq M_i y_i, \ x_i \geq m_i y_i, \ \sum_{i=1}^{n} y_i = k, \ \sum_{i=1}^{n} x_i = 1, \ x_i \geq 0\}. \quad (1)$$

Weighted portfolio selection model formulation

The portfolio selection model is thus formulated as

$$\begin{cases} \text{Max} & \text{Re} \\ \text{Min} & \text{En} \\ \text{Max} & \text{Sk} \\ x \in X. & \end{cases} \quad (2)$$

To convert the above tri-objective optimization problem into a preference-based single-objective optimization problem, let us consider three single-objective optimization problems optimizing separately the three objectives of the model subject to the constraints of the problem. The optimum values as well as the values of the remaining objective functions in each of the three cases are calculated. Considering all the three problems, let the minimum values of the three objectives be $\text{Re}^{\min}$, $\text{En}^{\min}$, and $\text{Sk}^{\min}$, respectively. Also, let the maximum values of the three objectives be $\text{Re}^{\max}$, $\text{En}^{\max}$, and $\text{Sk}^{\max}$, respectively. Then the above tri-objective portfolio selection model is transformed into the following model:

$$\begin{cases} \text{Max} \left[w_1 \dfrac{\text{Re}(x) - \text{Re}^{\min}}{\text{Re}^{\max} - \text{Re}^{\min}} + w_2 \dfrac{\text{En}^{\max} - \text{En}(x)}{\text{En}^{\max} - \text{En}^{\min}} + w_3 \dfrac{\text{Sk}(x) - \text{Sk}^{\min}}{\text{Sk}^{\max} - \text{Sk}^{\min}} \right] \\ \text{subject to} \\ \quad x \in X \\ \quad w_1 + w_2 + w_3 = 1 \end{cases} \quad (3)$$

where w_1, w_2, and w_3 are weights or preferences to the objectives *Re(x)*, *En(x)*, and *Sk(x)*, respectively. w_1, w_2, and w_3 will be allocated by the investor.

Genetic algorithm

After development of the genetic algorithm (GA) by Holland in 1975, it has been extensively used/modified to solve complex decision making problems in different fields of science and technology. A GA normally starts with a set of potential solutions (called initial population) of the decision making problem under consideration. Individual solutions are called chromosome. Crossover and mutation operations happen among the potential solutions to get a new set of solutions, and the process continues until terminating conditions are encountered. The following functions and values are adopted in the proposed GA to solve the problem [36]. The different parameters on which this GA depends are the number of generation (*MAXGEN*), population size (*POPSIZE*), probability of crossover (*PCROS*), and probability of mutation (*PMUTE*).

Chromosome representation

An important issue in applying a GA is to design an appropriate chromosome representation of solutions of the problem together with genetic operators. Traditional binary vectors used to represent the chromosome are not effective in many nonlinear problems. Since the proposed model is highly nonlinear, hence, to overcome the difficulty, a real-number representation is used. In this representation, each chromosome V_i is a string of n number of genes G_{ij} (i = 1, 2,..., POPSIZE, j = 1, 2,..., n) where these n number of genes respectively denote n number of decision variables x_j.

Initial population production

For each chromosome V_i, every gene G_{ij} is randomly generated between its boundary (LB_j, UB_j) where LB_j and UB_j are the lower and upper bounds of the variables x_j (j = 1, 2,..., n and i = 1, 2,..., POPSIZE), respectively.

Evaluation

Evaluation function plays the same role in GA as that the environment plays in natural evolution. Now, evaluation function (EVAL) for chromosome V_i is equivalent to the objective function $f(x_1, x_2,..., x_n)$. The following are the steps of evaluation:

1. Find EVAL(V_i) = $f(x_1, x_2,..., x_n)$, where the genes G_{ij} represent the decision variable x_j, j = 1, 2,..., n and f is the objective function.
2. Find total fitness of the population: $F = \sum_{i=1}^{\text{POPSIZE}} \text{EVAL}(V_i)$.
3. The probability p_i of selection for each chromosome V_i is determined by the formula $p_i = \frac{1}{F}\text{EVAL}(V_i)$.
4. Calculate the cumulative probability Y_i of selection for each chromosome V_i by the formula $Y_i = \sum_{j=1}^{i} p_i$.

Selection

The selection scheme in GA determines which solutions in the current population are to be selected for recombination. Many selection schemes, such as stochastic random sampling roulette wheel selection, have been proposed for various problems. In this paper, we adopt the roulette wheel selection process. This roulette wheel selection process is based on spinning the roulette wheel *POPSIZE* times each time we select a single chromosome for the new population in the following way:

(a) Generate a random (float) number r between 0 and 1.
(b) If $r < Y_1$, then the first chromosome is V_1; otherwise, select the ith chromosome V_i (2 ≤ i ≤ POPSIZE) such that $Y_{i-1} \le r < Y_i$.

Crossover

A crossover operator is mainly responsible for the search of new strings. Crossover operates on two parent solutions at a time and generates offspring solutions by recombining both parent solution features. After selection of chromosomes for new population, the crossover operator is applied. Here, the arithmetic crossover operation is used. It is defined as a linear combination of two consecutive selected chromosomes V_m and V_n, and resulting offspring's $V_m^{'}$ and $V_n^{'}$ are calculated as

$$V_m^{'} = cV_m + (1-c)V_n,$$
$$V_n^{'} = cV_n + (1-c)V_m,$$

where c is a random number between 0 and 1.

Mutation

A mutation operator is used to prevent the search process from converging to local optima rapidly. It is applied to each single chromosome V_i. The selection of a chromosome for mutation is performed in the following way:

1. Set $i \leftarrow 1$.
2. Generate a random number u from the range [0, 1].
3. If $u <$ PMUTE, then we select the chromosome V_i.
4. Set $i \leftarrow i + 1$.
5. If $i \leq$ POPSIZE, then go to step 2. Then the particular gene G_{ij} of the chromosome V_i selected by the abovementioned steps is randomly selected. In this problem, the mutation is defined as G_{ij}^{mut} random number from the range (LB_j, UB_j).

Termination

If the number of iteration is less than or equal to MAXGEN, then the process goes on; otherwise, it terminates.

Proposed GA procedure

```
Start
{
t ←0
while (all constraints are not satisfied)
{
initialize Population (t)
}
evaluate Population (t)
while(not terminate - condition)
{
t ← t + 1
select Population (t) from Population (t – 1)
crossover and mutate Population (t)
```

```
evaluate Population (t)
}
print optimum result
}.
```

Case study: Bombay Stock Exchange (BSE)

Bombay Stock Exchange is the oldest stock exchange in Asia with a rich heritage of over 133 years of existence. What is now popularly known as BSE was established as 'The Native Share & Stock Brokers' Association' in 1875. It is the first stock exchange in India which obtained permanent recognition (in 1956) from the Government of India under the Securities Contracts (Regulation) Act (SCRA) 1956. With demutualization, the stock exchange has two of world's prominent exchanges, Deutsche Borse and Singapore Exchange, as its strategic partners. Today, BSE is the world's number one exchange in terms of the number of listed companies and the world's fifth in handling of transactions through its electronic trading system. The companies listed on BSE command a total market capitalization of US$1.06 trillion as of July 2009.

The BSE index, SENSEX, is India's first and most popular stock market benchmark index. SENSEX is tracked worldwide. It constitutes 30 stocks representing 12 major sectors. It is constructed on a 'free-float' methodology, and is sensitive to market movements and market realities. Apart from SENSEX, BSE offers 23 indices, including 13 sectoral indices.

Case study

We have taken monthly share price data for 60 months (March 2003 to February 2008) of just five companies which are included in the BSE index. Though any finite number of stocks can be considered, we have taken only five stocks to reduce the complexity of representation.

The Table 1 shows the stocks along with their returns in the form of trapezoidal uncertain numbers, the average short-term returns, the average long-term returns, and the dividends. We also have $k_i = 0.001$. We consider, $x_i^0 = 0$ for $i = 1, 2, 3, 4, 5$.

Example

With respect to the above data, we consider the following tri-objective portfolio selection model:

Table 1 Stocks information

Stock	Return ($\tilde{r}_i$)	Short-term returns ($R_i^{(12)}$)	Long-term return ($R_i^{(36)}$)	Dividends (d_i, %)
Reliance energy	(−0.008, 0.020, 0.042,0.067)	0.0324	0.031	63
L&T	(−0.003, 0.029, 0.057, 0.087)	0.0524	0.044	85
Bhel	(−0.002,0.021, 0.051, 0.083)	0.0510	0.037	125
Tata steel	(0.009, 0.023, 0.038, 0.052)	0.0307	0.032	155
SBI	(−0.010, 0.022, 0.045, 0.079)	0.0387	0.035	140

Table 2 Solution

	Case 1	Case 2	Case 3	Case 4
w_1	1/3	0.6	0.2	0.2
w_2	1/3	0.2	0.6	0.2
w_3	1/3	0.2	0.2	0.6
Re(*x*)	0.03830	0.03878251	0.03787975	0.0389781
En(*x*)	0.0667776	0.06823548	0.06466012	0.06811745
Sk(*x*)	0.000000321	0.000002439285	0.0000002561417	0.000001260730
Dividend (%)	110.2186	113.1965	112.4341	110.2661
x_1	0	0	0	0
x_2	0.4922520	0.3969638	0.4552814	0.4865646
x_3	0.1805016	0.3313681	0.3565393	0
x_4	0	0	0.1881792	0.1981891
x_5	0.3272464	0.2716681	0	0.3152462

$$\begin{cases} \text{Max } \ \text{Re}(x) \\ \text{Min } \ \text{En}(x) \\ \text{Max } \text{Sk}(x) \\ \text{subject to} \\ R_{\text{st}}(x) \ \geq \ 0.039, \ \ R_{\text{lt}}(x) \ \geq \ 0.038, \ \ D(x) \geq 1.1, \\ x = (x_1, \ \ x_2, \ \ x_3, \ \ x_4, \ \ x_5), y_i \in \{0, 1\} \\ x_i \leq 0.6 y_i, x_i \geq 0.1 y_i, \ \ \sum_{i=1}^{5} x_i = 1, \ \ \sum_{i=1}^{5} y_i = 3. \end{cases} \tag{4}$$

Solution

To solve the above example, the GA is used with the parameters POPSIZE = 50, PCROS = 0.2, PMUTE = 0.2, *and* MAXGEN = 100. A real-number presentation is used here. In this representation, each chromosome x is a string of m (here, m = 5) number of genes; these represent decision variables. For each chromosome x, every gene (here, x_1, x_2, x_3, x_4, x_5) is randomly generated between its boundaries until it is feasible. In this

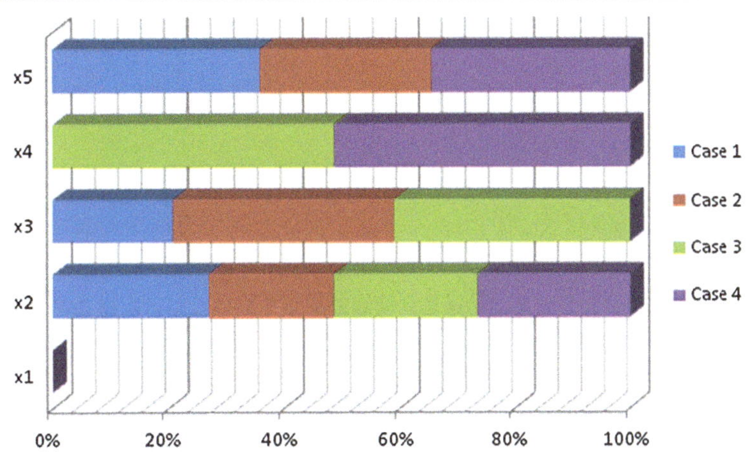

Figure 1 Portfolios.

Table 3 Individual mean, entropy, and skewness of the stocks

Stock	Return	Entropy	Skewness
Reliance energy	0.03025	0.0417492	-4.8197×10^{-7}
L&T	0.04250	0.0504081	-1.4632×10^{-5}
Bhel	0.03825	0.0384751	2.0318×10^{-6}
Tata steel	0.03050	0.0243972	0
SBI	0.0340	0.0489424	4.62×10^{-7}

problem, arithmetic crossover and random mutation are applied to generate new offsprings.

As discussed in the 'Weighted portfolio selection model formulation' section, optimizing the three single objectives Re(x), En(x), and Sk(x) separately subject to the constraints in (4), we obtain the minimum and maximum values of the objectives with the same parameters. In each case, only the best solution is considered.

With reference to model (3), the problem (4) is transformed into the following model:

$$\left\{\begin{array}{l} \text{Max} \left[w_1 \dfrac{\text{Re}(x)-0.0346375}{0.03957812-0.0346375} + w_2 \dfrac{0.06896848-\text{En}(x)}{0.06896848-0.05826382} + w_3 \dfrac{\text{Sk}(x)-(-0.000000199502)}{0.000000384282-(-0.000000199502)} \right] \\ \text{subject to} \\ R_{st}(x)\geq 0.039,\ R_{lt}(x)\geq 0.038,\ D(x)\geq 1.1,\ x=(x_1,\ x_2,\ x_3,\ x_4,\ x_5) \\ \sum_{i=1}^{5} x_i = 1,\ x_i \leq 0.6\ y_i,\ x_i \geq 0.1 y_i,\ \sum_{i=1}^{5} y_i = 3,\ y_i \in \{0,1\}, w_1 + w_2 + w_3 = 1. \end{array}\right. \tag{5}$$

For different preassigned values of w_1, w_2, and w_3, the above problem is solved. We have considered only the best solutions. The solutions obtained are shown in Table 2.

In case 1, where an investor gives same importance to all the three objectives, the portfolio states that the investor should invest 45%, 45%, and 10% of the money to the second, third, and fourth stocks, respectively. In case 3, where the importance is given towards minimization of risk, the investor should invest 39.6%, 18.8%, and 41.6% of the total money to the second, fourth, and fifth stocks, respectively. Similarly, we can explain the other two cases.

In case 2, where more importance is given to return, the investor gets a return of 0.03957813 which is higher than that of the other three cases {0.03893750, 0.03586275, and 0.03830000}. In case 3, where more importance is given to risk, the investors' risk (0.09320499) is lower than in all other cases {0.09551172, 0.09846223, and 0.09487761}. Similarly, in case 4, we get the best result for skewness. In case 1, where equal importance is given to all objectives, the outputs are intermediate. We represent the portfolios obtained in cases 1, 2, 3, and 4 graphically in Figure 1.

Table 4 Solution of the model in (5.4.1)

x_1	x_2	x_3	x_4	x_5	E
0.000000	0.3840023	0.5159977	0.000000	0.1000000	0.03945701

Table 5 Solution of the model of Ning et al

x_1	x_2	x_3	x_4	x_5	E
0.000000	0.4125000	0.4875000	0.000000	0.1000000	0.03957813

Some questions may arise on the appropriateness of the portfolios obtained in Table 2 under different circumstances (cases 1, 2, 3 and 4). For example, question may arise on the absence of reliance energy in all the obtained portfolios. To explain that, the individual mean, entropy and skewness of the stocks are calculated by Example 2 and are shown in Table 3. It is seen that reliance energy has the lowest rerun among the five stocks. It is also possessing negative skewness. So, the absences of reliance energy on the portfolios in cases 1, 2 and 4 are obvious. In case 3, where more importance is given to entropy, the selected portfolio contains L&T, Tata steel and Bhel. Tata steel and Bhel are the two stocks with lowest risks. Again, though L&T has a higher risk, it also has very high return. So, the portfolio in case 3 is not compromising too much towards entropy and is maintaining the characteristic of multiobjective optimization. This is also to note that if the constraint $x_i \geq 0.1y_i$ is not considered, then some of the portfolios would contain non-zero x_1.

Comparative study

We compare the results in Table 2 with other relevant literature to demonstrate how the results from the proposed technique compare with the literatures of uncertainty theory in the portfolio selection problem. Thus, the models in [33], [37], and [38] which apply uncertainty theory in portfolio selection are considered with the same data set as that in Table 1. We also used the following set of constraints (X) for each case:

We also used the following set of constraints (X) for each case:

$$X = \{R_{st}(x) \geq 0.039,\ R_{lt}(x) \geq 0.038,\ D(x) \geq 1.1,\ x_i \leq 0.6y_i,\ x_i \geq 0.1y_i,\ \sum_{i=1}^{5} x_i = 1,\ \sum_{i=1}^{5} y_i = 3\}.$$

Model of Yan

We considered the following model [33]:

$$\left.\begin{array}{l} \text{Maximize } E[\breve{r}_1x_1 + \breve{r}_2x_2 + \breve{r}_3x_3 + \breve{r}_4x_4 + \breve{r}_5x_5] \\ \text{Subject to the constraints} \\ V[\breve{r}_1x_1 + \breve{r}_2x_2 + \breve{r}_3x_3 + \breve{r}_4x_4 + \breve{r}_5x_5] \leq 0.002 \\ x \in X \end{array}\right\}.$$

Here E stands for mean (return) and V stands for variance (risk). The solution is shown in Table 4.

Table 6 Solution of the model of Liu and Qin

x_1	x_2	x_3	x_4	x_5	E
0.000000	0.2333333	0.6000000	0.000000	0.1666667	0.03853333

Model of Ning et al

We considered the following model [37]:

$$\left.\begin{array}{l}\text{Maximize } E[\breve{r}_1x_1+\breve{r}_2x_2+\breve{r}_3x_3+\breve{r}_4x_4+\breve{r}_5x_5]\\ \text{Subject to the constraints}\\ \text{TVaR}[\breve{r}_1x_1+\breve{r}_2x_2+\breve{r}_3x_3+\breve{r}_4x_4+\breve{r}_5x_5]\le -0.008\\ x\in X\end{array}\right\}.$$

Here E stands for mean (return) and $TvaR$ stands for tail value at risk. The solution is shown in Table 5.

Model of Liu and Qin

We considered the following model [38]:

$$\left.\begin{array}{l}\text{Maximize } E[\breve{r}_1x_1+\breve{r}_2x_2+\breve{r}_3x_3+\breve{r}_4x_4+\breve{r}_5x_5]\\ \text{Subject to the constraints}\\ \text{SAD}[\breve{r}_1x_1+\breve{r}_2x_2+\breve{r}_3x_3+\breve{r}_4x_4+\breve{r}_5x_5]\le 0.0144\\ x\in X\end{array}\right\}.$$

Here E stands for mean (return) and SAD stands for semi-absolute deviation (risk). The solution is shown in Table 6.

In the discussions done in the first and second sections, we see that using entropy as a measure of risk/uncertainty is analytically better than the other conventional measures. Again, if we compare Tables 4, 5, and 6 with Table 2, we see that the performance of the proposed model is clearly at par or better than the established models.

Conclusions

This paper has introduced a new framework of mean-entropy-skewness portfolio selection problem with transaction cost under the constrains on short-and long-term returns with transaction costs, dividends, number of assets in the portfolio, and the maximum and minimum allowable capital invested in stocks. Uncertainties of future return of stocks are characterized by uncertain variables. The efficiency of the portfolios is evaluated by looking for risk contraction on one hand and expected return and skewness augmentation on the other hand. An empirical application has served to illustrate the computational tractability of the approach and the effectiveness of the proposed algorithm. A comparative study with other relevant literatures proves the usefulness of the proposed model. In addition to the GA, some other meta-heuristic algorithms such as tabu search, simulated annealing, ant colony optimization, and particle swam optimization may be employed to solve the nonlinear programming problem.

Author details

[1]Department of Applied Science & Humanities, Global Institute of Management & Technology, Krishnagar, West Bengal 741102, India. [2]Department of Mathematics, National Institute of Technology, Durgapur, West Bengal 713209, India.

References

1. Markowitz, H: Portfolio selection. J. Finance. **7**, 77–91 (1952)
2. Markowitz, H: Portfolio Selection: Efficient Diversification of Investments. Wiley, New York (1959)
3. Roy, AD: Safety first and the holding of assets. Econometrics. **20**, 431–449 (1952)
4. Castellacci, G, Siclari, MJ: The practice of Delta-Gamma VaR: implementing the quadratic portfolio model. Eur. J. Oper. Res. **150**, 529–545 (2003)
5. Philippe, J: Value at risk: the new benchmark for controlling market risk. Irwin Professional, Chicago (1996)
6. Philippatos, GC, Wilson, CJ: Entropy, market risk and selection of efficient portfolios. Appl. Econ. **4**, 209–220 (1972)

7. Philippatos, GC, Gressis, N: Conditions of equivalence among E-V, SSD, and E-H portfolio selection criteria: the case for uniform, normal and lognormal distributions. Manag. Sci. **21**, 617–625 (1975)
8. Nawrocki, DN, Harding, WH: State-value weighted entropy as a measure of investment risk. Appl. Econ. **18**, 411–419 (1986)
9. Simonelli, MR: Indeterminacy in portfolio selection. Eur. J. Oper. Res. **163**, 170–176 (2005)
10. Huang, X: Mean-entropy models for fuzzy portfolio selection. IEEE Trans. Fuzzy Syst. **16**, 1096–1101 (2008)
11. Qin, Z, Li, X, Ji, X: Portfolio selection based on fuzzy cross-entropy. J. Comput. Appl. Math. **228**, 188–196 (2009)
12. Bhattacharyya, R, Kar, MB, Kar, S, Dutta Majumder, D: Mean-entropy-skewness fuzzy portfolio selection by credibility theory approach. In: Chaudhury, S, Mitra, S, Murthy, CA, Sastry, PS, Pal, SK (eds.) Pattern Recognition and Machine Intelligence, Proceedings of the Third International Conference, PReMI 2009, New Delhi, India, December 16–20, 2009, Lecture Notes in Computer Science, vol. 5909, pp. 603–608. (2009)
13. Samuelson, P: The fundamental approximation theorem of portfolio analysis in terms of means, variances an higher moments. Rev. Econ. Stud. **25**, 65–86 (1958)
14. Lai, T: Portfolio selection with skewness: a multiple-objective approach. Rev. Quant. Finance. Account. **1**, 293–305 (1991)
15. Konno, H, Suzuki, K: A mean-variance-skewness optimization model. J. Oper. Res. Soc. Jpn. **38**, 137–187 (1995)
16. Chunhachinda, P, Dandapani, P, Hamid, S, Prakash, AJ: Portfolio selection and skewness: evidence from international stock markets. J. Bank.Finance. **21**, 143–167 (1997)
17. Liu, SC, Wang, SY, Qiu, WH: A mean-variance-skewness model for portfolio selection with transaction costs. Int. J. Comput. Sci. Manag. Syst. **34**, 255–262 (2003)
18. Prakash, AJ, Chang, C, Pactwa, TE: Selecting a portfolio with skewness: recent evidence from US, European, and Latin American equity markets. J. Bank. Finance. **27**, 1375–1390 (2003)
19. Briec, W, Kerstens, K, Jokung, O: Mean-variance- skewness portfolio performance gauging: a general shortage function and dual approach. Manag. Sci. **53**, 135–149 (2007)
20. Yu, L, Wang, SY, Lai, K: Neural network based mean- variance-skewness model for portfolio selection. Comput. Oper. Res. **35**, 34–46 (2008)
21. Li, X, Qin, Z, Kar, S: Mean-variance-skewness model for portfolio selection with fuzzy returns. Eur. J. Oper. Res. **202**, 239–247 (2010)
22. Bhattacharyya, R, Kar, S, Dutta Majumder, D: Fuzzy mean-variance-skewness portfolio selection models by interval analysis. Comput. Math. Appl. **61**, 126–137 (2011)
23. Bhattacharyya, R, Kar, S: Possibilistic mean-variance-skewness portfolio selection models. Int. J. Oper. Res. **8**, 44–56 (2011)
24. Bhattacharyya, R, Kar, S: Multi-objective fuzzy optimization for portfolio selection: an embedding theorem approach. Turkish J. Fuzzy. Syst. **2**, 14–35 (2011)
25. Bhattacharyya, R: Possibilistic Sharpe ratio based novice portfolio selection models. In: Bhattacharyya, R, Bhaumik, AK (eds.) National Conference on Advancement of Computing in Engineering Research (ACER 13) Krishnagar, West Bengal, India, 2013. CS & IT-CSCP vol. 3, pp. 33–45. AIRCC, Chennai (2013)
26. Leon, T, Liern, V, Vercher, E: Viability of infeasible portfolio selection problems: a fuzzy approach. Eur. J. Oper. Res. **139**, 178–189 (2002)
27. Vercher, E, Bermudez, JD, Segura, JV: Fuzzy portfolio optimization under downside risk measures. Fuzzy Set. Syst. **158**, 769–782 (2007)
28. Dey, M, Bhattacharyya, R: Entropy-cost ratio maximization model for efficient stock portfolio selection using interval analysis. In: Bhattacharyya, R, Bhaumik, AK (eds.) National Conference on Advancement of Computing in Engineering Research (ACER 13) Krishnagar, West Bengal, India, 2013. CS & IT-CSCP vol. 3, pp. 119–134. AIRCC, Chenna (2013)
29. Huang, X: Fuzzy chance-constrained portfolio selection. Appl. Math. Comput. **177**, 500–507 (2006)
30. Huang, X: Mean-semivariance models for fuzzy portfolio selection. J. Comput. Appl. Math. **217**, 1–8 (2008)
31. Liu, B: Uncertainty Theory. Springer-Verlag, Berlin (2007)
32. Huang, X: Mean-risk model for uncertain portfolio selection. Fuzzy Optim. Decis. Making **10**, 71–89 (2011)
33. Yan, L: Optimal portfolio selection models with uncertain returns. Mod. Appl. Sci. **3**, 76–81 (2009)
34. Ehrgott, M, Klamroth, K, Schwehm, C: An MCDM approach to portfolio optimization. Eur. J. Oper. Res. **155**, 752–770 (2004)
35. Fang, Y, Lai, KK, Wang, SY: Portfolio rebalancing model with transaction costs based on fuzzy decision theory. Eur. J. Oper. Res. **175**, 879–893 (2006)
36. Roy, A, Maity, K, Kar, S, Maiti, M: A production-inventory model with remanufacturing for defective and usable items in fuzzy-environment. Comput. Ind. Eng. **56**, 87–96 (2009)
37. Ning, Y, Yan, L, Xie, Y: Mean-TVaR model for portfolio selection with uncertain returns. Information **15**, 129–137 (2012)
38. Liu, Y, Qin, Z: Mean semi-absolute deviation model for uncertain portfolio optimization problem. Journal of Uncertain Systems **6**, 299–307 (2012)

Computational modeling for parallel grid-based recursive Bayesian estimation: parallel computation using graphics processing unit

Xianqiao Tong[1*], Tomonari Furukawa[1] and Hugh Durrant-Whyte[2]

*Correspondence: txq1986@vt.edu
[1] Department of Mechanical Engineering, Virginia Tech, 800 Drillfield Dr, Blacksburg, VA 24061, USA
Full list of author information is available at the end of the article

Abstract

This paper presents the performance modeling of the real-time grid-based recursive Bayesian estimation (RBE), particularly the parallel computation using graphics processing unit (GPU). The proposed modeling formulates data transmission between the central processing unit (CPU) and the GPU as well as floating point operations to be carried out in each CPU and GPU necessary for one iteration of the real-time grid-based RBE. Given the specifications of the computer hardware, the proposed modeling can thus estimate the total amount of time cost for performing the grid-based RBE in a real-time environment. A new prediction formulation, which adopted separable convolution, is proposed to further accelerate the real-time grid-based RBE. The performance of the proposed modeling was investigated, and parametric studies have first demonstrated its validity in various conditions by showing that the average error of estimation in computational performance stays below 6% to 7%. Utilizing the prediction with separable convolution, the grid-based RBE has also been found to perform within 1 ms, although the size of the problem was relatively large.

Keywords: RBE; Bayesian; GPU; Real-time; Grid-base; Parallel

Introduction

Recursive Bayesian estimation (RBE) allows the estimation of belief of a dynamically moving target by updating the belief both in time and observation [1]. There are two fundamental processes for the RBE: prediction process and correction process. The prediction process updates the belief by the motion model of the target, whereas the correction process updates the belief through the current observation. If the target is observable, the accuracy of the RBE can be maintained by the correction process using the valid observations. When the target is not observable, the accuracy of the RBE heavily relies on the prediction process and the error accumulates due to the lack of the valid observation for the correction process. In order for an accurate estimation, the RBE has to be performed fast enough to catch the motion of the target with a well-defined target motion model, which requires a good synchronization between its discrete representation and the RBE. Recent years, as a result, have seen many real-time enhanced RBE techniques that help improve the speed of the RBE.

One of such techniques is the modified ensemble Kalman filter (EnKF). The EnKF allows non-Gaussian estimation by minimizing a cost function defined by a non-Gaussian observation error with a pre-conditioned conjugate gradient method [2]. Langevin-Markov Chain Monte Carlo (MCMC) method, which represents the non-Gaussian belief by sampling it using a Markov chain and Langevin equation, could be a non-Gaussian RBE technique [3]. Another sampling method is the interactive particle filter (IPF), which is able to flexibly mitigate the belief space complexity [4]. An ensemble Kalman-particle predictor-corrector filter is a hybrid method that combines the advantages of EnKF and IPF and is able to effectively deal with high-dimensional non-Gaussian problems [5]. A tree-based estimator approximates the posterior belief distribution at multiple resolutions to be effective for high-dimensional problems [6], whereas maximum likelihood state estimation method could also achieve non-Gaussian RBE [7] by using a finite Gaussian mixture model.

Grid-based RBE technique is able to maintain a good accuracy for the belief since the entire target space is spatially discretized [8]. The good accuracy is obtained by the subtle discretization of the target space but leads to an inefficient computation at the same time. Furukawa et al. [9,10] refined the grid-based RBE by developing a more general element-based RBE. The generalized element can help accurately represent the arbitrary target space with only the small number of elements compared with the grid-based RBE so as to reduce the computation of the RBE. Lavis et al. proposed an enhanced grid-based RBE that allows the update of not only the belief but also the target space [11]. Because of the dynamic adjustment of the target space, the computation of the RBE is additionally reduced. Further, the parallel grid-based RBE has been proposed, and it significantly accelerated the computation of the RBE and made its real-time implementation possible by utilizing the GPU's strong parallel computational capability [12]. Despite that these efforts successfully reduce the computation of the RBE to achieve the fast RBE, the accuracy of the RBE is not well kept when the prediction process dominates the RBE during the no-observation period. The time cost of one iteration of the RBE becomes critical for overcoming this issue because that only if it matches the time increment of the discrete target motion model, the RBE can maintain the accuracy during the no-observation period.

This paper presents a performance modeling for the parallel grid-based RBE, particularly the parallel computation using the GPU, and it is able to determine the time cost of one iteration of the RBE. The proposed modeling formulates the total amount of data transmission between the CPU and the GPU and the total number of floating point operations to be carried out in each CPU and GPU necessary for one iteration of the parallel grid-based RBE. Given the specifications of the computer hardware, it is thus possible to estimate the time cost for one iteration of the parallel grid-based RBE. In order to perform the parallel grid-based RBE at maximum speed, the proposed modeling also reformulates and implements the prediction process with separable convolution.

The paper is organized as follows. The following section reviews the recursive Bayesian estimation as well as the parallel grid-based RBE. Section presents the proposed reformulation of the prediction process for the parallel grid-based RBE and its computational performance modeling. Section demonstrates the validation and efficacy of the proposed modeling through numerical examples, and the Conclusion and future work are summarized in the final section.

Parallel grid-based RBE

Problem statement

The motion of an object, o, is deterministically given by the following equation:

$$\dot{\mathrm{x}}^o = \mathbf{f}^o(\mathrm{x}^o, \mathbf{u}^o, \mathrm{w}^o, t), \tag{1}$$

where x^o represents the state of the object, u^o represents the object control input, w^o represents the system noise, which includes environmental influences on the target, and t represents the time. In general, the state of the object describes its two-dimensional location but may also include other variables such as velocity. Let the time interval between the consecutive time steps be defined as Δt. By integrating Equation (1), the state of the object at the time step k is given by

$$\mathrm{x}_k^o = \mathrm{x}_{k-1}^o + \int_{t_{k-1}}^{t_{k-1}+\Delta t} \mathbf{f}^o(\mathbf{x}^o, \mathbf{u}^o, \mathbf{w}^o, t)dt, \tag{2}$$

where t_{k-1} is the time which corresponds to the time step $k-1$.

Recursive Bayesian estimation

Prediction

The prediction process starts with the numerical implementation of the object motion model defined in Equation (2). For simplicity, the numerical integration is carried out by Riemann left sum algorithm. By dividing the time interval Δt between the consecutive time steps into n subintervals. The state of the object at the time step k is given by

$$\mathrm{x}_k^o = \mathrm{x}_{k-1}^o + \sum_{i=0}^{n-1} \mathbf{f}^o\left(\mathbf{x}^o, \mathbf{u}^o, \mathbf{w}^o, t_{k-1} + i\frac{\Delta t}{n}\right)\frac{\Delta t}{n}. \tag{3}$$

Let a sequence of the observations of the object from time step 1 to $k-1$ be defined as ${}^s\tilde{\mathbf{z}}_{1:k-1} \equiv \{{}^s\tilde{\mathbf{z}}_i | \forall i \in \{1, \ldots, k-1\}\}$. Notice here that $\tilde{(\cdot)}$ represents an instance of variable $(\cdot)$. The prediction process computes the belief of the current state $p(\mathbf{x}_k^o|{}^s\tilde{\mathbf{z}}_{1:k-1})$ from the belief in the previous time step $p(\mathbf{x}_{k-1}^o|{}^s\tilde{\mathbf{z}}_{1:k-1})$. The prediction is iteratively carried out by Chapman-Kolmogorov equation and given by

$$\begin{aligned} &p(\mathbf{x}_k^o|{}^s\tilde{\mathbf{z}}_{1:k-1}) \\ &= \int_{\mathcal{X}^o} p(\mathbf{x}_k^o|\mathbf{x}_{k-1}^o)p(\mathbf{x}_{k-1}^o|{}^s\tilde{\mathbf{z}}_{1:k-1})d\mathbf{x}_{k-1}^o, \end{aligned} \tag{4}$$

where $p(\mathbf{x}_k^o|\mathbf{x}_{k-1}^o)$ is the probabilistic representation of the object motion model defined in Equation (3), which maps the probability of transition from the previous state $\mathbf{x}_{k-1}^o$ to the current state $\mathbf{x}_k^o$. The prediction process at $k=1$ is carried out by letting $p(\mathbf{x}_{k-1}^o|{}^s\tilde{\mathbf{z}}_{1:k-1}) = p(\tilde{\mathbf{x}}_0^o)$, where $p(\tilde{\mathbf{x}}_0^o)$ is defined as a prior belief of the object in terms of the probability density function. Equation (4) indicates that the performance of the prediction process relies on the object motion model $p(\mathbf{x}_k^o|\mathbf{x}_{k-1}^o)$. Due to the fact that the object motion model is usually non-Gaussian when only prediction process applies to the RBE, the belief could eventually become heavily non-Gaussian.

Correction

The correction process is associated with the definition of the observation model. Let the probability of detection (PoD) be $0 \leq P_{\mathrm{d}}(\mathbf{x}_k^o) \leq 1$ as a reliable measure for detecting the object in terms of the object state. Observable region ${}^s\mathcal{X}_k^o$ is defined as

$$ {}^s\mathcal{X}_k^o = \{\mathbf{x}_k^o | 0 < P_\mathrm{d}(\mathbf{x}_k^o) \leq 1\}. \quad (5)$$

The observation ${}^s\mathbf{z}_k$ at the time step k is given by

$$ {}^s\mathbf{z}_k = \begin{cases} \mathbf{h}^s(\mathbf{x}_k^o, \mathbf{v}_k^s) & \mathbf{x}_k^o \in {}^s\mathcal{X}_k^o \\ \varnothing & \mathbf{x}_k^o \notin {}^s\mathcal{X}_k^o, \end{cases} \quad (6)$$

where $\mathbf{v}_k^s$ represents the observation noise at the time step k, and $\varnothing$ represents an empty element, indicating that the observation contained no information on the object or that the target is unobservable when it is not within the observable region.

The correction process then computes the belief $p(\mathbf{x}_k^o | {}^s\tilde{\mathbf{z}}_{1:k})$ given the corresponding observations up to the previous time step $p(\mathbf{x}_k^o | {}^s\tilde{\mathbf{z}}_{1:k-1})$ and a new observation ${}^s\tilde{\mathbf{z}}_k$. The equation is derived by applying formulas for marginal distribution and conditional independence and given by

$$ p(\mathbf{x}_k^o | {}^s\tilde{\mathbf{z}}_{1:k}) = \frac{l(\mathbf{x}_k^o | {}^s\tilde{\mathbf{z}}_k) p(\mathbf{x}_k^o | {}^s\tilde{\mathbf{z}}_{1:k-1})}{\int_{\mathcal{X}^o} l(\mathbf{x}_k^o | {}^s\tilde{\mathbf{z}}_k) p(\mathbf{x}_k^o | {}^s\tilde{\mathbf{z}}_{1:k-1}) d\mathbf{x}_k^o}, \quad (7)$$

where $l(\mathbf{x}_k^o | {}^s\tilde{\mathbf{z}}_k)$ represents the observation likelihood of $\mathbf{x}_k^o$. The observation likelihood is defined with reference to the PoD and is given by

$$ l(\mathbf{x}_k^o | {}^s\tilde{\mathbf{z}}_k) = \begin{cases} p(\mathbf{x}_k^o | {}^s\tilde{\mathbf{z}}_k) & {}^s\tilde{\mathbf{z}}_k \in {}^s\mathcal{X}_k^o \\ 1 - P_\mathrm{d}(\mathbf{x}_k^o) & {}^s\tilde{\mathbf{z}}_k \notin {}^s\mathcal{X}_k^o, \end{cases} \quad (8)$$

where $p(\mathbf{x}_k^o | {}^s\tilde{\mathbf{z}}_k)$ is the probabilistic representation of the observation model defined in Equation (6). When the object is within the observable region, a positive observation is obtained and the observation likelihood is a probability density function given the current of the object observation. When the object is out of the observable region, the negative observation is defined with respect to the PoD as the observation likelihood. Due to the fact that the observation likelihood of the negative observation is non-Gaussian, when the negative observation occurs in the RBE, the object belief would immediately become heavily non-Gaussian.

Parallel grid-based RBE

Representation of target space and belief

The grid-based RBE achieves non-Gaussian belief estimation by first representing the arbitrary target space $\mathcal{X}^t$ in terms of a set of grid cells by constructing a rectangular space $\mathcal{X}^r$ that covers the target space. For simplicity, let us consider a two-dimensional target space, and it is represented as $\mathbf{m}^t = [x^t, y^t] \in \mathcal{X}^t$. The creation of a rectangular space $\mathcal{X}^r$ is achieved then by defining the minimum and maximum values of the target space

$$ x_{\mathrm{min}}^t = \min\{x^t\}, x_{\mathrm{max}}^t = \max\{x^t\}$$
$$ y_{\mathrm{min}}^t = \min\{y^t\}, y_{\mathrm{max}}^t = \max\{y^t\}$$

and subsequently creating a rectangular space as $\mathcal{X}^r = \{\mathbf{m} | \forall x \in [x_{\mathrm{min}}^t, x_{\mathrm{max}}^t], \forall y \in [y_{\mathrm{min}}^t, y_{\mathrm{max}}^t]\} \supseteq \mathcal{X}^t$, where $\mathbf{m} = [x, y]$. The grid space is further introduced by discretizing the rectangular space by n_x and n_y grid cells in two directions, respectively. The dimensions of a grid cell are defined as $\Delta x^r = (x_{\mathrm{max}}^t - x_{\mathrm{min}}^t)/n_x$ and $\Delta y^r = (y_{\mathrm{max}}^t - y_{\mathrm{min}}^t)/n_y$.

This results in introducing the center of each grid cell as

$$\bar{\mathbf{m}}^r_{i_x,i_y} = [\bar{x}^r_{i_x}, \bar{y}^r_{i_y}] = [(i_x - 0.5)\Delta x^r + x^t_{\min}, (i_y - 0.5)\Delta y^r + y^t_{\min}], \tag{9}$$

where $\forall i_x \in \{1, \ldots, n_x\}$ and $\forall i_y \in \{1, \ldots, n_y\}$. Each grid cell is defined as

$$\mathcal{X}^r_{i_x,i_y} = \{\mathbf{m} \,|\, |x - \bar{x}^r_{i_x}| < \frac{1}{2}\Delta x^r, |y - \bar{y}^r_{i_y}| < \frac{1}{2}\Delta y^r\}. \tag{10}$$

Note that $\bigcup_{i_x=1}^{n_x} \bigcup_{i_y=1}^{n_y} \mathcal{X}^r_{i_x,i_y} = \mathcal{X}^r$ and $\bigcap_{i_x=1}^{n_x} \bigcap_{i_y=1}^{n_y} \mathcal{X}^r_{i_x,i_y} = \varnothing$. Finally, the selection of grid cells that represent the target space is performed by selecting a grid cell when its center is located in the target space, $\mathcal{X}^r_{i_x,i_y} \subset \mathcal{X}^t$ if $\bar{\mathbf{x}}^r_{i_x,i_y} \in \mathcal{X}^t$. The approximate target space derived by the processes described above is $\mathcal{X}^t \approx \{\mathcal{X}^r_1, \mathcal{X}^r_2, \ldots, \mathcal{X}^r_{n_g}\}$, where n_g is the number of grid cells approximating the target space.

The belief is usually represented by a probability density function over the target space. Similar to the discretization of the target space, the belief could also be represented discretely by grid cells. The position of each grid cell can be described in the two-dimensional integer space as $[i_x, i_y]$, where $i_x \in 1, \ldots, n_x$ and $i_y \in 1, \ldots, n_y$. With the integer representation, the belief at the grid cell $[i_x, i_y]$ can be represented as $p^{i_x,i_y}(\cdot)$.

Prediction

The prediction process requires the numerical evaluation of Equation (4). Given the belief of the previous state $p^{i_x,i_y}(\mathbf{x}^t_k|^s\tilde{\mathbf{z}}_{1:k-1})$ at the grid cell $[i_x, i_y]$ and the target motion model $p^{I_x,I_y}(\mathbf{x}^t_k|\mathbf{x}^t_{k-1})$ constructed in the matrix of size $I_x \times I_y$ as a convolution kernel, the predicted belief of the current state can be numerically computed as

$$\begin{aligned} &p^{i_x,i_y}(\mathbf{x}^t_k|^s\tilde{\mathbf{z}}_{1:k-1}) \\ &= p^{i_x,i_y}(\mathbf{x}^t_{k-1}|^s\tilde{\mathbf{z}}_{1:k-1}) \otimes p^{I_x,I_y}(\mathbf{x}^t_k|\mathbf{x}^t_{k-1}), \end{aligned} \tag{11}$$

where $\otimes$ indicates the two-dimensional convolution of the belief of the previous state with the probabilistic target motion model. Therefore, the belief of the current state is given by

$$\begin{aligned} &p^{i_x,i_y}(\mathbf{x}^t_k|^s\tilde{\mathbf{z}}_{1:k-1}) \\ &= \sum_{\beta=1}^{I_y} \sum_{\alpha=1}^{I_x} p^{\alpha,\beta}(\mathbf{x}^t_k|\mathbf{x}^t_{k-1}) p^{i_x-\alpha+1, i_y-\beta+1}(\mathbf{x}^t_{k-1}|^s\tilde{\mathbf{z}}_{1:k-1}). \end{aligned} \tag{12}$$

The parallelization of the prediction process is straightforward. Since the prediction at each grid cell, given by Equation (12), can be performed independently, the parallelization of the prediction corresponds to the parallelization of the equation and achieves a parallel efficiency of 100% in an ideal environment. However, this equation also shows that the computation for the prediction process is largely dominated by the size of the convolution kernel. In order for real-time performance, it is important that the convolution kernel of an appropriate size, which needs to be big enough to capture the motion of the target as well as small enough to perform fast computation, is utilized.

Correction

The correction process corresponds to the numerical computation of Equation (7). Given the predicted belief $p(\mathbf{x}^t_k|^s\tilde{\mathbf{z}}_{1:k-1})$ and the new observation likelihood $l^{i_x,i_y}(\mathbf{x}^t_k|^s\tilde{\mathbf{z}}_k)$ at the grid cell $[i_x, i_y]$, the corrected belief is computed by

$$p^{i_x,i_y}(\mathbf{x}_k^t|^s\tilde{\mathbf{z}}_{1:k}) = \frac{q^{i_x,i_y}(\mathbf{x}_k^t|^s\tilde{\mathbf{z}}_{1:k})}{A_c \sum_{\alpha=1}^{n_g} q^{\alpha}(\mathbf{x}_k^t|^s\tilde{\mathbf{z}}_{1:k})}, \tag{13}$$

where A_c is the area of a grid cell, and

$$q^{i_x,i_y}(\mathbf{x}_k^t|^s\tilde{\mathbf{z}}_{1:k}) = l^{i_x,i_y}(\mathbf{x}_k^t|^s\tilde{\mathbf{z}}_k)p^{i_x,i_y}(\mathbf{x}_k^t|^s\tilde{\mathbf{z}}_{1:k-1}). \tag{14}$$

The parallelization of the correction process requires the breakdown of the process as it identifies which subprocesses are parallelizable. By observing the mathematical operations, the correction process can be broken down into three steps:

1. Calculate $q^{i_x,i_y}(\mathbf{x}_k^t|^s\tilde{\mathbf{z}}_{1:k})$ by multiplying the predicted belief $p^{i_x,i_y}(\mathbf{x}_k^t|^s\tilde{\mathbf{z}}_{1:k-1})$ with the observation likelihood $l^{i_x,i_y}(\mathbf{x}_k^t|^s\tilde{\mathbf{z}}_k)$;
2. Sum $\sum_{\alpha=1}^{n_g} q^{\alpha}(\mathbf{x}_k^t|^s\tilde{\mathbf{z}}_{1:k})$ and multiply the sum by A_c;
3. Calculate $p^{i_x,i_y}(\mathbf{x}_k^t|^s\tilde{\mathbf{z}}_{1:k})$ by dividing $q^{i_x,i_y}(\mathbf{x}_k^t|^s\tilde{\mathbf{z}}_{1:k})$ by $A_c \sum_{\alpha=1}^{n_g} q^{\alpha}(\mathbf{x}_k^t|^s\tilde{\mathbf{z}}_{1:k})$.

The breakdown indicates that steps 1 and 3 are grid-wise sub-processes, which can be conducted independently. Therefore, for the correction process, steps 1 and 3 can be computed in parallel, whereas step 2 is not parallelizable.

Target state evaluation

In the parallel grid-based RBE, the state of the target is evaluated by Equation (2) in the integral form at each time step. For an accurate evaluation of the target state an appropriate choice of the time interval Δt is necessary. Given a specific computer hardware configuration, each iteration of the parallel grid-based RBE requires the certain amount of time Δt_c to perform the computation, including both the prediction and correction processes. In order to achieve an accurate evaluation of the target state, the time interval Δt needs to be chosen such that it matches the Δt_c. As shown in Figure 1, only when the Δt is identical with the Δt_c the evaluated target states could match the real target states. When the Δt is smaller or larger than the Δt_c, the evaluation of the target states fails and eventually leads to large accumulated errors. The Δt_c is determined by not only the parallel grid-based RBE itself but also its computational performance for the specific computer hardware configuration.

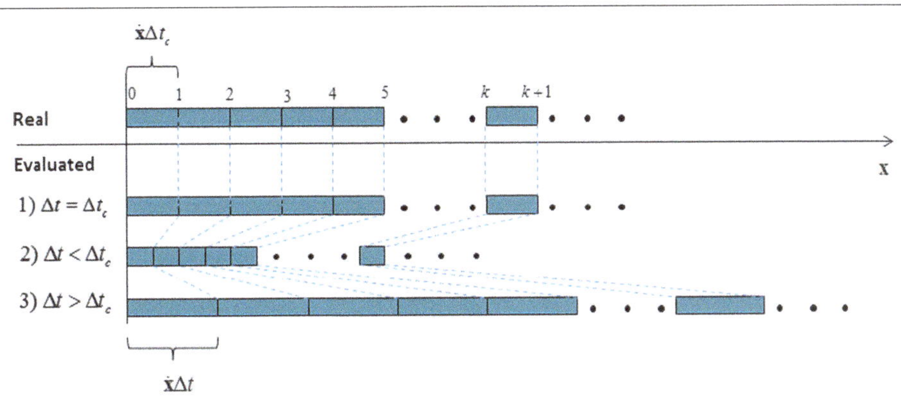

Figure 1 Target states evaluation.

Computational performance modeling

Acceleration of prediction process

Since the RBE designed with high frequency results in using the Markovian target motion model well approximated by a Gaussian probability density, the proposed modeling first reformulates the prediction process with the Gaussian assumption as a pre-process and accelerates the parallel grid-based RBE to achieve the maximum performance. With the Gaussian assumption, the convolution kernel in the matrix of size $I_x \times I_y$ can be separated into two vector kernels in the name of separable convolution: a column kernel of length I_x and a row kernel of length I_y. Therefore, the target motion model matrix is separated as

$$p^{I_x,I_y}(\mathbf{x}_k^t|\mathbf{x}_{k-1}^t) = {}^c p^{I_x}(\mathbf{x}_k^t|\mathbf{x}_{k-1}^t)\,{}^r p^{I_y}(\mathbf{x}_k^t|\mathbf{x}_{k-1}^t), \tag{15}$$

where ${}^c p^{I_x}(\mathbf{x}_k^t|\mathbf{x}_{k-1}^t)$ and ${}^r p^{I_y}(\mathbf{x}_k^t|\mathbf{x}_{k-1}^t)$ are the column kernel and row kernel, respectively, with the size of a vector of $I_x + I_y$. Substituting Equation (15) into Equation (11), the predicted belief of the current state can be computed as

$$\begin{aligned} &p^{i_x,i_y}(\mathbf{x}_k^t|{}^s\tilde{\mathbf{z}}_{1:k-1}) \\ &= p^{i_x,i_y}(\mathbf{x}_{k-1}^t|{}^s\tilde{\mathbf{z}}_{1:k-1}) \otimes {}^c p^{I_x}(\mathbf{x}_k^t|\mathbf{x}_{k-1}^t) \otimes {}^r p^{I_y}(\mathbf{x}_k^t|\mathbf{x}_{k-1}^t), \end{aligned} \tag{16}$$

which means that the prediction process can be broken down into two steps:

$$\begin{aligned} &u^{i_x,i_y}(\mathbf{x}_k^t|{}^s\tilde{\mathbf{z}}_{1:k-1}) \\ &= p^{i_x,i_y}(\mathbf{x}_{k-1}^t|{}^s\tilde{\mathbf{z}}_{1:k-1}) \otimes {}^c p^{I_x}(\mathbf{x}_k^t|\mathbf{x}_{k-1}^t) \\ &= \sum_{\alpha=1}^{I_x} {}^c p^{\alpha}(\mathbf{x}_k^t|\mathbf{x}_{k-1}^t)\, p^{i_x-\alpha+1,i_y}(\mathbf{x}_{k-1}^t|{}^s\tilde{\mathbf{z}}_{1:k-1}), \end{aligned} \tag{17}$$

and

$$\begin{aligned} &p^{i_x,i_y}(\mathbf{x}_k^t|{}^s\tilde{\mathbf{z}}_{1:k-1}) \\ &= u^{i_x,i_y}(\mathbf{x}_k^t|{}^s\tilde{\mathbf{z}}_{1:k-1}) \otimes {}^r p^{I_y}(\mathbf{x}_k^t|\mathbf{x}_{k-1}^t) \\ &= \sum_{\beta=1}^{I_y} {}^r p^{\beta}(\mathbf{x}_k^t|\mathbf{x}_{k-1}^t)\, u^{i_x,i_y-\beta+1}(\mathbf{x}_{k-1}^t|{}^s\tilde{\mathbf{z}}_{1:k-1}). \end{aligned} \tag{18}$$

These equations show that the prediction process at each grid cell is carried out by performing two one-dimensional convolutions, each in horizontal and vertical directions instead of the original one two-dimensional convolution while remaining complete parallelizability. For Equation (17), the number of floating point operations for each grid cell is seen $2I_x$ since I_x times of one multiplication and one summation are necessary, whereas the number of floating point operations for Equation (18) is $2I_y$ via the similar observation. Having a total of n_g grid cells, the total number of floating point operations for the prediction process is thus given by

$$N_{\mathrm{p}} = 2n_g I_x + I_y. \tag{19}$$

This is considerably small compared to that of the original formulation which is derived as $2n_g I_x I_y$ via Equation (12) since $I_x + I_y \ll I_x I_y$ for an appropriate prediction process.

Parallel computation using GPU

Following Equations (16) and (13) for the prediction and correction process, respectively, Figure 2 shows the schematic diagram of the proposed accelerated parallel grid-based RBE using GPU. For efficiency, the GPU stores the entire data for RBE in the global memory and performs RBE using local memories. As a result, the data transmission between the CPU's memory and the GPU's local memories is carried out via the GPU's global memory, and all the parallelizable floating point operations are executed using the local memories. For the prediction process, the data to be transmitted from the CPU's memory to the GPU's local memories are the previous belief $p(\mathbf{x}_{k-1}^t|^s\tilde{\mathbf{z}}_{1:k-1})$ and the target motion model $p(\mathbf{x}_k^t|\mathbf{x}_{k-1}^t)$. Since the predicted belief is in the local memories, the correction needs only the observation likelihood to be initially transmitted in addition. After performing the multiplication of $p(\mathbf{x}_k^t|^s\tilde{\mathbf{z}}_{1:k-1})$ and the observation likelihood $l(\mathbf{x}_k^t|^s\tilde{\mathbf{z}}_k)$ using GPU's local memories, the result $q(\mathbf{x}_k^t|^s\tilde{\mathbf{z}}_{1:k})$ is transmitted to the CPU's memory to calculate the sum $A_c \sum_{\alpha=1}^{n_g} q^\alpha(\mathbf{x}_k^t|^s\tilde{\mathbf{z}}_{1:k})$. The sum is then transmitted back to the GPU's local memories to perform divisions in parallel and update the belief $p(\mathbf{x}_k^t|^s\tilde{\mathbf{z}}_{1:k})$. Finally, the belief is transmitted back to the CPU's memory for the next iteration of the accelerated parallel grid-based RBE.

Modeling of computational performance

The computational performance of the accelerated parallel grid-based RBE using GPU is determined not only by the performance of the CPU but also by the performance of the GPU and that of data transmission. As a result, the time cost of one iteration of the accelerated parallel grid-based RBE is given by

$$\Delta t_c = \Delta t_{trans} + \Delta t_G + \Delta t_C, \tag{20}$$

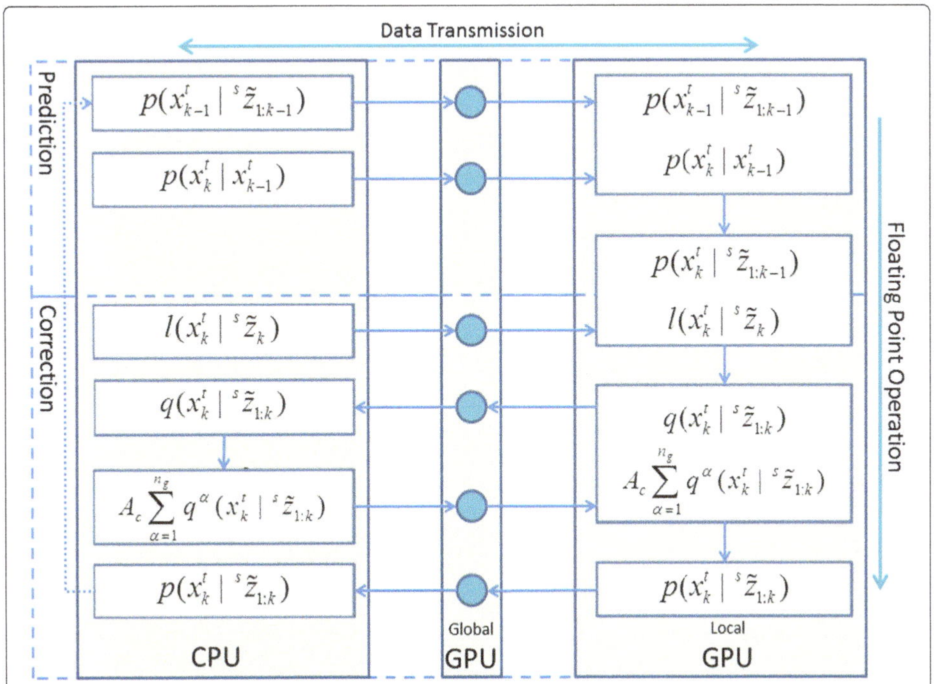

Figure 2 Procedures in the parallel grid-based RBE using GPU.

where Δt_{trans} represents the data transmission time cost between the CPU's memory and the GPU's global memory as well as that between the local and the global memory inside the GPU, Δt_{G} represents the time cost of the parallel computation performed on the GPU, and Δt_{C} represents the time cost of the computation performed on the CPU.

Data transmission

In order to determine the data transmission time cost Δt_{trans} for one iteration of the accelerated parallel grid-based RBE, the data transmitted among the CPU's memory, GPU's global memory, and GPU's local memory need to be evaluated in both the prediction and correction processes. Let the amount of data transmitted in the unit of bytes be defined as

$$A = PN, \tag{21}$$

where P is the precision of the numerical representation, and N is defined as the number of data transmitted. Since the precision is usually constant, the amount of data transmitted could be derived in terms of the number of data transmitted. The numbers of data of the belief and the target motion model for the prediction process are n_g and $I_x + I_y$, respectively. The same numbers of data, n_g and $I_x + I_y$, are transmitted to the GPU's local memory to perform parallel calculation. In the correction process, the number of data of the likelihood to be transmitted from the CPU's memory to the GPU's local memory through the GPU's global memory is n_g, whereas the number of data of the result $q(\mathbf{x}_k^t|^s\tilde{\mathbf{z}}_{1:k})$ to be transmitted from the GPU's local memory to the CPU's memory through the GPU's global memory is similarly n_g. The number of data of the sum, $A_{\text{c}} \sum_{\alpha=1}^{n_g} q^{\alpha}(\mathbf{x}_k^t|^s\tilde{\mathbf{z}}_{1:k})$, to be then transmitted to the GPU's local memory to perform parallel divisions is 1, and finally, the number of data to be transmitted back to the CPU's memory for the next RBE is n_g.

By observing the data transmission for one iteration of the accelerated parallel grid-based RBE, the total number of data transmitted from the CPU's memory to the GPU's global memory is given by

$$\begin{aligned} N_{\text{CG}} &= n_g + I_x + I_y + 1 + n_g \\ &= 2n_g + I_x + I_y + 1, \end{aligned} \tag{22}$$

and all the data are transmitted continuously from the GPU's global memory to the GPU's local memory

$$N_{\text{GL}} = N_{\text{CG}} = 2n_g + I_x + I_y + 1. \tag{23}$$

The total number of data transmitted from the GPU's local memory to the GPU's global memory is

$$N_{\text{LG}} = n_g + n_g = 2n_g, \tag{24}$$

and that from the GPU's global memory to the CPU's memory similarly becomes

$$N_{\text{GC}} = N_{\text{LG}} = 2n_g. \tag{25}$$

The data transmission time cost Δt_{trans} for one iteration of the accelerated parallel grid-based RBE is given by

$$\Delta t_{\text{trans}} = P\left(\frac{N_{\text{CG}}}{B_{\text{CG}}} + \frac{N_{\text{GC}}}{B_{\text{GC}}} + \frac{N_{\text{GG}}}{B_{\text{GG}}}\right), \tag{26}$$

where N_{CG} and B_{CG} are the total number of data transmitted and the copy bandwidth with the unit of bytes per second from the CPU's memory to the GPU's global memory, respectively, N_{GC} and B_{GC} are those from the GPU's global memory to the CPU's memory, respectively, and N_{GG} and B_{GG} represent those between the GPU's global memory and the GPU's local memory. Due to the fact that the copy bandwidth from the GPU's global memory to the GPU's local memory and the one in opposite direction are the same, the number of data transmitted inside the GPU is given by

$$N_{GG} = N_{GL} + N_{LG} = 4n_g + I_x + I_y + 1. \tag{27}$$

Substitute Equations (22), (25), and (27) into Equation (26), the data transmission time cost for one iteration of the accelerated parallel grid-based RBE is given by

$$\Delta t_{\text{trans}} = P\left(\frac{2n_g + I_x + I_y + 1}{B_{CG}} + \frac{2n_g}{B_{GC}} + \frac{4n_g + I_x + I_y + 1}{B_{GG}}\right). \tag{28}$$

It is to be noted here that these parameters of copy bandwidths are inherent for a specific computer hardware configuration and can be determined experimentally.

Floating point operations

In order to determine the GPU computation time cost Δt_G and CPU computation time cost Δt_C for one iteration of the accelerated parallel grid-based RBE, the number of floating point operations performed on both CPU and GPU needs to be evaluated. The number of floating point operations performed on the GPU for the prediction process is seen $2n_g(I_x + I_y)$ as the Equation (19) indicated. The number of floating point operations performed on the GPU for the correction process is identified as $2n_g$ in total since n_g parallel multiplications and n_g parallel divisions are performed for steps 1 and 3 in Subsection 2, respectively. Meanwhile, the number of floating point operations performed on the CPU is n_g by n_g summations in step 2 of the Subsection 2. As a consequence, the total number of floating point operations performed on the GPU and the CPU for one iteration of the accelerated parallel grid-based RBE is given, respectively, by

$$N_G = 2n_g(I_x + I_y) + 2n_g = 2n_g(I_x + I_y + 1),$$

$$N_C = n_g.$$

The GPU computation time cost for one iteration of the accelerated parallel grid-based RBE is given by

$$\Delta t_G = \frac{N_G}{V_G}, \tag{29}$$

where N_G is the number of floating point operations performed on the GPU, and V_G is the computational rate of GPU with the unit of FLOPS. Substituting Equation (29) into Equation (29), the GPU computation time cost is given by

$$\Delta t_G = 2n_g\frac{I_x + I_y + 1}{V_G}. \tag{30}$$

Similarly, the CPU computation time cost for one iteration of the accelerated parallel grid-based RBE is given by

$$\Delta t_C = \frac{N_C}{V_C}, \tag{31}$$

where N_C represents the number of floating point operations performed on the CPU, and V_C is the computational rate of CPU with the unit of FLOPS. In the same way, by substituting Equation (29) into Equation (31), the CPU computation time cost is given by

$$\Delta t_C = \frac{n_g}{V_C}. \tag{32}$$

It is to be noted here that the computational rates, V_G and V_C, are also inherent for a specific CPU and GPU configuration and can be determined experimentally.

Experimental validation

Table 1 shows the setup specifications which have been available for the validation and other investigations. Setup 1 is the fastest in both CPU and GPU, whereas setup 3 is the slowest. This section firstly shows the improvement of the parallel grid-based RBE using GPU by adopting the separable convolution in the prediction process with the specification listed in setup 1. Moreover, the proposed computational modeling for the parallel grid-based RBE is validated via setups 1 to 3. In the end, a simulated target searching task is introduced to further evaluate the efficacy of the proposed modeling.

Improvement in prediction process

The efficiency of the prediction process accelerated by separable convolution was evaluated with a problem having a fixed grid space size of $1,000 \times 1,000$ and varying the convolution kernel size from 1 to 50 on the computer setup 1. The result of the time cost by GPU is shown in Figure 3 together with the corresponding result by the original prediction. Even when the convolution kernel size is 50, the accelerated prediction is seen to require the time cost of only 1 ms. Its superiority can also be understood by comparing it to the original prediction, which needs the time cost 25 times as much as that of the accelerated prediction process when the convolution kernel size is 50.

Validation

This set of tests was aimed at validating the proposed modeling of computer performance by estimating the total iteration time cost Δt of the parallel grid-based RBE using GPU and comparing it with the actual iteration time cost experimentally measured in three different computer setups. Each component, Δt_{trans}, Δt_G, or Δt_C, is also compared with the actual performance, respectively. All the time cost results are measured by averaging the time cost of $10,000$ iterations. Needless to say, the convolution kernel size $I_x + I_y$ and grid space size n_g are the two major factors in the proposed modeling. Two tests were thus conducted by each, changing the convolution kernel size and the grid space size.

Table 1 Test computer system specifications

Setup	Processor	Memory (GB)	GPU
1	Intel Dual-Core, 2.70 GHz	4.0	Nvidia GeForce GT220
2	Intel Dual-Core, 2.40 GHz	4.0	Nvidia GeForce GT320M
3	Intel Dual-Core, 2.40 GHz	4.0	Nvidia GeForce GS8400

Nvidia GeForce (Santa Clara, CA, USA), Intel (Santa Clara, CA, USA).

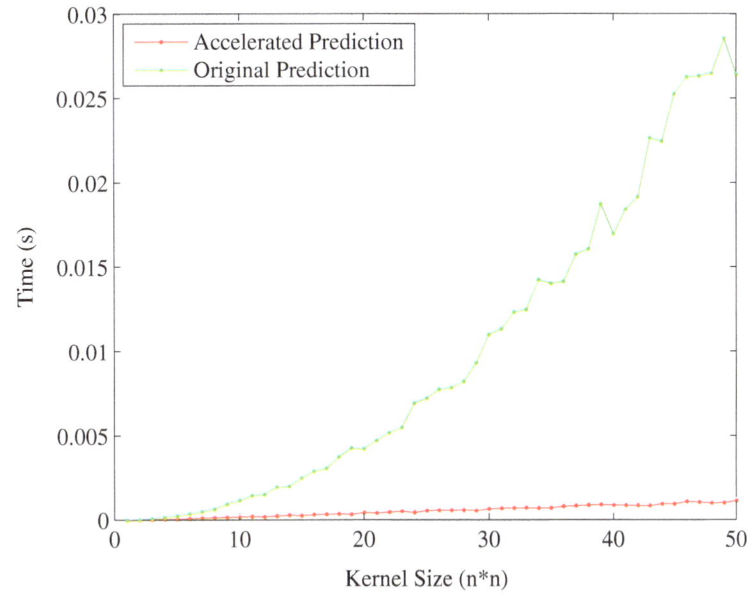

Figure 3 GPU calculation time cost.

Test 1

Test 1 was performed by fixing the grid space size of the parallel grid-based RBE to $1,000 \times 1,000$ and varying the convolution kernel size $I_x = I_y = i$ from 1 to 200. A convolution kernel size over 200 was not explored since it is unlikely that the target motion model requires such a large convolution kernel. The square convolution kernel was because of the insignificance in changing size in both x and y directions, and this additionally allows visualization of results in two-dimensional space.

The results of all the components of the time cost for the three computer setups are shown in Figures 4, 5, and 6. Each solid line represents the estimated total and component time costs, whereas each solid dot line with the same color represents the corresponding actual performance. These figures primarily show that the total and component time costs estimated by the proposed modeling well match to the actual performance. Values listed in Table 2 also support this and indicate the effectiveness of the proposed modeling since the average and the maximum relative errors are below 7% and 12%, respectively. While the time cost of data transmission is seen to contribute most, it is also seen that the time cost by GPU increases the total time cost with increase in convolution kernel size particularly when the GPU is of low quality. It is thus important to use a high-performance GPU if fast RBE with large convolution kernel size is necessary.

Test 2

Test 2 was performed by fixing the convolution kernel size of the parallel grid-based RBE to 16×16 or 32×32 and varying grid space size $n_x = n_y = n$ from 100 to 1,000. These convolution kernel sizes often represent the target motion model with sufficient accuracy, and the grid space size $n = 1,000$, which creates $1,000,000$ grid cells, also provides good accuracy in many practical problems. Similarly to test 1, the square grid size enables two-dimensional visualization of results.

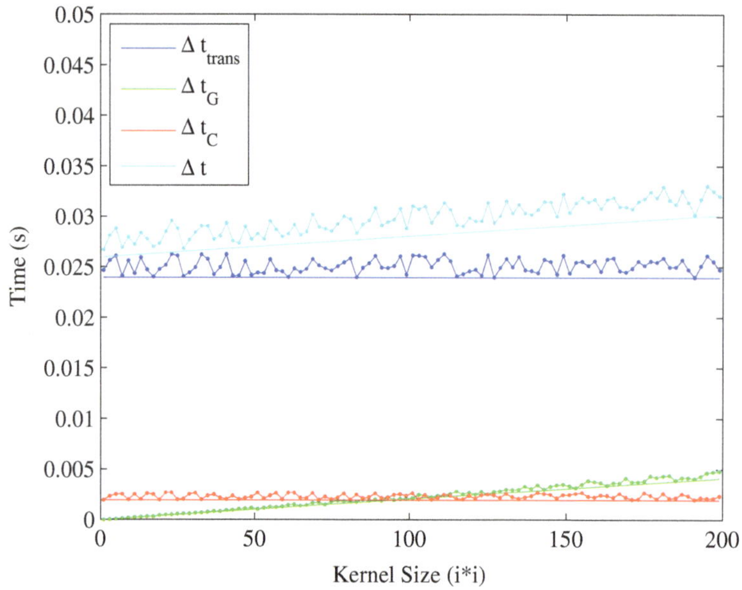

Figure 4 Time cost of all the components for setup 1 with fixed grid space.

The results for all the components of the time cost for the three computer setups are shown in Figures 7, 8, and 9, respectively. These figures firstly show that the proposed modeling is also able to well estimate the actual performance of the parallel grid-based RBE regardless of different grid space sizes. Similarly to test 1, Table 3 shows small average and maximum relative errors, which are below 6% and 11%, respectively. Secondly, from these results, it is seen that the total time cost is dominated by the time cost of data transmission particularly when the ratio of the grid space size to the convolution kernel

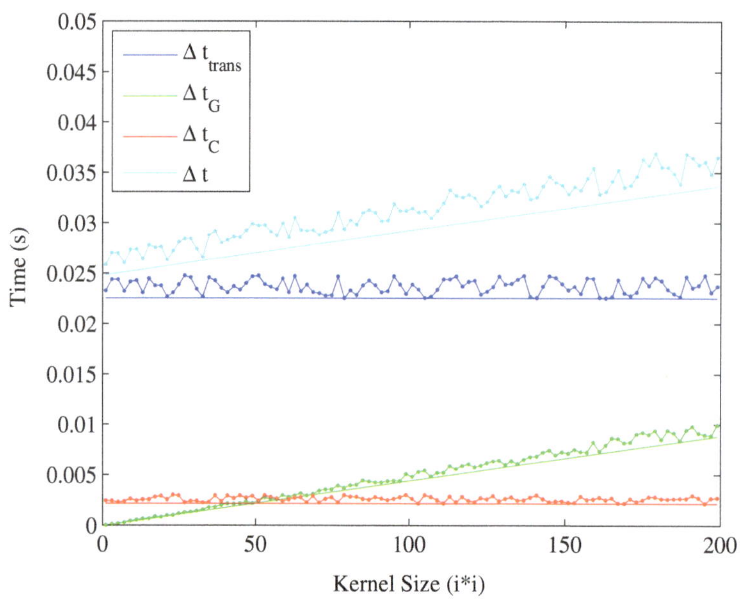

Figure 5 Time cost of all the components for setup 2 with fixed grid space.

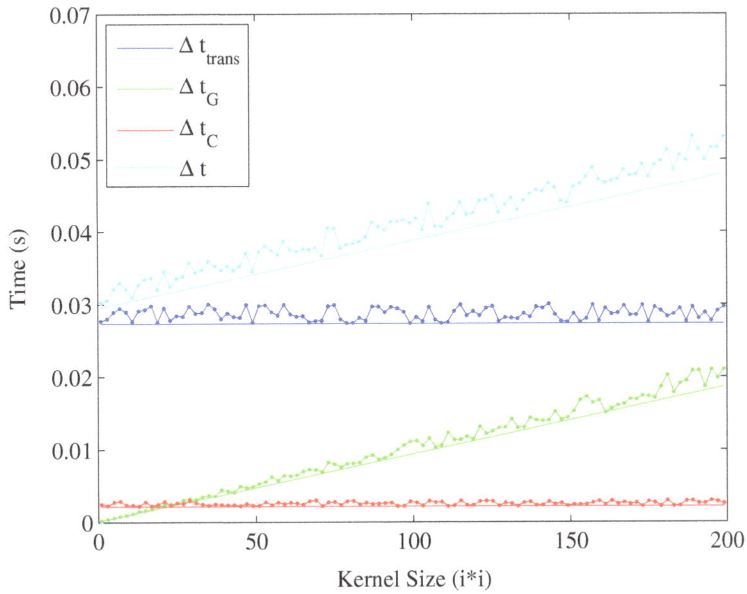

Figure 6 Time cost of all the components for setup 3 with fixed grid space.

size is large. Since the data transmission rate is determined by the quality of the memory, the utilization of a high-quality memory is the first priority for fast RBE.

Simulated target searching task

The performance of the prediction process dominates the accuracy of the RBE when no valid observations are obtained. The aim of this test is to evaluate how well the proposed modeling help the prediction process keep the accuracy during the no observation period. A simplified target searching task is described in this subsection. The motion model of the simulated target is given by

$$\begin{aligned} x_{k+1}^t &= x_k^t + \Delta t \cdot v_k^t \cos \gamma_k^t \\ y_{k+1}^t &= y_k^t + \Delta t \cdot v_k^t \sin \gamma_k^t, \end{aligned} \tag{33}$$

Table 2 Quantitative results for test 1

	Time cost	Setup 1	2	3
Average relative error	Δt_{trans}	1.159 ms	1.165 ms	1.305 ms
	Δt_G	0.216 ms	0.462 ms	0.856 ms
	Δt_C	0.402 ms	0.446 ms	0.382 ms
	Δt	1.777 ms	2.073 ms	2.543 ms
		(5.88%)	(6.55%)	(6.05%)
Maximum relative error	Δt_{trans}	2.351 ms	2.254 ms	2.670 ms
	Δt_G	0.716 ms	1.464 ms	3.259 ms
	Δt_C	0.779 ms	0.857 ms	0.818 ms
	Δt	3.228 ms	4.149 ms	6.081 ms
		(10.63%)	(11.24%)	(11.45%)

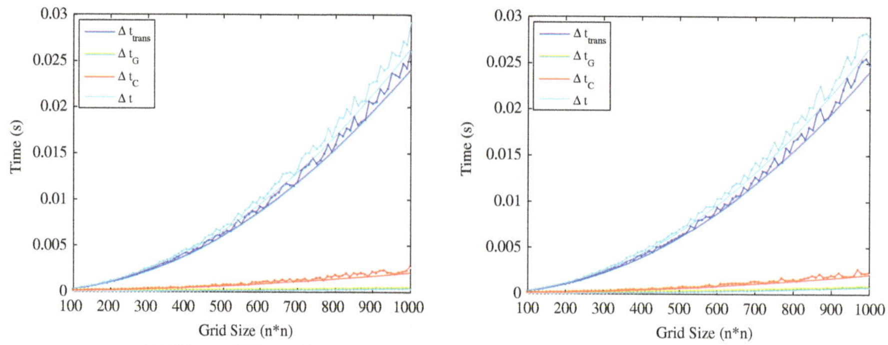

(a)KernelSize=16 (b)KernelSize=32

Figure 7 Time cost of all the components for setup 1 with fixed kernel.

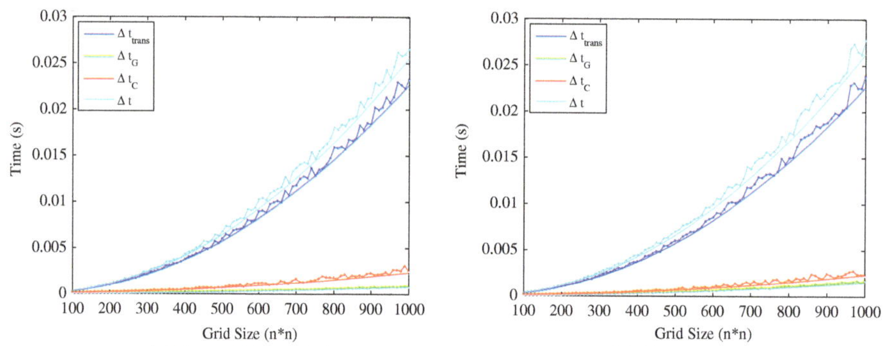

(a)KernelSize=16 (b)KernelSize=32

Figure 8 Time cost of all the components for setup 2 with fixed kernel.

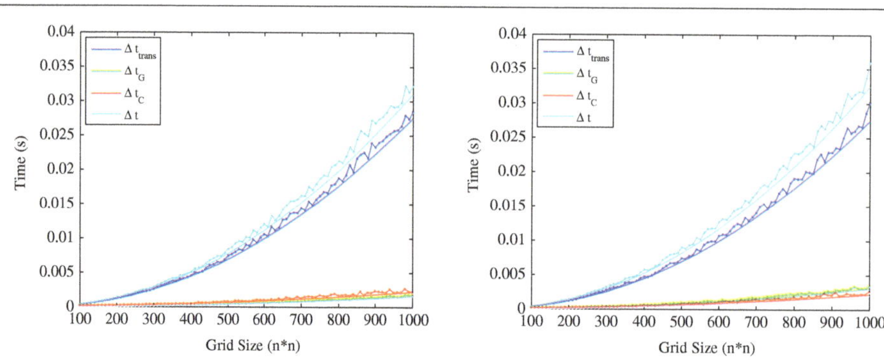

(a)KernelSize=16 (b)KernelSize=32

Figure 9 Time cost of all the components for setup 3 with fixed kernel.

Table 3 Quantitative results for test 2

	Total time cost	Setup		
		1	2	3
Average relative error	Δt	0.513 ms (5.59%)	0.530 ms (5.68%)	0.617 ms (5.90%)
Maximum relative error	Δt	2.140 ms (10.08%)	2.491 ms (10.64%)	2.835 ms (10.26%)

where v^t and γ^t are the velocity and direction of the target motion, respectively, each subject to a Gaussian noise, and Δt is the time increment. The prior belief on the target is also Gaussian. The autonomous sensor platforms are assumed to move on a horizontal plane and given by

$$\begin{aligned} x_{k+1}^{s_i} &= x_k^{s_i} + \Delta t \cdot v_k^{s_i} \cos \gamma_k^{s_i} \\ y_{k+1}^{s_i} &= y_k^{s_i} + \Delta t \cdot v_k^{s_i} \sin \gamma_k^{s_i} \\ \theta_{k+1}^{s_i} &= \theta_k^{s_i} + \Delta t \cdot \alpha^{s_i} \gamma_k^{s_i}, \end{aligned} \tag{34}$$

where v^{s_i} and γ^{s_i} are the velocity and turn of the sensor platform (s_i) respectively, and α^{s_i} is a coefficient governing the rate of turn. The probability of detection $P_{\mathrm{d}}(\mathbf{x}_k^t|\mathbf{x}_k^{s_i})$ is given by a Gaussian distribution, whereas the likelihood $l(\mathbf{x}_k^t|^{s_i}\tilde{\mathbf{z}}_k^t, \tilde{\mathbf{x}}_k^{s_i})$ when the target is detected is given by a Gaussian distribution with variances proportional to the distance between the sensor platform s_i and the target. Table 4 shows the major parameters of this simulated target searching task. The convolution kernel constructed by the target motion model is represented by a 32×32 matrix, and the grid space size is set as $1,000 \times 1,000$. The computer specifications followed the setup 3 in the Table 1. With the proposed approach, the time increment Δt was chosen as 0.032 s, the time cost of one iteration of the RBE estimated by the proposed modeling. For the case without the proposed approach, the time increment Δt was chosen as 0.02 s randomly in order to show the comparison.

Figure 10 shows the initial and final states of four sensor platforms without and with proposed prediction reformulation, respectively. Without the proposed prediction improvement, all the sensor platforms lost the target, whereas all of them successfully found the target under the condition of utilizing the proposed prediction reformulation. The reason is that the proposed prediction process made the grid-based RBE to update the belief much faster than the original one, resulting in a reliable tracking on the moving target. The evaluation of the proposed modeling for this simulated search and rescue task

Table 4 Major parameters of the target searching task

	Parameter	Value
Sensor platform, s_i	Velocity $v_k^{s_i}$	0.05
	Turn coef. α^{s_i}	0.8
	PoD var.	[0.2, 0.2]
Target, t	Velocity v_k^t	N(0.1, 0.02)
	Direction γ_k^t	N(0rad, 0.7rad)
	Prior $[x_0^t, y_0^t]$	N([20, 25], diag{200, 200 })

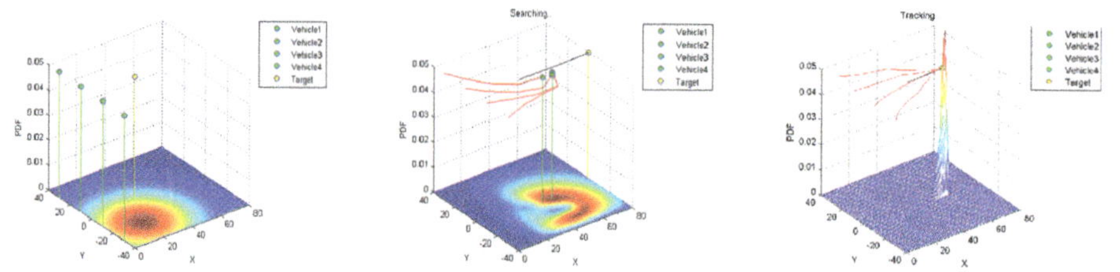

(a) Initial state (b) Final state (original approach) (c) Final state (propose approach)

Figure 10 Search and rescue task.

is conducted, and the corresponding quantitative results were concluded in the Table 5. The result shows small average and maximum relative errors as well, which are below 7% and 10% respectively, and indicates that the proposed modeling is able to estimate the actual time cost for the grid-based RBE using GPU.

Conclusion and future work

The performance modeling for the real-time grid-based RBE, especially parallel computation using GPU, has been proposed to identify the best resolution of the RBE with given computer hardware. The modeling allows the estimation of time costs necessary within CPU and GPU and that of data transmission between CPU and GPU for the real-time grid-based RBE. In order to speed up the RBE, the prediction has been additionally reformulated with the separable convolution.

The proposed modeling was experimentally investigated by varying its major parameters. The result of the first test with varying convolution kernel size shows that the average error of the estimation by the proposed modeling stays below 7% regardless of the convolution kernel size and that a high-performance GPU is necessary if the convolution kernel size is large. In the second test with varying grid space size, it is found that the proposed modeling estimates within the average error of 6%, irrespective of the grid space size, and that a high-quality memory is necessary if fast RBE is required for large grid space. Utilizing prediction with separable convolution, the RBE has also been found to perform within 1 ms, although the size of the problem was relatively large.

The current study is still the first step for achieving high-fidelity RBE in a real-time environment. The project is further planned to utilize the best resolution of the RBE identified by the proposed modeling and investigate its efficacy.

Table 5 Quantitative results for simulated search and rescue task

	Total time	Sensor platform			
		1	2	3	4
Average relative error	Δt	0.618 ms (5.78%)	0.633 ms (6.21%)	0.626 ms (5.82%)	0.618 ms (5.93%)
Maximum relative error	Δt	2.856 ms (9.89%)	2.823 ms (9.56%)	2.892 ms (9.25%)	2.854 ms (9.68%)

Author details
[1]Department of Mechanical Engineering, Virginia Tech, 800 Drillfield Dr, Blacksburg, VA 24061, USA. [2]Australian Centre for Field Robotics (ACFR), Rose St, Sydney 2006, Australia.

References

1. Tarantola, A: Inverse Problem Theory and Methods for Model Parameter Estimation. Society for Industrial and Applied Mathematics, Philadelphia (2005)
2. Harlim, J, Hunt, BR: A non-Gaussian ensemble filter for assimilating infrequent noisy observations. Tellus A. **59**, 225–237 (2007)
3. Apte, A, Hairer, M, Stuart, AM, Voss, J: Sampling the posterior: an approach to non-Gaussian data assimilation. Physica D. **230**, 50–64 (2007)
4. Doshi, P, Gmytrasiewicz, PJ: Monte Carlo sampling methods for approximating interactive POMDPs. J. Artif. Intell. Res. **34**, 297–337 (2009)
5. Mandel, J, Beezley, JD: An ensemble Kalman-particle predictor-corrector filter for non-Gaussian data assimilation. Comput. Sci. ICCS. **2009**, 470–478 (2009)
6. Stenger, B, Thayananthan, A, Torr, PHS, Cipolla, R: Filtering using a tree-based estimator. IEEE Int. Conf. Comput. Vis. **2**, 1063–1070 (2003)
7. Huang, D, Leung, H: Maximum likelihood state estimation of semi-Markovian switching system in non-Gaussian measurement noise. IEEE Trans. Aerosp. Electron. Syst. **46**, 133–146 (2010)
8. Bergman, N: Recursive Bayesian estimation navigation and tracking applications. PhD Dissertation, Linkopings University (1999)
9. Furukawa, T, Durrant-Whyte, HF, Lavis, B: The element-based method—theory and its application to Bayesian search and tracking. Paper presented at the IEEE/RSJ international conference on intelligent robots and systems, San Diego (29 Oct–2 Nov 2007)
10. Lavis, B, Furukawa, T: HyPE: Hybrid particle-element approach for recursive Bayesian searching and tracking. Proceedings of Robotic: Science and Systems IV, pp. 135–142. MIT Press, Zurich (2008)
11. Lavis, B, Furukawa, T, Durrant-Whyte, HF: Dynamic space reconfiguration for Bayesian search and tracking with moving targets. Auto. Robots. **24**(4), 387–399 (2008)
12. Furukawa, T, Lavis, B, Durrant-Whyte, HF: Parallel grid-based recursive Bayesian estimation using GPU for real-time autonomous navigation. Paper presented at the IEEE international conference on robotics and automation, Anchorage, AK, USA (3–7 May 2010)

15

A solid transportation problem with safety factor under different uncertainty environments

Abhijit Baidya[1*], Uttam Kumar Bera[1] and Manoranjan Maiti[2]

* Correspondence: abhijitnita@yahoo.in
[1]Department of Mathematics, National Institute of Technology, Agartala, Jirania 799055, West Tripura, India
Full list of author information is available at the end of the article

Abstract

In this paper, we introduce a new concept 'safety factor' in a transportation problem. When items are transported from plants to destinations through different conveyances, there are some difficulties/risks to transport the items due to bad road, insurgency, land slide, etc. in some routes. Due to these, a desired total safety factor is being introduced, and depending upon the nature of the safety factor, we develop five models. In this paper, a solid transportation problem (STP) with imprecise unit costs is considered. The sources' availabilities, destinations' demands, and capacities of conveyances are also represented by fuzzy numbers like trapezoidal and triangular numbers. The transportation problem has been formulated with and without a safety factor. To reduce the different models into its crisp equivalent, we introduce different methods as chance-constraint programming, an approach using interval approximation of fuzzy numbers and the application of the expected value model. Generalized reduced gradient technique is used to find the optimal solutions for a set of given numerical data. To illustrate the model, a numerical example has been presented and solved using LINGO.12 software. The effect of safety factors on transported amount is illustrated.

Keywords: Solid transportation problem; Safety factor; Fuzzy variable; Random variable; Hybrid variable; Expected value model; An approach using interval approximation of fuzzy number; Chance-constraint programming

Introduction

A transportation model plays a vital role in ensuring the efficient movement and in-time availability of raw materials and finished goods from sources to destinations. Transportation problem is a linear programming problem that stemmed from a network structure consisting of a finite number of nodes and arcs attached to them. Effective algorithms have been developed to solve the transportation problem when the cost coefficients and the supply and demand quantities are known exactly. The occurrence of randomness and imprecision in the real world is inevitable owing to some unexpected situations. There are cases that the cost coefficients and the supply and demand quantities of a transportation problem may be uncertain due to some unmanageable factors. To deal quantitatively with imprecise information in making decisions, Bellman and Zadeh [1] and Zadeh [2,3] introduced the impression of fuzziness. In many industrial problems, a homogeneous product is delivered from an origin to a destination by means of different modes of transport called conveyances, such as trucks, cargo flights, goods trains, and ships. As a generalization of traditional transportation problem, solid transportation problem (STP) was stated by Shell [4] in 1955, which he considered the three item properties in the constraint set instead of two items, namely source and destination. Haley [5] described a solution procedure of a solid transportation problem, which is an extension of the Modi method. In general, real-life problems are modeled with multi-objective functions which are measured in different respects, and they are noncommensurable and conflicting in nature. Furthermore, it is frequently difficult for the decision maker to combine the objective functions in one overall utility function. Due to insufficient information, lack of evidence, and fluctuating financial market, the available data of a transportation system such as resources, demands, and conveyance capacities are not always crisp or precise but are fuzzy, random, or both. Thus, the fuzziness and randomness can be present in the objective function as well as in the constraints of a STP. Since then, significant work has been done by researchers based on the uncertainty theory both in theoretical and practical aspects. If the cost parameters of the transportation problem are uncertain variables, we call the problem uncertain cost transportation problem. Along with global economic development, production and demand have much more importance. The importance of goods transportation is also increasingly reflected. Dealing with different types of uncertainty, Kundu et al. [6] solve solid transportation problems with budget constraint in an uncertain environment. Our aim is to formulate and solve solid transportation problems with safety constraints with different types of uncertain (fuzzy, random, and hybrid) parameters. Recently, Baidya et al. [7,8] solved two problems based on safety factors and uncertainty in transportation problem.

In this paper, we develop five models and consider three types of uncertainty (stochastic, fuzzy, and hybrid) in different models for unit transportation cost and safety factor. Also, we consider resources, demands, and conveyance capacities as fuzzy. Model 1 was solved without considering a safety factor, but other models were solved with safety factors that are crisp, random, fuzzy, and hybrid for their respective models. To derive the crisp equivalences of the models, appropriate methods, i.e., chance-constrained programming, an approach that uses interval approximation of fuzzy numbers, and expected value model techniques are applied in these five different models. The models are illustrated with specific numerical data. Finally, all the models are solved using generalized reduced gradient method using LINGO.12 software. A comparison of the different models is presented.

Literature review

Zadeh [2] first introduced the concept of fuzzy set theory. Later on, several authors such as Zadeh [3], Kaufmann [9], Zimmermann [10], Liu [11], and Dubois and Prade [12] developed and applied the fuzzy set theory. Chanas and Kuchta [13] studied transportation problem with fuzzy cost coefficients. Jimenez and Verdegay [14] considered two types of uncertain STP, one with interval numbers and the other with fuzzy numbers. Charnes and Cooper [15] introduced chance-constrained programming to deal with uncertain environment. Liu and Iwamura [16] presented chance-constrained programming with fuzzy parameters. Liu and Liu [17] presented expected value model for fuzzy programming. Yang and Liu [18] applied expected value model, chance-constrained programming model, and dependent-chance programming in a fixed charge solid transportation problem in a fuzzy environment. Mula, Poler, and Garcia Sabater [19] applied possibilistic programming approach to a material requirement planning problem with fuzzy constraints and fuzzy coefficients using the definition of possibility measure of fuzzy number. Chen and Tsai [20] considered time–cost trade-off problem with fuzzy parameters and constructed a two-level mathematical program for it using the concept α-cut of fuzzy number. Fegad, Jadhav, and Muley [21] obtained an optimal solution of a transportation problem using interval and triangular membership functions. Hulsurkar, Biswal, and Sinha [22] applied the fuzzy programming approach to multi-objective stochastic programming problems after transforming them into crisp forms. Some methodologies to deal with objective function having random parameters are highlighted in Rao's book [23].

Preliminaries

Definition 1. (Fuzzy number) [24] A fuzzy subset $\tilde{A}$ of real number $\mathfrak{R}$ with membership function $\mu_{\tilde{A}} : \mathfrak{R} \rightarrow [0,1]$ is said to be a fuzzy number if the following conditions are met:

1. $\mu_{\tilde{A}}(x)$ is an upper semicontinuous membership function;
2. $\tilde{A}$ is normal, i.e., $\exists$ an element x_0 s.t. $\mu_{\tilde{A}}(x_0) = 1$;
3. $\tilde{A}$ is fuzzy convex, i.e., $\mu_{\tilde{A}}(\lambda x_1 + (1-\lambda)x_2) \geq \mu_{\tilde{A}}(x_1) \wedge \mu_{\tilde{A}}(x_2) \forall x_1, x_2 \in \mathfrak{R}$ and $\lambda \in [0,1]$;
4. Support of $\tilde{A} = \{x \in \mathfrak{R} : \mu_{\tilde{A}}(x) > 0\}$ is bounded.

Definition 2. (Triangular fuzzy number) By a triangular fuzzy number η, we mean the fuzzy variable fully determined by the triplet $A = (a_1, a_2, a_3)$ of crisp numbers with $a_1 < a_2 < a_3$ whose membership function is given by

$$\mu_A(x) = \begin{cases} \dfrac{x-a_1}{a_2-a_1} & \text{if } a_1 \leq x < a_2 \\ \dfrac{x-a_3}{a_2-a_3} & \text{if } a_2 \leq x < a_3 \\ 0 & \text{otherwise.} \end{cases}$$

Definition 3. (Trapezoidal fuzzy number) A TrFN $\tilde{\xi}$ is a fuzzy number fully determined by quadruplet $\tilde{a} = (a_1, a_2, a_3, a_4)$ of crisp numbers with $r_1 \leq r_2 \leq r_3 \leq r_4$, whose membership function is given by

$$\mu_{\tilde{a}}(x) = \begin{cases} \dfrac{x-a_1}{a_2-a_1}, & \text{if } a_1 \leq x \leq a_2; \\ 1, & \text{if } a_2 \leq x \leq a_3; \\ \dfrac{a_4-x}{a_4-a_3}, & \text{if } a_3 \leq x \leq a_4; \\ 0, & \text{otherwise.} \end{cases}$$

Example. A man is driving along a highway where the speed limit is 55 mph. He tries to hold his speed at exactly 55 mph, but his car lacks 'cruise control,' so the speed varies from moment to moment. If a graph of his instantaneous speed over a period of several minutes is to be plotted and the result will be in rectangular coordinates, he will get a function that looks like the curves shown in Figures 1 and 2 which represent a triangular fuzzy number and trapezoidal fuzzy number,

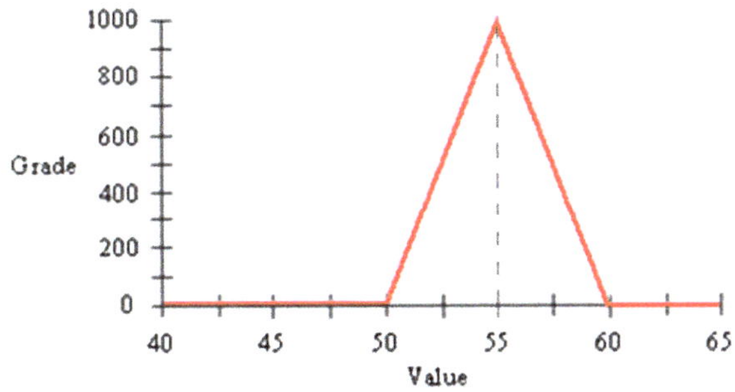

Figure 1 Triangular fuzzy number.

respectively. Crisp number b can be expressed as fuzzy number $\tilde{b}$, whose membership function $\mu_{\tilde{b}}$ (Figure 3) is given by

$$\mu_{\tilde{b}}(x) = \begin{cases} 1 \text{ for } x = b, \\ 0 \text{ for } x \neq b. \end{cases}$$

Definition 4. (Credibility measure) Credibility measure was presented by Liu and Liu [17]. For a fuzzy variable ξ with membership function $\mu_{\xi}(x)$ and for any set B of real numbers, credibility measure of fuzzy event $\{\xi \in B\}$ is defined as

$$\mathrm{Cr}\{\xi \in B\} = \frac{1}{2}(\mathrm{Pos}\{\xi \in B\} + \mathrm{Nec}\{\xi \in B\}),$$

where possibility and necessity of $\{\xi \in B\}$ are respectively defined as
and

$$\mathrm{Pos}\{\xi \in B\} = \sup_{x \in B} \mu_{\xi}(x)$$

and

$$\mathrm{Nec}\{\xi \in B\} = 1 - \sup_{x \in B^c} \mu_{\xi}(x).$$

Definition 5. [11,25]. Let ξ be a fuzzy variable. Then, the expected value of ξ is defined as

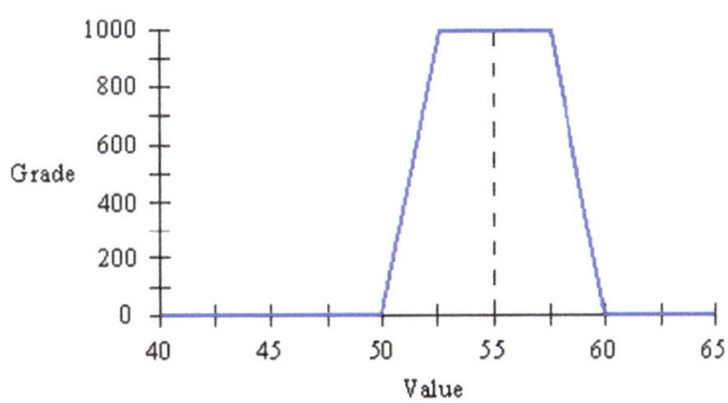

Figure 2 Trapezoidal fuzzy number.

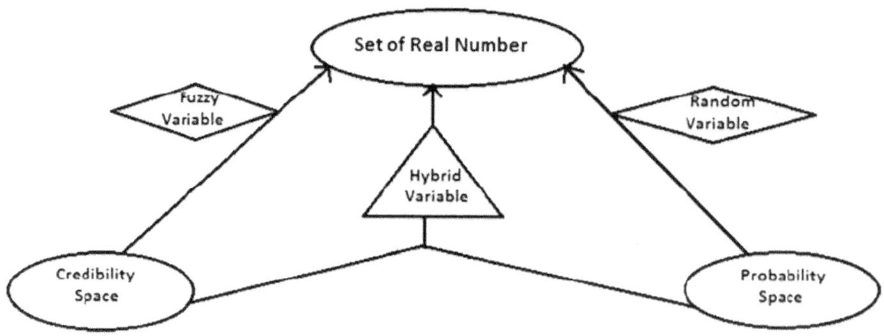

Figure 3 Geometrical representation of hybrid variable.

$$E(\xi) = \int_0^{\infty} \text{cr}\{\xi \geq r\}\text{dr} - \int_{-\infty}^{0} \text{cr}\{\xi \leq r\}\text{dr}$$

provided that at least one of the two integral is finite.

For example, the expected value of a triangular fuzzy variable (r_1, r_2, r_3) is $[\xi] = \frac{r_1+2r_2+r_3}{4}$.

Definition 6. [11]. Let ξ be a fuzzy variable and $\alpha \in [0, 1]$. Then

$\xi_{\sup}(\alpha) = \sup\{r: \text{cr}\{\xi \geq r\} \geq \alpha\}$ is called α-optimistic value to ξ, and

$\xi_{\inf}(\alpha) = \inf\{r: \text{cr}\{\xi \leq r\} \geq \alpha\}$ is called α-pessimistic value to ξ.

Example 1. Let $\xi = (r_1, r_2, r_3, r_4)$ be a trapezoidal fuzzy variable. Then, its α-optimistic and α-pessimistic values are

$$\xi_{\sup}(\alpha) = \begin{cases} 2\alpha r_3 + (1-2\alpha)r_4, \text{if } \alpha \leq 0.5; \\ (2\alpha-1)r_1 + 2(1-\alpha)r_2, \text{if } \alpha > 0.5. \end{cases}$$

$$\xi_{\inf}(\alpha) = \begin{cases} 2\alpha r_2 + (1-2\alpha)r_1, \text{if } \alpha \leq 0.5; \\ (2\alpha-1)r_4 + 2(1-\alpha)r_3, \text{if } \alpha > 0.5. \end{cases}$$

Nearest interval approximation

The nearest interval approximation of a fuzzy number [24] $\tilde{A}$ with distance metric d is given by $C_d(\tilde{A}) = [C_L - C_U]$, where $C_L = \int_0^1 A_L(\alpha)d\alpha$ and $C_U = \int_0^1 A_U(\alpha)d\alpha$, where distance metric d measure distance of $\tilde{A}$ from $C_d(\tilde{A})$ is given by

$$d(d, C_d(\tilde{A})) = \sqrt{\int_0^1 \{A_L(\alpha) - C_L\}^2 d\alpha + \int_0^1 \{A_U(\alpha) - C_U\}^2 d\alpha}$$

and $[A_L(\alpha), A_U(\alpha)$ is the α–cut$(0 \leq \alpha \leq 1)$ of $\tilde{A}$. For example, α – cut of a trapezoidal fuzzy number (r_1, r_2, r_3, r_4) is $[r_1 + \alpha(r_2 - r_1), r_4 - \alpha(r_4 - r_3)]$ and its interval approximation is $\left[\frac{r_1+r_2}{2}, \frac{r_3+r_4}{2}\right]$.

Theorem 1. [18] *Suppose that ξ is a fuzzy number with continuous membership function $\mu_\xi(x)$, and $r_0 = \sup\{r: \mu_\xi(x) = 1\}$, $g(x, \xi) = h(x) - \xi$. Then, we have $Cr\{g(x, \xi) \geq 0\} \geq \alpha$ if and only if $h(x) \geq F_\alpha$, where*

$$F_\alpha = \begin{cases} inf\{F|F = \mu_\xi^{-1}(2\alpha)\}, & if \ \alpha \le 0.5, \\ inf\{F|F = \mu_\xi^{-1}(2(1-\alpha)), F > r_0\} & if \ \alpha > 0.5. \end{cases}$$

Theorem 2. [18] *Suppose that ξ is a fuzzy number with continuous membership. Function $\mu_\xi(x)$ and $r_0 = inf\{r : \mu_\xi(r) = 1\}$, $g(x, \xi) = h(x) - \xi$. Then, we have $Cr\{g(x, \xi) \le 0\} \ge \alpha$ if and only if $h(x) \le F_\alpha$, where*

$$F_\alpha = \begin{cases} sup\{F|F = \mu_\xi^{-1}(2\alpha)\}, & if \ \alpha \le 0.5, \\ sup\{F|F = \mu_\xi^{-1}(2(1-\alpha)), F < r_0\} & if \ \alpha > 0.5. \end{cases}$$

Theorem 3. [11,25] *Assume that the function $g(x, \xi)$ can be written as*

$$g(x, \xi) = h_1(x)\xi_1 + h_2(x)\xi_2 + \cdots\cdots\cdots + h_t(x)\xi_t + h_0(x)$$

where ξ_k are trapezoidal fuzzy variables $(r_{k1}, r_{k2}, r_{k3}, r_{k4})$, $k = 1, 2, \ldots t$, respectively. We define two functions $h_k^+(x) = h_k(x) \vee 0$ and $h_k^-(x) = -(h_k(x) \vee 0)$ for $k = 1, 2, \ldots .t$. Then, we have the following:

1. *When $\alpha \le \frac{1}{2}$, $Cr\{g(x, \xi) \le 0\} \ge \alpha$ if and only if*

$$(1-2\alpha)\sum_{k=1}^{t}\left[r_{k1}h_k^+(x) - r_{k4}h_k^-(x)\right] + 2\alpha\sum_{k=1}^{t}\left[r_{k2}h_k^+(x) - r_{k3}h_k^-(x)\right] + h_0(x) \le 0;$$

2. *When $\alpha > \frac{1}{2}$, $Cr\{g(x, \xi) \le 0\} \ge \alpha$ if and only if*

$$(2-2\alpha)\sum_{k=1}^{t}\left[r_{k3}h_k^+(x) - r_{k2}h_k^-(x)\right] + (2\alpha-1)\sum_{k=1}^{t}\left[r_{k4}h_k^+(x) - r_{k1}h_k^-(x)\right] + h_0(x) \le 0;$$

Corollary 1. *If $\tilde{\xi} = (r_1, r_2, r_3, r_4)$ is a trapezoidal fuzzy variable and $h(x)$ is a function of x, then $Cr\{h(x) \le \tilde{\xi}\} \ge \alpha$ if and only if $h(x) \le F_\xi$, where*

$$F_\xi = \begin{cases} (1-2\alpha)r_4 + 2\alpha r_3, & if \ \alpha \le \dfrac{1}{2} \\ 2(1-\alpha)r_2 + (2\alpha-1)r_1, & if \ \alpha > \dfrac{1}{2} \end{cases}$$

Proof. $\mathrm{Cr}\{h(x) \le \tilde{\xi}\} \ge \alpha \leftrightarrow \mathrm{Cr}\left\{\tilde{\xi}' + h(x) \le 0\right\} \ge \alpha$, where $\tilde{\xi}' = -\tilde{\xi} = (-r_4, -r_3, -r_2, -r_1)$. Then, from the above theorem, it follows that this inequality holds if and only if

1. $(1-2\alpha)(-r_4) + 2\alpha(-r_3) + h(x) \le 0$, if $\alpha \le \frac{1}{2}$;
2. $(2-2\alpha)(-r_2) + (2\alpha-1)(-r_1) + h(x) \le 0$, if $\alpha > \frac{1}{2}$.

Hence, the corollary follows.

Corollary 2. *If $\tilde{\xi} = (r_1, r_2, r_3, r_4)$ is a trapezoidal fuzzy variable and $h(x)$ is a function of x, then $Cr\{h(x) \ge \tilde{\xi}\} \ge \alpha$ if and only if $h(x) \ge F_\xi$, where*

$$F_\xi = \begin{cases} (1-2\alpha)r_1 + 2\alpha r_2, \text{ if } \alpha \leq \frac{1}{2}; \\ (2-2\alpha)r_3 + (2\alpha-1)r_4, \text{ if } \alpha > \frac{1}{2}. \end{cases}$$

Proof. $\text{Cr}\{h(x) \geq \tilde{\xi}\} \geq \alpha \Leftrightarrow \text{Cr}\left\{\tilde{\xi}' - h(x) \leq 0\right\} \geq \alpha$, where $\tilde{\xi}' = -\tilde{\xi} = (-r_4, -r_3, -r_2, -r_1)$.

Then, from the above theorem, it follows that this inequality holds if and only if

1. $(1-2\alpha)r_1 + 2\alpha r_2 - h(x) \leq 0, \ \text{if } \alpha \leq \frac{1}{2}$;
2. $(2-2\alpha)r_3 + (2\alpha-1)r_4 - h(x) \leq 0, \ \text{if } \alpha > \frac{1}{2}$.

Hence, the corollary follows.

Corollary 3. *If $\tilde{\xi} = (r_1, r_2, r_3, r_4)$ and $\tilde{\eta} = (t_1, t_2, t_3, t_4)$ are trapezoidal fuzzy variable and $h(x) \geq 0 \ \forall x$, then $Cr\{h(x)\tilde{\xi} \leq \tilde{\eta}\} \geq \alpha$ if and only if*

1. $(1-2\alpha)h(x)r_1 + 2\alpha h(x)r_2 - h(x) \leq (1-2\alpha)t_4 + 2\alpha t_3, \ \text{if } \alpha \leq \frac{1}{2}$;
2. $(2-2\alpha)h(x)r_3 + (2\alpha-1)h(x)r_4 - h(x) \leq (2-2\alpha)t_2 + (2\alpha-1)t_1, \ \text{if } \alpha > \frac{1}{2}$.

Proof. $\text{Cr}\{h(x)\tilde{\xi} \leq \tilde{\eta}\} \geq \alpha \Leftrightarrow \text{Cr}\{h(x)\tilde{\xi} - \tilde{\eta} \leq 0\} \geq \alpha \Leftrightarrow \text{Cr}\{h(x)\tilde{\xi} + \tilde{\eta}' \leq 0\} \geq \alpha$, where $\tilde{\eta}' = -\tilde{\eta} = (-t_4, -t_3, -t_2, -t_1)$. Then, from the above theorem, it follows that this inequality holds if and only if

1. $(1-2\alpha)(h(x)r_1 - t_4) + 2\alpha(h(x)r_2 - t_3) \leq 0, \ \text{if } \alpha \leq \frac{1}{2}$;
2. $(2-2\alpha)(h(x)r_3 - t_2) + (2\alpha-1)(h(x)r_4 - t_1) + h(x) \leq 0, \ \text{if } \alpha > \frac{1}{2}$

Hence, the corollary follows.

It is obvious that these three corollaries help us to determine crisp equivalences of various inequalities with fuzzy parameters.

Random variable

For the probability spaces (Ω, S, P) where Ω is a set of elementary events, S is a set of all events ($a\ \sigma$ – field of events) and P: $S \rightarrow [0, 1]$ is a probability function, the mapping $\widehat{X} : (\Omega, S, P) \rightarrow \mathfrak{R}$ is called a random variable.

By the probability distribution function of the random variable $\widehat{X}$, we mean a function $F(x) = P\{\omega : \hat{X}(\omega) \leq X\}$ for all $x \in N$ with $F(-\infty) = 0$ and $F(\infty) = 1$.

Hybrid variable

Recall that a random variable is a measurable function from a probability space to the set of real number and that a fuzzy variable is a function from a credibility space to the set of real numbers. In order to describe a quantity with both fuzziness and randomness, we introduce the concept of hybrid variable as follows:

Definition 7. [11] A hybrid variable is a measurable function from a chance space (Θ, P, Cr) × (Ω, S, Pr) to real numbers, i.e., for any Borel set B of real numbers, the set

$$\{\xi \in B\} = \{(\theta, \omega) \in \Theta \times \Omega \colon \xi(\theta, \omega) \in B\}$$

is an event.

Remark 1. [11] A hybrid variable degenerates to a fuzzy variable if the value of $\xi(\theta, \omega)$ does not vary with ω. For example, $\xi(\theta, \omega) = \theta, \xi(\theta, \omega) = \theta^2 + 1, \xi(\theta, \omega) = \sin\theta$.

Remark 2. [11] A hybrid variable degenerate to a random variable if the value of $\xi(\theta, \omega)$ does not vary with θ. For example, $\xi(\theta, \omega) = \omega, \xi(\theta, \omega) = \omega^2 + 1, \xi(\theta, \omega) = \sin\omega$.

Remark 3. A hybrid variable $\xi(\theta, \omega)$ may also be regarded as a function from a credibility space $(\Theta, \mathcal{P}, \mathrm{Cr})$ to the set $\{\xi(\theta,.) \mid \theta \in \Theta\}$ of random variable. Thus, ξ is a random fuzzy variable defined by Liu [11].

Remark 4. A hybrid variable $\xi(\theta, \omega)$ may also be regarded as a function from a probability space $(\Omega, \mathcal{A}, \mathrm{Pr})$ to the set $\{\xi(., \omega) \mid \omega \in \Omega\}$ of fuzzy variables. If $\mathrm{Cr}\{\xi(., \omega) \in B\}$ is a measurable function of ω for any Borel set B of real number, then ξ is a fuzzy random in the sense of [17].

Definition 8. [17] Then, a chance measure of an event $\wedge$ is defined as

$$\mathrm{ch}(\wedge) = \begin{cases} \sup_{\theta\in\Theta}(\mathrm{Cr}\{\theta\}\wedge \mathrm{Pr}\{\wedge(\theta)\}), & \textit{if } \sup_{\theta\in\Theta}(\mathrm{Cr}\{\theta\}\wedge \mathrm{Cr}\{\wedge(\theta)\}) < 0.5; \\ 1-\sup_{\theta\in\Theta}(\mathrm{Cr}\{\theta\}\wedge \mathrm{Pr}\{\wedge(\theta)\}) & \textit{if } \sup_{\theta\in\Theta}(\mathrm{Cr}\{\theta\}\wedge \mathrm{Pr}\{\wedge(\theta)\}) \geq 0.5. \end{cases}$$

Then $\mathrm{ch}(\phi) = 0$, $\mathrm{ch}\{\Theta \times \Omega\} = 1$, $0 \leq \mathrm{ch}\{\wedge\} \leq 1$ for any event $\wedge$.

Example 2. If $\tilde{\mathbf{a}}$ is a fuzzy variable and $\hat{\eta}$ is a random variable, then the sum $\hat{\tilde{\xi}} = \tilde{\mathbf{a}} + \hat{\eta}$ is a hybrid variable, i.e., if $f\colon \Re^2 \to \Re$ is a measurable function, then $\hat{\tilde{\xi}} = f(\tilde{\mathbf{a}}, \hat{\eta})$ is a hybrid variable. Now, suppose that $\tilde{\mathbf{a}}$ has a membership function μ, and $\hat{\eta}$ has a probability density function ϕ. Then, for any Borel set B of real numbers, we have

$$\mathrm{ch}\{f(\tilde{a}, \hat{\eta}) \in B\} = \begin{cases} \sup_x \left[\frac{\mu(x)}{2} \wedge \int_{f(x,y)\in B} \phi(y)\mathrm{d}y\right], & \textit{if } \sup_x \left[\frac{\mu(x)}{2} \wedge \int_{f(x,y)\in B} \phi(y)dy\right] < 0.5; \\ 1-\sup_x \sup_x \left[\frac{\mu(x)}{2} \wedge \int_{f(x,y)\in B} \phi(y)dy\right], & \textit{if } \sup_x \left[\frac{\mu(x)}{2} \wedge \int_{f(x,y)\in B} \phi(y)dy\right] \geq 0.5. \end{cases}$$

Definition 9. [17] Let $\hat{\tilde{\xi}}$ be a hybrid variable. Then, the expected value of $\hat{\tilde{\xi}}$ is defined by

$$E\left[\hat{\tilde{\xi}}\right] = \int_0^{+\infty} \mathrm{ch}\left\{\hat{\tilde{\xi}} \geq r\right\} dr - \int_{-\infty}^0 \mathrm{ch}\left\{\hat{\tilde{\xi}} \geq r\right\} dr$$

provided that at least one of the integrals is finite.

Example 3. For the hybrid variable $\hat{\tilde{\xi}} = \tilde{\mathbf{a}} + \hat{\eta}$, expected value of $\hat{\tilde{\xi}}$ is $E\left(\hat{\tilde{\xi}}\right) = E(\tilde{\mathbf{a}}) + \mathbf{E}(\hat{\eta})$. For example, if $\tilde{\mathbf{a}} = (t_1, t_2, t_3)$ is the triangular fuzzy number and $\hat{\eta} = \mathcal{N}(\mu, \sigma^2)$ is normally distributed as random variable. Then, $E\left(\hat{\tilde{\xi}}\right) = \frac{t_1+2t_2+t_3}{4} + \mu$.

Description of the problem

STP is a problem of transporting goods from some sources to some destinations through some conveyances (modes of transportation), and the main objective is to find the optimal transportation plan so that the total transportation cost is minimum. Also, the goods transported through each source cannot exceed its supply capacity, the requirements of each destination must be satisfied, and the total transported amount must not exceed the capacity of conveyances. In this manuscript, we impose a desired total safety factor for the whole transportation system. The decision of future transportation planning is generally founded upon the past record. However, the available data from previous experiments are not always precise; those are often imprecise due to uncertainty in judgment, fluctuating financial market, linguistic information, imperfect statistical analysis, insufficient information, etc. For example, transportation cost depends on the fuel price, labor charges, tax charges, etc., each of which are fluctuating from time to time. Similarly, the supply of a source cannot be always exact because it depends upon the availability of manpower, raw materials, market competition, product demands, etc. Fuzzy set theory and random set theory are most widely used and are successfully applied tools to deal with uncertainty. In the next section, we formulate five STPs with safety constraints and different uncertain (fuzzy, random, hybrid) parameters.

Model formulation

Sometimes, information about supplies at plants, demands at destination, and conveyance capacity are not known precisely, i.e., these amounts are erstwhile vague in nature. For this reason, we consider here that the supplies at plants, demands at destination, and conveyance capacity are all fuzzy in nature.

Model 1: solid transportation problem with hybrid penalties, fuzzy resources, demands, conveyance capacities, and without safety factor

To formulate the model, we assume that there are no risks to transport the commodities from plants to destination by different conveyances, i.e., all routes are totally safe for the transport of goods, and unit transportation cost is a hybrid variable.

$$\mathrm{Min}(Z) = \sum_{i=1}^{m}\sum_{j=1}^{n}\sum_{k=1}^{K} \tilde{\hat{C}}_{ijk}\, x_{ijk} \tag{1}$$

subject to constraints

$$\sum_{j=1}^{n}\sum_{k=1}^{k} x_{ijk} \le \tilde{a}_i, \quad i = 1, 2, \ldots, m, \tag{2}$$

$$\sum_{i=1}^{m}\sum_{k=1}^{k} x_{ijk} \ge \tilde{b}_j, \quad j = 1, 2, \ldots, n, \tag{3}$$

$$\sum_{j=1}^{n}\sum_{k=1}^{k} x_{ijk} \le \tilde{e}_k, \quad k = 1, 2, \ldots, K, \tag{4}$$

$$x_{ijk} \ge 0, \quad \forall i, j, k,$$

where $\tilde{a}_i$ denotes the fuzzy amount of the product available at the ith origin, $\tilde{b}_j$ denotes the fuzzy demand of the product of the jth destination, $\tilde{e}_k$ denotes the fuzzy transportation capacity of conveyance k, $\tilde{\hat{C}}_{ijk}$ denotes the hybrid unit transportation penalty from the ith origin to the jth destination via the kth conveyance for the objective function.

The penalty could represent transportation cost or quantity of goods delivered or underused capacity, etc. Here, the penalties represent per unit transportation costs.

Solution methodology (expected value model)

Liu and Liu [17] presented a spectrum of expected value model of fuzzy programming to obtain the optimum expected value of objective function under some expected constraints. Considering $\tilde{\hat{C}}_{ijk}$ as $\tilde{\hat{C}}_{ijk} = C_{ijk} + \hat{C}_{ijk}$ and constructing the expected value model [17,18,26] for the model I, we have the corresponding crisp form as

$$\text{Min } E(Z) = \sum_{i=1}^{m} \sum_{j=1}^{n} \sum_{k=1}^{K} E\left(\tilde{\hat{C}}_{ijk}\right) x_{ijk} \tag{5}$$

subject to constraints

$$\sum_{j=1}^{n} \sum_{k=1}^{k} x_{ijk} \leq E(\tilde{a}_i), \quad i = 1, 2, ..., m, \tag{6}$$

$$\sum_{i=1}^{m} \sum_{k=1}^{k} x_{ijk} \geq E(\tilde{b}_j), \quad j = 1, 2, ..., n, \tag{7}$$

$$\sum_{j=1}^{n} \sum_{k=1}^{k} x_{ijk} \leq E(\tilde{e}_k), \quad k = 1, 2, ..., K, \tag{8}$$

$$x_{ijk} \geq 0, \quad \forall i, j, k.$$

Model 2: solid transportation problem with crisp penalties, fuzzy resources, demands, conveyance capacities, and desired safety measure as crisp

To formulate the model, we assume that all transportation routes are not equally safe to transport the commodity from the source to the destination by different conveyances, and we consider a crisp safety factor for each transportation:

$$\text{Min}(Z) = \sum_{i=1}^{m} \sum_{j=1}^{n} \sum_{k=1}^{K} C_{ijk}\, x_{ijk}$$

subject to the constraint

$$\sum_{i=1}^{m} \sum_{j=1}^{n} \sum_{k=1}^{K} S_{ijk}\, y_{ijk} \geq B, \tag{9}$$

with constraints (2), (3), and (4),

$$x_{ijk} \geq 0, \quad \forall i, j, k,$$

where S_{ijk} is the safety factor when an item is transformed from the ith origin to the jth destination by the kth conveyance. If an item is transported from source i to destination j by conveyance k, then the safety factor S_{ijk} is considered. This implies that if $x_{ijk} > 0$, then we consider the safety factor for this route as a part of the safety constraint. Thus, for the convenience of modeling, the following notation is introduced:

$$y_{ijk} = \begin{cases} 1 & \textit{for} \quad x_{ijk} > 0 \\ 0 & \textit{otherwise} \end{cases}$$

and B is the desired safety measure for the whole transportation system.

Solution methodology

$$\text{Min}(Z) = \sum_{i=1}^{m}\sum_{j=1}^{n}\sum_{k=1}^{K} C_{ijk}\, x_{ijk}$$

subject to constraints

$$\text{Cr}\left(\sum_{j=1}^{n}\sum_{k=1}^{k} x_{ijk} \le \tilde{a}_i\right) \ge \alpha_i, \quad i = 1, 2, ..., m, \tag{10}$$

$$\text{Cr}\left(\sum_{i=1}^{m}\sum_{k=1}^{k} x_{ijk} \ge \tilde{b}_j\right) \ge \beta_j, \quad j = 1, 2, ..., n, \tag{11}$$

$$\text{Cr}\left(\sum_{j=1}^{n}\sum_{k=1}^{k} x_{ijk} \le \tilde{e}_k\right) \ge \gamma_k, \quad i = 1, 2, ..., k, \tag{12}$$

and (9)

$$x_{ijk} \ge 0, \quad \forall i, j, k.$$

Crisp equivalences

$$\text{Min}(Z) = \sum_{i=1}^{m}\sum_{j=1}^{n}\sum_{k=1}^{K} C_{ijk}\, x_{ijk}$$

subject to constraints

$$\sum_{j=1}^{n}\sum_{k=1}^{k} x_{ijk} \le F_{\alpha_i}, \quad i = 1, 2, ..., m, \tag{13}$$

$$\sum_{i=1}^{m}\sum_{k=1}^{k} x_{ijk} \ge F_{\beta_j}, \quad j = 1, 2, ..., n, \tag{14}$$

$$\sum_{j=1}^{n}\sum_{k=1}^{k} x_{ijk} \le F_{\gamma_k}, \quad k = 1, 2, ..., K, \tag{15}$$

and (9)

$$x_{ijk} \ge 0, \quad \forall i, j, k.$$

where $\tilde{a}_i = (a_i^1, a_i^2, a_i^3, a_i^4), \tilde{b}_j = \left(b_j^1, b_j^2, b_j^3, b_j^4\right), \tilde{e}_k = (e_k^1, e_k^2, e_k^3, e_k^4)$ and

$$F_{\alpha_i} = a_{i_{\text{sup}}}(\alpha_i) = \begin{cases} (1-2\alpha_i)a_i^4 + 2\alpha_i a_i^3 \quad , & \text{if } \alpha_i \le 0.5; \\ 2(1-\alpha_i)a_i^2 + (2\alpha_i - 1)a_i^1, & \alpha_i > 0.5. \end{cases}$$

$$F_{\beta_j} = a_{j_{\text{inf}}}\left(\beta_j\right) = \begin{cases} (1-2\beta_j)b_j^1 + 2\beta_j b_j^2 \quad , & \text{if } \alpha_i \le 0.5; \\ 2(1-\beta_j)b_j^3 + (2\beta_j - 1)b_j^4, \alpha_i > 0.5. \end{cases}$$

$$F_{\gamma_k} = a_{k_{\text{sup}}}(\alpha_i) = \begin{cases} (1-2\gamma_k)e_k^4 + 2\gamma_k e_i^3 \quad , & \text{if } \alpha_i \le 0.5; \\ 2(1-\gamma_k)e_k^2 + (2\gamma_k - 1)e_i^1, & \alpha_i > 0.5. \end{cases}$$

Model 3: solid transportation problem with random penalties, fuzzy resources, demands, conveyance capacities, and DSM as random

Here, we formulate the respective model by taking unit transportation cost, desired total safety factor, and safety factor as a random variables because it may happen that the safety factor is uncertain, not precisely known, but some past data about it is available:

$$\text{Min}(Z) = \sum_{i=1}^{m}\sum_{j=1}^{n}\sum_{k=1}^{K} \hat{C}_{ijk}\, x_{ijk}$$

subject to the constraint

$$\sum_{i=1}^{m}\sum_{j=1}^{n}\sum_{k=1}^{K}\hat{S}_{ijk}\,y_{ijk} \geq \hat{B} \tag{16}$$

with constraints (2), (3), and (4),

$$x_{ijk} \geq 0, \quad \forall i, j, k,$$

where $\hat{C}_{ijk}$ are random unit transportation cost from the ith origin to the jth destination by the kth conveyance, $\hat{S}_{ijk}$ is the random safety factor for a particular route, and $\hat{B}$ is the desired total random safety factor for the whole transportation system.

Solution methodology (an approach using interval approximation of fuzzy number)

Let us denote the interval approximations of $\tilde{a}_i, \tilde{b}_j, \tilde{e}_k$, and $\tilde{B}$ obtained using α - cut of these fuzzy numbers are $C_d(\tilde{a}_i) = [a_{iL}, a_{iU}], C_d(\tilde{b}_j) = [b_{jL}, b_{jU}], C_d(e_k) = [e_{kL}, e_{kU}]$, respectively. Then, using these interval approximations and predetermined probability confidence level δ for the constraint (11), the above model becomes

$$\text{Min}(Z) = \sum_{i=1}^{m}\sum_{j=1}^{n}\sum_{k=1}^{K}\hat{C}_{ijk}\,x_{ijk} \tag{17}$$

subject to constraints

$$\sum_{j=1}^{n}\sum_{k=1}^{k}x_{ijk} \leq C_d(\tilde{a}_i) \tag{18}$$

$$\sum_{i=1}^{m}\sum_{k=1}^{k}x_{ijk} \geq C_d(\tilde{b}_j) \tag{19}$$

$$\sum_{j=1}^{n}\sum_{k=1}^{k}x_{ijk} \leq C_d(\tilde{e}_k) \tag{20}$$

$$\text{Prob}\left[\sum_{i=1}^{m}\sum_{j=1}^{n}\sum_{k=1}^{K}\hat{S}_{ijk}\,y_{ijk} \geq \hat{B}\right] \geq \epsilon, \tag{21}$$

$$x_{ijk} \geq 0, \quad \forall i, j, k.$$

Crisp equivalences

We assumed that all $\hat{C}_{ijk}$ are mutually independent normally distributed random variables with known mean $E(\hat{C}_{ijk}) = \bar{C}_{ijk}$ and variance $\text{var}(\hat{C}_{ijk})$. Then, Z will also be a normally distributed random variable with mean $\bar{z} = \sum_{i=1}^{m}\sum_{j=1}^{n}\sum_{k=1}^{K}\bar{C}_{ijk}\,x_{ijk}$ and var $(z) = X^T VX$, where V is the covariance matrix of $\hat{C}_{ijk}$, where

$$V_{ij} = \begin{pmatrix} \text{var}(\hat{C}_{ij1}) & \text{cov}(\hat{C}_{ij1}, \hat{C}_{ij2}) & \cdots & \text{cov}(\hat{C}_{ij1}, \hat{C}_{ijk}) \\ \text{cov}(\hat{C}_{ij2}, \hat{C}_{ij1}) & \text{var}(\hat{C}_{ij2}) & \cdots & \text{cov}(\hat{C}_{ij2}, \hat{C}_{ijk}) \\ \cdots & \cdots & \cdots & \cdots \\ \text{cov}(\hat{C}_{ijk}, \hat{C}_{ij1}) & \text{cov}(\hat{C}_{ijk}, \hat{C}_{ij2}) & \cdots & \text{var}(\hat{C}_{ijk}) \end{pmatrix}$$

For $i = 1, 2, \ldots, m; j = 1, 2, \ldots, n;$ and $k = 1, 2, \ldots, K.$

Then, a new deterministic nonlinear objective function for minimization can be formulated as

$$z(X) = \lambda_1\bar{z} + \lambda_2\sqrt{X^T VX} \tag{22}$$

where λ_1 and λ_2 are nonnegative constants whose values indicate the relative importance of mean and standard deviation of z for minimization. Thus, $\lambda_2 = 0$(i. e., $\lambda_1 = 1$) indicates that only the expected value (mean value) of the objective function is to be minimized without caring for the standard deviation (SD) of the objective function. Similarly, $\lambda_1 = 0\ \lambda$ indicates that only the SD, i.e., the variability of the objective function about its mean is to be minimized without caring of for its mean. $\lambda_1 = \lambda_2 = 1$ indicates that equal importance is given to the minimization of both the mean and SD of the objective function.

If the r.v. $\hat{C}_{ijk}$ are independent, then $Z(X)$ reduces to

$$Z(X) = \lambda_1 \sum_{i=1}^{m} \sum_{j=1}^{n} \sum_{k=1}^{K} \bar{C}_{ijk} x_{ijk} + \lambda_2 \sqrt{\sum_{i=1}^{m} \sum_{j=1}^{n} \sum_{k=1}^{K} \mathrm{var}(\hat{C}_{ijk}) x_{ijk}^2} \tag{23}$$

If ε are the probabilities of nonviolation of the constraint (16) then the constraint can be written as

$$\mathrm{Prob}\left[\sum_{i=1}^{m} \sum_{j=1}^{n} \sum_{k=1}^{K} \hat{S}_{ijk} y_{ijk} \geq \hat{B}\right] \geq \epsilon,$$

$$\Rightarrow \mathrm{Prob}\left[\sum_{i=1}^{m} \sum_{j=1}^{n} \sum_{k=1}^{K} \hat{S}_{ijk} y_{ijk} - \hat{B} \geq 0\right] \geq \epsilon,$$

$$\Rightarrow \mathrm{Prob}\left[\hat{Q} \geq 0\right] \geq \epsilon, \text{ where } \hat{Q} = \sum_{i=1}^{m} \sum_{j=1}^{n} \sum_{k=1}^{K} \hat{S}\, ijk\, y_{ijk} - \hat{B}$$

$$\Rightarrow \mathrm{Prob}\left[\frac{\hat{Q} - E(\hat{Q})}{\mathrm{Var}(\hat{Q})} \geq -\frac{E(\hat{Q})}{\mathrm{Var}(\hat{Q})}\right] \geq \epsilon,$$

$$\Rightarrow \mathrm{Prob}\left[\hat{T} \geq -K\right] \geq \epsilon,$$

$$\Rightarrow \mathrm{Prob}\left[\hat{T} \leq K\right] \geq \epsilon,$$

where $\hat{T} = \frac{\hat{Q} - E(\hat{Q})}{\mathrm{Var}(\hat{Q})}$ are the standard normal variate and $K = \frac{E(\hat{Q})}{\mathrm{Var}(\hat{Q})}$

$$\Rightarrow E(\hat{Q}) \geq \lambda\ \mathrm{Var}(\hat{Q}).$$

$$\Rightarrow \sum_{i=1}^{m} \sum_{j=1}^{n} \sum_{k=1}^{K} E(\hat{S}_{ijk}) y_{ijk} - E(\hat{B}) \geq \lambda \left[\sum_{i=1}^{m} \sum_{j=1}^{n} \sum_{k=1}^{K} \mathrm{Var}(\hat{S}_{ijk}) y_{ijk} - Var(\hat{B})\right]. \tag{24}$$

where λ be the real number such that $\mathrm{Prob}\left[\hat{T} \geq \lambda\right] = \epsilon$.

Now, denote the left-hand side expressions of the constraints (19), (20), (21), and (23) by S_i, D_j, E_k. The right-hand side expressions of the constraints are interval numbers. Now, using the idea of possibility degree of interval numbers [27] that represent a certain degree by which one interval is larger or smaller than the other, we define the possibility degree of satisfaction of these constraints as follows:

$$P_{S_i \leq [a_{iL}, a_{iU}]} = \begin{cases} 1 & S_i \leq a_{iL}; \\ \dfrac{a_{iU} - S_i}{a_{iU} - a_{iL}} & a_{iL} \leq S_i \leq a_{iU}; \\ 0 & S_i > a_{iU}. \end{cases}$$

$$P_{D_j \geq [b_{jL}, b_{jU}]} = \begin{cases} 0 & D_j \leq b_{jL}; \\ \dfrac{D_j - b_{iL}}{b_{jU} - a_{jL}} & a_{jL} \leq S_j \leq a_{jU}; \\ 1 & D_j > a_{jU}. \end{cases}$$

$$P_{E_k \le [e_{kL}, e_{kU}]} = \begin{cases} 1 & E_k \le e_{kL}; \\ \dfrac{e_{kU} - e_k}{e_{kU} - e_{kL}} & e_{kL} \le E_k \le e_{kU}; \\ 0 & E_i > e_{kU}. \end{cases}$$

Now, for a predetermined possibility degree of satisfaction, $\alpha_i, \beta_j, \gamma_k$ and $\epsilon(0 \le \alpha_i, \beta_j, \gamma_k \le 1)$, respectively, for the constraints, i.e., $P_{S_i \le [a_{iL}, a_{iU}]} \ge \alpha_i, P_{D_j \ge [b_{jL}, b_{jU}]} \ge \beta_j, P_{E_k \le [e_{kL}, e_{kU}]} \ge \gamma_k \forall i, j, k$, then the equivalent deterministic inequalities of the respective constraints are obtained as follows:

$$S_i \le a_{iU} - \alpha_i (a_{iU} - a_{iL}) = a_{ic}(\text{say}),$$

$$D_j \ge b_{jL} + \beta_j (b_{jU} - b_{jL}) = b_{jc},$$

$$E_k \le e_{kU} - \gamma_k (e_{kU} - e_{kL}) = e_{kc}, \tag{25}$$

Now, using the deterministic form (23) of the objective function (17) and the constraints (27), the deterministic form of problems (17) to (21) becomes

$$\text{Min } z(x) = \lambda_1 \sum_{i=1}^{m} \sum_{j=1}^{n} \sum_{k=1}^{K} \bar{C}_{ijk} x_{ijk} + \lambda_2 \sqrt{\sum_{i=1}^{m} \sum_{j=1}^{n} \sum_{k=1}^{K} var(\hat{C}_{ijk}) x_{ijk}^2}, \tag{26}$$

subject to constraints (25) and (24)

$$x_{ijk} \ge 0, \quad \forall i, j, k.$$

Model 4: solid transportation problem with fuzzy penalties, resources, demands, conveyance capacities, and DSM as fuzzy

In this section, we formulate the model by taking the unit transportation cost, desired total safety factor, and safety factor as fuzzy numbers because it may happen that the safety factor is vague in nature, i.e., imprecisely known:

$$\text{Min}(Z) = \sum_{i=1}^{m} \sum_{j=1}^{n} \sum_{k=1}^{K} \tilde{C}_{ijk} x_{ijk}$$

subject to the constraint

$$\sum_{i=1}^{m} \sum_{j=1}^{n} \sum_{k=1}^{K} \tilde{S}_{ijk} y_{ijk} \ge \tilde{B} \tag{27}$$

with constraints (2), (3), and (4),

$$x_{ijk} \ge 0, \quad \forall i, j, k,$$

where $\tilde{C}_{ijk}$ is the fuzzy unit transportation cost from the ith origin to the jth destination by the kth conveyance, $\tilde{S}_{ijk}$ is the fuzzy safety factor for a particular route, and $\tilde{B}$ is the desired total fuzzy safety measure for the whole transportation system.

Solution methodology (chance-constrained programming)

Chance-constrained programming technique was introduced by Charnes and Cooper [15] for stochastic programming. Chance-constrained programming in a fuzzy case was developed by Liu and Iwamura [16], Liu [25], Yang and Liu [18], and many more authors. This method is used to solve the problems with chance constraints. In this method, the uncertain constraints are allowed to be violated such that constraints must be satisfied at some chance (or confidence) level. Applying this idea, we reformulate the above problem as follows:

Min $\bar{z}$, subject to constraints,

$$\mathrm{Cr}\left\{\sum_{i=1}^{m}\sum_{j=1}^{n}\sum_{k=1}^{K}\tilde{C}_{ijk}\, x_{ijk} \le \bar{z}\right\} \ge \eta \tag{28}$$

$$\mathrm{Cr}\left\{\sum_{i=1}^{m}\sum_{j=1}^{n}\sum_{k=1}^{K}\tilde{S}_{ijk}\, y_{ijk} \ge \tilde{B}\right\} \ge \delta, \tag{29}$$

$$\mathrm{Cr}\left\{\sum_{j=1}^{n}\sum_{k=1}^{k} x_{ijk} \le \tilde{a}_i\right\} \ge \alpha_i \tag{30}$$

$$\mathrm{Cr}\left\{\sum_{i=1}^{m}\sum_{k=1}^{k} x_{ijk} \ge \tilde{b}_j\right\} \ge \beta_j \tag{31}$$

$$\mathrm{Cr}\left\{\sum_{j=1}^{n}\sum_{k=1}^{k} x_{ijk} \le \tilde{e}_k\right\} \ge \gamma_k \tag{32}$$

$x_{ijk} \ge 0,\ \ \forall i, j, k.$

where η indicates that we are going to optimize the η – critical value of the objective z, and δ indicates the credibility level of satisfaction of the safety constraint.

Crisp equivalence

$\tilde{C}_{ijk} = \left(C^1_{ijk}, C^2_{ijk}, C^3_{ijk}, C^4_{ijk}\right), \tilde{a}_i = \left(a^1_i, a^2_i, a^3_i, a^4_i\right), \tilde{b}_j = \left(b^1_j, b^2_j, b^3_j, b^4_j\right), \tilde{e}_k = \left(e^1_k, e^2_k, e^3_k, e^4_k\right),$ $\tilde{B} = (B_1, B_2, B_3, B_4)$ are trapezoidal fuzzy numbers for all i, j, and k.

Now, since $\tilde{C}_{ijk}$ are trapezoidal fuzzy numbers and x_{ijk} is 0 for all i, j, k, so $Z(x) = \sum_{i=1}^{m}\sum_{j=1}^{n}\sum_{k=1}^{K}\tilde{C}_{ijk}\, x_{ijk}$ are also trapezoidal fuzzy numbers for any feasible solution x and given by $Z(x) = (r_1(x), r_2(x), r_3(x), r_4(x))$. Then, the objective in the above model (18), i.e., Min $\bar{Z}$ s.t. $Cr\{Z(x) \le \bar{Z}\} \ge \eta$ can be equivalently computed as $\bar{Z} = \inf$ $\{\text{r: } Cr\{Z(x) \le \bar{Z}\} \ge \eta\}$ which is nothing but η – pessimistic value of $Z(Z_{\inf}(\eta))$ and so is equal to $z^{'}(x)$, z where

$$Z^{'}(x) = \begin{cases} (1-2\eta)r_1(x) + 2\eta\, r_2(x) & \mathit{if}\ \ \alpha \le 0.5, \\ 2(1-\eta)r_3(x) + (2\eta-1)4(x) & \mathit{if}\ \ \alpha > 0.5 \end{cases}$$

$$r_1(x) = \sum_{i=1}^{m}\sum_{j=1}^{n}\sum_{k=1}^{K} C^1_{ijk}\, x_{ijk}, r_2(x) = \sum_{i=1}^{m}\sum_{j=1}^{n}\sum_{k=1}^{K} C^2_{ijk}\, x_{ijk}, r_3(x)$$
$$= \sum_{i=1}^{m}\sum_{j=1}^{n}\sum_{k=1}^{K} C^3_{ijk}\, x_{ijk}, r_4(x) = \sum_{i=1}^{m}\sum_{j=1}^{n}\sum_{k=1}^{K} C^4_{ijk}\, x_{ijk}.$$

Now, the safety constraint in model 4 is in the form

$$\mathrm{Cr}\left\{\sum_{i=1}^{m}\sum_{j=1}^{n}\sum_{k=1}^{K}\left(S^1_{ijk}, S^2_{ijk}, S^3_{ijk}, S^4_{ijk}\right)x_{ijk} + (-1)(B_1, B_2, B_3, B_4) \ge 0\right\} \ge \delta$$

Since $x_{ijk} \ge 0$ x from Corollary 3 ('Nearest interval approximation' section), it is obvious that this constraint will be active if and only if $g \ge 0$, where

$$g = \begin{cases} (1-2\delta)\left(\sum_{i=1}^{m}\sum_{j=1}^{n}\sum_{k=1}^{K} S^1_{ijk}\, x_{ijk} - B_4\right) + 2\delta\left(\sum_{i=1}^{m}\sum_{j=1}^{n}\sum_{k=1}^{K} S^2_{ijk}\, x_{ijk} - B_3\right) & \mathit{if}\ \ \alpha \le 0.5, \\ 2(1-\delta)\left(\sum_{i=1}^{m}\sum_{j=1}^{n}\sum_{k=1}^{K} S^3_{ijk}\, x_{ijk} - B_2\right) + (2\delta-1)\left(\sum_{i=1}^{m}\sum_{j=1}^{n}\sum_{k=1}^{K} S^4_{ijk}\, x_{ijk} - B_1\right) & \alpha > 0.5 \end{cases}$$

Thus, finally an equivalent crisp form of the above model (17) can be written as

$$\text{Min}\left[Z'(x)\right]$$
$$\text{s.t. } g \geq 0, \tag{33}$$

with the constraints 13, 14, and 15,

$$x_{ijk} \geq 0, \quad \forall i, j, k.$$

Model 5: solid transportation problem with hybrid penalties, fuzzy resources, demands, conveyance capacities, and DSM as hybrid

The respective model is formulated by assuming the unit transportation cost, desired total safety factor, and safety factor as hybrid variables:

$$\text{Min}(Z) = \sum_{i=1}^{m}\sum_{j=1}^{n}\sum_{k=1}^{K}\tilde{\hat{C}}_{ijk}\, x_{ijk}$$

subject to the constraint

$$\sum_{i=1}^{m}\sum_{j=1}^{n}\sum_{k=1}^{K}\tilde{\hat{S}}_{ijk}\, y_{ijk} \geq \tilde{\hat{B}} \tag{34}$$

with constraints (2), (3), and (4),

$$x_{ijk} \geq 0, \quad \forall i, j, k,$$

where $\tilde{\hat{C}}$ is the hybrid unit transportation cost from the ith origin to the jth destination by the kth conveyance and $\tilde{\hat{B}}$ is the desired total hybrid risk factor for the whole transportation system.

Solution methodology (expected value model)

$$\text{Min } E(Z) = \sum_{i=1}^{m}\sum_{j=1}^{n}\sum_{k=1}^{K} E\left(\tilde{\hat{C}}_{ijk}\right) x_{ijk}$$

Subject to the constraint

$$\sum_{j=1}^{n}\sum_{k=1}^{k} x_{ijk} \leq E(\tilde{a}_i), \quad i = 1, 2, \ldots, m, \tag{35}$$

$$\sum_{i=1}^{m}\sum_{k=1}^{k} x_{ijk} \geq E(\tilde{b}_j), \quad j = 1, 2, \ldots, n, \tag{36}$$

$$\sum_{j=1}^{n}\sum_{k=1}^{k} x_{ijk} \leq E(\tilde{e}_k), \quad k = 1, 2, \ldots, K, \tag{37}$$

$$\sum_{i=1}^{m}\sum_{j=1}^{n}\sum_{k=1}^{K} E\left(\tilde{\hat{S}}_{ijk}\right) y_{ijk} \geq E\left(\tilde{\hat{B}}\right) \tag{38}$$

$$x_{ijk} \geq 0, \quad \forall i, j, k.$$

Numerical experiments

Identical products are produced in three factories and sent to three warehouses for delivery through the two different types of conveyances, i.e., we considered the following $(3 \times 3 \times 2)$ solid transportation problem.

Input data

For model 1, see Table 1 for the hybrid unit transport cost.

For model 2, see Table 2 for the crisp unit transportation cost and safety factor.

$$\tilde{a}_1 = (10, 15, 16, 17), \tilde{a}_2 = (14, 15, 18, 19), \tilde{a}_3 = (22, 23, 24, 26), \tilde{b}_1 = (9, 10, 11, 12),$$
$$\tilde{b}_2 = (12, 13, 14, 16),$$

$$\tilde{b}_3 = (5, 7, 8, 9), \tilde{e}_1 = (17, 19, 20, 21), \tilde{e}_2 = (13, 14, 16, 17), \alpha_i = \beta_j = \gamma_k = 0.7,$$
$$F_{\alpha_1} = 13, F_{\alpha_2} = 14.6,$$

$$F_{\alpha_3} = 22.6, F_{\beta_1} = 9.6, F_{\beta_2} = 12.6, F_{\beta_3} = 6.2, F_{\gamma_1} = 18.2, F_{\gamma_2} = 13.6. \quad B = 2.1.$$

For model 3, see Table 3 for the assumed mean and variance for random unit transportation cost and safety

$$\alpha_i = \beta_j = \gamma_k = 0.7, \text{and equal priority to mean and variance, i.e.,} \lambda_1 = \lambda_2 = 1,$$
$$a_{1c} = 14.35, a_{2c} = 13.7,$$

$$a_{3c} = 21.7, b_{1c} = 13.3, b_{2c} = 11.27, b_{3c} = 4.7, e_{1c} = 17.8, e_{2c} = 12.55, \; E(\tilde{B}) = 2,$$
$$\text{Var}(\tilde{B}) = 2, \lambda = 0.5.$$

For model 4, see Tables 4 and 5 for fuzzy unit transport cost and fuzzy safety factor, respectively.

$$\eta = 0.7, \delta = 0.6, \tilde{B} = (2, 3, 5, 6)$$

For model 5, see Table 6 for the hybrid safety factor.

$$\tilde{\hat{B}} = (1, 1.5, 1.6) + (1, 0.7)$$

Table 1 Hybrid unit transportation cost

Unit transportation cost	Hybrid value
$\hat{C}_{111}$	(7, 8, 9) + (2, 0.2)
$\tilde{\hat{C}}_{211}$	(5, 8, 10) + (12, 0.3)
$\tilde{\hat{C}}_{311}$	(6, 9, 10) + (13, 0.7)
$\tilde{\hat{C}}_{121}$	(10, 12, 13) + (14, 1.6)
$\tilde{\hat{C}}_{221}$	(15, 16, 17) + (11, 2.3)
$\tilde{\hat{C}}_{321}$	(1, 2, 3) + (1, 0.1)
$\tilde{\hat{C}}_{131}$	(10, 12, 14) + (5, 1.5)
$\tilde{\hat{C}}_{231}$	(15, 17, 19) + (8, 0.7)
$\tilde{\hat{C}}_{331}$	(7, 10, 12) + (12, 1.3)
$\tilde{\hat{C}}_{112}$	(8, 12, 14) + (14, 2.33)
$\tilde{\hat{C}}_{212}$	(5, 7, 10) + (7, 1)
$\tilde{\hat{C}}_{312}$	(7, 8, 9) + (14, 0.7)
$\tilde{\hat{C}}_{122}$	(11, 12, 14) + (9, 1.2)
$\tilde{\hat{C}}_{222}$	(17, 19, 21) + (19, 0.9)
$\tilde{\hat{C}}_{322}$	(19, 20, 21) + (11, 1.4)
$\tilde{\hat{C}}_{132}$	(17, 18, 22) + (12, 1.7)
$\tilde{\hat{C}}_{232}$	(9, 13, 15) + (15, 0.75)
$\tilde{\hat{C}}_{332}$	(6, 10, 11) + (8, 1.35)

Table 2 Crisp unit transportation cost and safety factor

Unit transportation cost	Crisp value	Safety factor	Crisp value
c_{111}	20	s_{111}	0.25
c_{211}	19	s_{211}	0.30
c_{311}	21	s_{311}	0.82
c_{121}	25	s_{121}	0.35
c_{221}	26	s_{221}	0.30
c_{321}	11	s_{321}	0.50
c_{131}	24	s_{131}	0.80
c_{231}	33	s_{231}	1
c_{331}	21	s_{331}	0.35
c_{112}	25	s_{112}	0.42
c_{212}	14	s_{212}	0.20
c_{312}	26	s_{312}	0.20
c_{122}	26	s_{122}	0.98
c_{222}	37	s_{222}	0.82
c_{322}	30	s_{322}	0.70
c_{132}	30	s_{132}	0.50
c_{232}	27	s_{232}	0.50
c_{332}	17	s_{332}	0.20

Optimal results

The optimum results of the different models using various methods are shown in Table 7.

Overview of the results of the five models

It may be noted that model 1 was solved without having to take the safety constraint, but the remaining models were solved by taking the safety constraint where safety factors were crisp, random, fuzzy, and hybrid for the respective models. After solving these five models, we have seen that the total transportation cost of model 1 is minimum compared with the other models. This is as per expectation as the nature of the four variables in safety factors for different models are different (crisp, random, fuzzy, and hybrid). Moreover, as the impreciseness for the source, demand, and conveyance capacity constraints in both models 2 and 4 are removed using credibility measure, the total satisfied demands (i.e., total transported amount) are the same, but the objective and safety values are different because of the different types of unit transportation costs (i.e., hybrid in models 1 and 5, crisp in model 2, random in model 3, and fuzzy in model 4) and safety factor (i.e., crisp in model 2, random in model 3, fuzzy in model 4, hybrid in model 5).

If we think practically the meaning of safety factor in the transportation problem, it is very large because the total transportation cost varies if the safety factor increases or decreases in some transportation problem. In this manuscript, we formulated five models with numerical example and solved these models using three methods, namely expected value modeling, chance-constrained programming, and an approach using interval approximation of fuzzy number. We have seen that if we introduce a safety factor in a transportation problem, the transportation cost for the whole transportation system increases, which was found in these five models. The transportation cost for the second, third, fourth, and fifth models were more than the transportation cost of the first model due to the safety factor.

Table 3 Assumed mean and variance for random unit transportation cost and safety factor

		Mean	Variance
Random safety factor	$\hat{S}_{111}$	0.22	0.11
	$\hat{S}_{211}$	0.19	0.1
	$\hat{S}_{311}$	0.4	0.2
	$\hat{S}_{121}$	0.5	0.14
	$\hat{S}_{221}$	0.6	0.2
	$\hat{S}_{321}$	0.24	0.3
	S_{131}	0.7	0.23
	$\hat{S}_{231}$	0.8	0.15
	$\hat{S}_{331}$	0.2	0.27
	$\hat{S}_{112}$	0.3	0.37
	$\hat{S}_{212}$	0.6	0.19
	$\hat{S}_{312}$	0.5	0.16
	$\hat{S}_{121}$	0.8	0.25
	$\hat{S}_{222}$	0.6	0.2
	$\hat{S}_{322}$	0.55	0.3
	$\hat{S}_{132}$	0.61	0.32
	$\hat{S}_{232}$	0.45	0.26
	$\hat{S}_{332}$	0.4	0.27
Random unit transportation cost	$\hat{C}_{111}$	19	2
	$\hat{C}_{211}$	19	1
	$\hat{C}_{311}$	17	0.4
	$\hat{C}_{121}$	22	0.75
	$\hat{C}_{221}$	22	0.9
	$\hat{C}_{321}$	11	0.5
	C_{131}	21	0.23
	$\hat{C}_{231}$	27	1.56
	$\hat{C}_{331}$	2	2.7
	$\hat{C}_{112}$	22	1.56
	$\hat{C}_{212}$	15	0.65
	$\hat{C}_{312}$	11	1
	$\hat{C}_{121}$	22	2.25
	$\hat{C}_{222}$	30	1.2
	$\hat{C}_{322}$	26	1.3
	$\hat{C}_{132}$	25	3.5
	$\hat{C}_{232}$	22	2.5
	$\hat{C}_{332}$	17	1.25

Conclusions

To deliver the goods through different routes with different modes of conveyances is not equally safe due to insurgency and bad road. For this reason, safety constraints are introduced, and in using this concept, we have seen that the transportation cost increased when we introduced safety constraint in transportation modeling. The main objective of this paper is to present a solution procedure of a solid transportation problem under various precise and imprecise environments. In this paper, we formulated

Table 4 Fuzzy unit transportation cost

Unit transportation cost	TrFN
$\tilde{C}_{111}$	(4, 5, 6, 11)
$\tilde{C}_{211}$	(7, 10, 11, 17)
$\tilde{C}_{311}$	(6, 8, 9, 10)
$\tilde{C}_{121}$	(5, 6, 8, 12)
$\tilde{C}_{221}$	(10, 11, 12, 14)
$\tilde{C}_{321}$	(12, 14, 17, 18)
$\tilde{C}_{131}$	(13, 14, 15, 20)
$\tilde{C}_{231}$	(15, 17, 19, 21)
$\tilde{C}_{331}$	(15, 17, 19, 21)
$\tilde{C}_{112}$	(3, 5, 7, 9)
$\tilde{C}_{212}$	(13, 15, 19, 20)
$\tilde{C}_{312}$	(11, 13, 14, 22)
$\tilde{C}_{122}$	(9, 10, 13, 14)
$\tilde{C}_{222}$	(12, 17, 18, 20)
$\tilde{C}_{322}$	(7, 8, 11, 12)
$\tilde{C}_{132}$	(6, 9, 13, 18)
$\tilde{C}_{232}$	(17, 18, 19, 21)
$\tilde{C}_{332}$	(18, 20, 21, 22)

Table 5 Fuzzy safety factor

Safety factor	TrFN
$\tilde{S}_{111}$	(0.18, 0.22, 0.25, 0.3)
$\tilde{S}_{211}$	(0.25, 0.3, 0.35, 0.4)
$\tilde{S}_{311}$	(0.39, 0.44, 0.49, 0.55)
$\tilde{S}_{121}$	(0.5, 0.55, 0.6, 0.65)
$\tilde{S}_{221}$	(0.65, 0.7, 0.75, 0.8)
$\tilde{S}_{321}$	(0.1, 0.15, 0.2, 0.25)
$\tilde{S}_{131}$	(0.73, 0.78, 0.83, 0.88)
$\tilde{S}_{231}$	(0.83, 0.88, 0.93, 0.98)
$\tilde{S}_{331}$	(0.25, 0.3, 0.35, 0.4)
$\tilde{S}_{112}$	(0.3, 0.35, 0.4, 0.45)
$\tilde{S}_{212}$	(0.72, 0.77, 0.82, 0.87)
$\tilde{S}_{312}$	(0.62, 0.67, 0.73, 0.77)
$\tilde{S}_{122}$	(0.84, 0.89, 0.94, 0.99)
$\tilde{S}_{222}$	(0.75, 0.8, 0.85, 0.9)
$\tilde{S}_{322}$	(0.65, 0.7, 0.75, 0.8)
$\tilde{S}_{132}$	(0.6, 0.65, 0.7, 0.75)
$\tilde{S}_{232}$	(0.50, 0.55, 0.6, 0.65)
$\tilde{S}_{332}$	(0.45, 0.5, 0.55, 0.6)

Table 6 Hybrid safety factor

Safety factor	Hybrid value
$\tilde{\hat{S}}_{111}$	(0.17, 0.18, 0.19) + (0.1, 0.3)
$\tilde{\hat{S}}_{211}$	(0.15, 0.16, 0.17) + (0.19, 0.11)
$\tilde{\hat{S}}_{311}$	(0.4, 0.42, 0.43) + (0.35, 0.12)
$\tilde{\hat{S}}_{121}$	(0.20, 0.21, 0.22) + (0.25, 0.14)
$\tilde{\hat{S}}_{221}$	(0.34, 0.35, 0.36) + (0.35, 0.2)
$\tilde{\hat{S}}_{321}$	(0.05, 0.06, 0.07) + (0.11, 0.1)
$\tilde{\hat{S}}_{131}$	(0.6, 0.61, 0.62) + (0.15, 0.11)
$\tilde{\hat{S}}_{231}$	(0.7, 0.71, 0.73) + (0.2, 0.13)
$\tilde{\hat{S}}_{331}$	(0.14, 0.15, 0.17) + (0.17, 0.16)
$\tilde{\hat{S}}_{112}$	(0.16, 0.17, 0.18) + (0.19, 0.18)
$\tilde{\hat{S}}_{212}$	(0.55, 0.56, 0.57) + (0.21, 0.15)
$\tilde{\hat{S}}_{312}$	(0.2, 0.21, 0.22) + (0.5, 0.19)
$\tilde{\hat{S}}_{122}$	(0.45, 0.46, 0.47) + (0.41, 0.7)
$\tilde{\hat{S}}_{222}$	(0.35, 0.36, 0.37) + (0.44, 0.17)
$\tilde{\hat{S}}_{322}$	(0.31, 0.32, 0.33) + (0.37, 0.3)
$\tilde{\hat{S}}_{132}$	(0.39, 0.4, 0.41) + (0.2, 0.13)
$\tilde{\hat{S}}_{232}$	(0.3, 0.31, 0.32) + (0.15, 0.14)
$\tilde{\hat{S}}_{332}$	(0.15, 0.17, 0.18) + (0.31, 0.2)

Table 7 Optimum results of different models using various methods

Optimal solution	Expected value model		Chance-constrained programming		Approach using interval approximation of fuzzy number
	Model 1	Model 5	Model 2	Model 4	Model 3
Min (Z)	267.75	443.06	456.80	278.50	326.61
x_{111}	6.25	0	5.60	0	0.25
x_{211}	0	0	0	0	0.50
x_{311}	0	0	0	4.30	0
x_{121}	0	0	0	7.70	0
x_{221}	0			0	0
x_{321}	11	11	12.6	0	11.27
x_{131}	0	0	0	0	0
x_{231}	0	0	0	0	0
x_{331}	0	2.50	0	6.20	4.7
x_{112}	0	0	0	5.30	0
x_{212}	3.25	9.5	4	0	6.82
x_{312}	0	0	0	0	5.73
x_{122}	0	0	0	0	0
x_{222}	0	0	0	0	0
x_{322}	0	0	0	4.90	0
x_{132}	0	0	0	0	0
x_{232}	0	0	0	0	0
x_{332}	4.75	2.25	6.20	0	0

five models where penalties are crisp, random, fuzzy, and hybrid variables. Also, supplies, demands, and conveyance capacities are all fuzzy in nature, but the safety factor for models 2, 3, 4, and 5 were crisp, random, fuzzy, and hybrid variables, respectively. The present investigation is one of the few transportation models with hybrid penalties available in the literature, and for the first time, safety factors have been introduced in solid transportation problem. In this paper, we solved all mathematical problems using LINGO 12.0 software. The models could be extended to include breakable/deteriorating items, space constraints, price discount, etc. The methods, used for solutions here are quite general in nature, and these can be applied to other similar uncertain/imprecise models in other areas such as inventory control, ecology, and sustainable farm management.

Author details
[1]Department of Mathematics, National Institute of Technology, Agartala, Jirania 799055, West Tripura, India.
[2]Department of Applied Mathematics, Vidyasagar University, Midnapore 721102, West Bengal, India.

References

1. Bellman, RE, Zadeh, LA: Decision-making in a fuzzy environment. Manag. Sci. **17**, B141–B164 (1970)
2. Zadeh, LA: Fuzzy sets. Inf. Control **8**, 338–353 (1965)
3. Zadeh, LA: Fuzzy sets as a basis for a theory of possibility. Fuzzy Set. Syst. **1**, 3–28 (1978)
4. Schell, ED: Distribution of a product by several properties. Proceedings of the 2nd Symposium in Linear Programming, Washington, D.C., 27–29 January 1955, pp. 615–642. National Bureau of Standards, Washington, D.C (1955)
5. Haley, KB: The solid transportation problem. Oper. Res. **10**, 448–463 (1962)
6. Kundu, P, Kar, S, Maiti, M: Multi-objective solid transportation problems with budget constraint in uncertain environment. Int. J. Syst. Sci (2013). doi:10.1080/00207721.2012.748944
7. Baidya, A, Bera, UK, Maiti, M: Multi-item interval valued solid transportation problem with safety measure under fuzzy-stochastic environment. J. Transp. Secur. **6**, 151–174 (2013)
8. Baidya, A, Bera, UK, Maiti, M: Solution of multi-item interval valued solid transportation problem with safety measure using different methods. OPSEARCH, Springer (2013). doi:10.1007/s12597 013 0129–2
9. Kaufmann, A: Introduction to the Theory of Fuzzy Subsets. Academic Press, New York (1975)
10. Zimmermann, HJ: Fuzzy programming and linear programming with several objective functions. Fuzzy Set. Syst. **1**, 45–55 (1978)
11. Liu, B: A survey of credibility theory. Fuzzy Optim. Decis. Making **5**, 387–408 (2006)
12. Dubois, D, Prade, H: Possibility Theory: An Approach to Computerized Processing of Uncertainty. Plenum, New York (1998)
13. Chanas, S, Kuchta, D: A concept of the optimal solution of the transportation problem with fuzzy cost coefficients. Fuzzy Set. Syst. **82**, 299–305 (1996)
14. Jimenez, F, Verdegay, JL: Solving fuzzy solid transportation problems by an evolutionary algorithm based parametric approach. Eur. J. Oper. Res. **117**, 485–510 (1999)
15. Charnes, A, Cooper, AA: Chance-constrained programming. Manag. Sci. **6**, 73–79 (1959)
16. Liu, B, Iwamura, K: Chance constrained programming with fuzzy parameters. Fuzzy Set. Syst. **94**, 227–237 (1998)
17. Liu, B, Liu, YK: Expected value of fuzzy variable and fuzzy expected value models. IEEE Trans. Fuzzy Syst. **10**, 445–450 (2002)
18. Yang, L, Liu, L: Fuzzy fixed charge solid transportation problem and algorithm. Appl. Soft Comput. **7**, 879–889 (2007)
19. Mula, J, Poler, R, Garcia Sabater, JP: Material requirement planning with fuzzy constraints and fuzzy coefficients. Fuzzy Set. Syst. **158**, 783–793 (2007)
20. Chen, SP, Tsai, MJ: Time-cost trade-off analysis of the project network in fuzzy environment. Eur. J. Oper. Res. **212**, 386–397 (2011)
21. Fegad, MR, Jadhav, VA, Muley, AA: Finding an optimal solution of transportation problem using interval and triangular membership functions. Eur. J. Sci. Res. **60**, 415–421 (2011)
22. Hulsurkar, S, Biswal, MP, Sinha, SB: Fuzzy programming approach to multi-objective stochastic linear programming problem. Fuzzy Set. Syst. **88**, 173–181 (1997)
23. Rao, SS: Engineering Optimization-Theory and Practice, 3rd edn. New Age International Publishers, New Delhi (2006)
24. Grzegorzewski, P: Nearest interval approximation of a fuzzy number. Fuzzy Set. Syst. **130**, 321–330 (2002)
25. Liu, B: Minimax chance constrained programming model for fuzzy decision systems. Inform. Sci. **112**, 25–38 (1998)
26. Yang, L, Feng, Y: A bicriteria solid transportation problem with fixed charge under stochastic environment. Appl. Math. Model. **31**, 2668–2683 (2007)
27. Zhang, Q, Fan, Z, Pan, D: A ranking approach for interval numbers in uncertain multiple attribute decision making problems. Syst. Eng. - Theory Pract. **5**, 129–133 (1999)

16

Extreme values and integral of solution of uncertain differential equation

Kai Yao

Correspondence:
yaok09@mails.tsinghua.edu.cn
Department of Mathematical Sciences, Tsinghua University, Beijing 100084, China

Abstract

Uncertain differential equation is a type of differential equation involving uncertain process. This paper will give uncertainty distributions of the extreme values, first hitting time, and integral of the solution of uncertain differential equation. Some solution methods are also documented in this paper.

Keywords: Uncertainty theory; Uncertain process; Uncertain differential equation

Background

Probability theory, since it was founded by Kolmogorov in 1933, has been a crucial tool to model indeterminacy phenomena when probability distributions of the possible events are available. However, due to economical or technological reasons, we often cannot obtain sample data, based on which we estimate the probability distribution via statistics. In this case, we have to invite domain experts to evaluate their belief degree. Since human beings tend to overweight unlikely events, the belief degree usually has a much larger range than the real frequency (Kahneman and Tversky [1]). As a result, it cannot be treated as probability, otherwise some counterintuitive results may happen. An extreme counterexample was proposed by Liu [2].

In order to model the belief degree, an uncertainty theory was founded by Liu [3], and refined by Liu [4] based on normality, duality, subadditivity and product axioms. So far, it has been applied to many areas, and has brought many branches such as uncertain programming (Liu [5]), uncertain risk analysis (Liu [6]), uncertain inference (Liu [7]), uncertain logic (Liu [8]), and uncertain statistics (Liu [4]).

In order to describe the evolution of an uncertain phenomenon, Liu [9] proposed a concept of uncertain process. Then Liu [10] designed a Liu process that is an uncertain process with stationary and independent normal uncertain increments, and founded uncertain calculus to deal with the integral and differential of an uncertain process with respect to Liu process. Then Liu and Yao [11] extended uncertain integral from single Liu process to multiple ones. Besides, Chen and Ralescu [12] founded uncertain calculus with respect to general Liu process. As a complementary, Yao [13] founded uncertain calculus with respect to uncertain renewal process.

Uncertain differential equation was first proposed by Liu [9] as a type of differential equation driven by Liu process. Chen and Liu [14] gave an analytic solution for linear uncertain differential equation. Following that, Liu [15] and Yao [16] gave some methods to solve two types of nonlinear uncertain differential equations. Then Yao and Chen [17] proposed a numerical method for solving uncertain differential equation. As extensions of uncertain differential equation, uncertain differential equation with jumps was proposed by Yao [13], and uncertain delayed differential equation was studied among others by Barbacioru [18], Liu and Fei [19], and Ge and Zhu [20]. In addition, backward uncertain differential equation was proposed by Ge and Zhu [21].

Due to the paradox of stochastic finance theory (Liu [22]), Liu [10] presented an uncertain stock model via uncertain differential equation, and gave its European option pricing formulas, opening the door of uncertain finance theory. Then Chen [23] derived the American option pricing formulas of the stock model. After that, Peng and Yao [24] presented a mean-reverting uncertain stock model, and Chen et al [25] proposed an uncertain stock model with periodic dividends. Besides, Chen and Gao [26] proposed an uncertain interest rate model, and Liu et al [27] proposed an uncertain currency model. In addition, Zhu [28] applied uncertain differential equation to optimal control problems.

With many applications of uncertain differential equation, the study on properties of the solutions was also developed well. Chen and Liu [14] gave a sufficient condition for an uncertain differential equation having a unique solution. Then Gao [29] weakened the condition. After that, Yao et al [30] gave a sufficient condition for an uncertain differential equation being stable.

In this paper, we will consider the extreme values, first hitting time, and integral of the solution of an uncertain diffusion process. The rest of this paper is organized as follows. In the section of Preliminary, we will review some basic concepts about uncertain variable, uncertain process and uncertain differential equation. After that, we study the extreme values of the solution of an uncertain differential equation, and give their uncertainty distributions in the section of Extreme values. Then by the relationship between first hitting time and extreme value, we give an uncertainty distribution of the first hitting time of the solution of an uncertain differential equation in the section of First hitting time. Following that, we consider the integral of the solution of an uncertain differential equation, and give its inverse uncertainty distribution in the section of Integral. At last, some remarks are given in the section of Conclusions.

Preliminary

In this section, we will first review some basic concepts and results in uncertainty theory. Then we introduce the concept of uncertain process, uncertain calculus and uncertain differential equation.

Uncertainty theory

Definition 1. (Liu [3]) Let $\mathcal{L}$ be a σ-algebra on a nonempty set Γ. A set function $\mathcal{M}$: $\mathcal{L} \to [0,1]$ is called an uncertain measure if it satisfies the following axioms:

Axiom 1: (Normality Axiom) $\mathcal{M}\{\Gamma\} = 1$ for the universal set Γ.

Axiom 2: (Duality Axiom) $\mathcal{M}\{\Lambda\} + \mathcal{M}\{\Lambda^c\} = 1$ for any event Λ.

Axiom 3: (Subadditivity Axiom) For every countable sequence of events $\Lambda_1, \Lambda_2, \cdots$, we have

$$\mathcal{M}\left\{\bigcup_{i=1}^{\infty} \Lambda_i\right\} \leq \sum_{i=1}^{\infty} \mathcal{M}\{\Lambda_i\}.$$

Besides, in order to provide the operational law, Liu [10] defined the product uncertain measure on the product σ-algebre $\mathcal{L}$ as follows.

Axiom 4: (Product Axiom) Let $(\Gamma_k, \mathcal{L}_k, \mathcal{M}_k)$ be uncertainty spaces for $k = 1, 2, \cdots$ Then the product uncertain measure $\mathcal{M}$ is an uncertain measure satisfying

$$\mathcal{M}\left\{\prod_{i=1}^{\infty} \Lambda_k\right\} = \bigwedge_{k=1}^{\infty} \mathcal{M}_k\{\Lambda_k\}$$

where Λ_k are arbitrarily chosen events from $\mathcal{L}_k$ for $k = 1, 2, \cdots$, respectively.

An uncertain variable is essentially a measurable function from an uncertainty space to the set of real numbers. In order to describe an uncertain variable, a concept of uncertainty distribution is defined as follows.

Definition 2. (Liu [3]) The uncertainty distribution of an uncertain variable ξ is defined by

$$\Phi(x) = \mathcal{M}\{\xi \le x\}$$

for any $x \in \Re$.

Expected value is regarded as the average value of an uncertain variable in the sense of uncertain measure.

Definition 3. (Liu [3]) The expected value of an uncertain variable ξ is defined by

$$E[\xi] = \int_0^{+\infty} \mathcal{M}\{\xi \ge x\}dx - \int_{-\infty}^{0} \mathcal{M}\{\xi \le x\}dx$$

provided that at least one of the two integrals exists.

Assuming ξ has an uncertainty distribution Φ, Liu [3] proved that the expected value of ξ is

$$E[\xi] = \int_0^{+\infty} (1 - \Phi(x))\, dx - \int_{-\infty}^{0} \Phi(x)dx.$$

The inverse function Φ^{-1} of the uncertainty distribution Φ of uncertain variable ξ is called the inverse uncertainty distribution of ξ if it exists and is unique for each $\alpha \in (0, 1)$. Inverse uncertainty distribution plays a crucial role in operations of independent uncertain variables.

Definition 4. (Liu[10]) The uncertain variables $\xi_1, \xi_2, \cdots, \xi_n$ are said to be independent if

$$\mathcal{M}\left\{\bigcap_{i=1}^{n}(\xi_i \in B_i)\right\} = \bigwedge_{i=1}^{n} \mathcal{M}\{\xi_i \in B_i\}$$

for any Borel set $B_1, B_2, \cdots, B_n$ of real numbers.

Theorem 1. *(Liu [4]) Let $\xi_1, \xi_2, \cdots, \xi_n$ be independent uncertain variables with uncertainty distributions $\Phi_1, \Phi_2, \cdots, \Phi_n$, respectively. If $f(x_1, x_2, \cdots, x_n)$ is strictly increasing* *with respect to* $x_1, x_2, \cdots, x_m$ *and strictly decreasing with respect to* $x_{m+1}, x_{m+2}, \cdots, x_n$, *then* $\xi = f(\xi_1, \xi_2, \cdots, \xi_n)$ *is an uncertain variable with an inverse uncertainty distribution*

$$\Phi^{-1}(\alpha) = f\left(\Phi_1^{-1}(\alpha), \cdots, \Phi_m^{-1}(\alpha), \Phi_{m+1}^{-1}(1-\alpha), \cdots, \Phi_n^{-1}(1-\alpha)\right).$$

Theorem 2. *(Liu and Ha [31]) Let* $\xi_1, \xi_2, \cdots, \xi_n$ *be independent uncertain variables with uncertainty distributions* $\Phi_1, \Phi_2, \cdots, \Phi_n$, *respectively. If* $f(x_1, x_2, \cdots, x_n)$ *is strictly increasing with respect to* $x_1, x_2, \cdots, x_m$ *and strictly decreasing with respect to* $x_{m+1}, x_{m+2}, \cdots, x_n$, *then the expected value of uncertain variable* $\xi = f(\xi_1, \xi_2, \cdots, \xi_n)$ *is*

$$E[\xi] = \int_0^1 f\left(\Phi_1^{-1}(\alpha), \cdots, \Phi_m^{-1}(\alpha), \Phi_{m+1}^{-1}(1-\alpha), \cdots, \Phi_n^{-1}(1-\alpha)\right) d\alpha.$$

Uncertain process

In order to model the evolution of uncertain phenomena, an uncertain process was proposed by Liu [9] as a sequence of uncertain variables driven by time or space.

Definition 5. (Liu [9]) Let T be an index set, and let $(\Gamma, \mathcal{L}, \mathcal{M})$ be an uncertainty space. An uncertain process is a measurable function from $T \times (\Gamma, \mathcal{L}, \mathcal{M})$ to the set of real numbers, i.e., for each $t \in T$ and any Borel set B of real numbers, the set

$$\{X_t \in B\} = \{\gamma | X_t(\gamma) \in B\}$$

is an event.

Definition 6. Let X_t be an uncertain process and let z be a given level. Then the uncertain variable

$$\tau_z = \inf\left\{t \geq 0 \mid X_t = z\right\}$$

is called the first hitting time that X_t reaches the level z.

Independent increment uncertain process is an important type of uncertain processes. Its formal definition is given below.

Definition 7. (Liu [9]) An uncertain process is said to have independent increments if

$$X_{t_0}, X_{t_1} - X_{t_0}, X_{t_2} - X_{t_1}, \cdots, X_{t_k} - X_{t_{k-1}}$$

are independent uncertain variables where t_0 is the initial time and $t_1, t_2, \cdots, t_k$ are any times with $t_0 < t_1 < \cdots < t_k$.

For a sample-continuous independent increment process X_t, Liu [32] proved the following *extreme value theorem*,

$$\mathcal{M}\left\{\sup_{0\le t\le s} X_t \le x\right\} = \inf_{0\le t\le s} \mathcal{M}\{X_t \le x\},$$

$$\mathcal{M}\left\{\inf_{0\le t\le s} X_t \le x\right\} = \sup_{0\le t\le s} \mathcal{M}\{X_t \le x\}.$$

Definition 8. (Liu [9]) An uncertain process is said to have stationary increments if for any given $t > 0$, the increments $X_{t+s} - X_s$ are identically distributed uncertain variables for all $s > 0$.

Uncertain calculus

Definition 9. (Liu [10]) An uncertain process C_t is said to be a canonical Liu process if

(i) $C_0 = 0$ and almost all sample paths are Lipschitz continuous,
(ii) C_t has stationary and independent increments,
(iii) every increment $C_{s+t} - C_s$ is a normal uncertain variable with expected value 0 and variance t^2, whose uncertainty distribution is

$$\Phi_t(x) = \left(1 + \exp\left(-\frac{\pi x}{\sqrt{3}t}\right)\right)^{-1}, \quad x \in \Re.$$

Definition 10. (Liu [10]) Let X_t be an uncertain process and C_t be a canonical Liu process. For any partition of closed interval $[a, b]$ with $a = t_1 < t_2 < \cdots < t_{k+1} = b$, the mesh is written as

$$\Delta = \max_{1\le i\le k} |t_{i+1} - t_i|.$$

Then Liu integral of X_t is defined by

$$\int_a^b X_t \mathrm{d}C_t = \lim_{\Delta\to 0} \sum_{i=1}^{k} X_{t_i} \cdot (C_{t_{i+1}} - C_{t_i})$$

provided that the limit exists almost surely and is finite. In this case, the uncertain process X_t is said to be Liu integrable.

Definition 11. (Liu [10]) Let C_t be a canonical Liu process, and μ_s and σ_s be two uncertain processes. Then the uncertain process

$$Z_t = Z_0 + \int_0^t \mu_s \mathrm{d}s + \int_0^t \sigma_s \mathrm{d}C_s$$

is called a Liu process with drift μ_t and diffusion σ_t. The differential form of Liu process is written as

$$\mathrm{d}Z_t = \mu_t \mathrm{d}t + \sigma_t \mathrm{d}C_t.$$

Theorem 3. *(Liu [10]) (Fundamental Theorem of Uncertain Calculus) Let C_t be a canonical Liu process, and $h(t, c)$ be a continuously differentiable function. Then $Z_t = h(t, C_t)$ is a Liu process with*

$$\mathrm{d}Z_t = \frac{\partial h}{\partial t}(t, C_t)\mathrm{d}t + \frac{\partial h}{\partial c}(t, C_t)\mathrm{d}C_t.$$

Uncertain differential equation

An uncertain differential equation is essentially a type of differential equation driven by Liu process.

Definition 12. (Liu [9]) Suppose C_t is a canonical Liu process, and f and g are two given functions. Then

$$\mathrm{d}X_t = f(t, X_t)\mathrm{d}t + g(t, X_t)\mathrm{d}C_t$$

is called an uncertain differential equation.

Yao and Chen [17] proposed a concept of α-path, and found a connection between an uncertain differential equation and a spectrum of ordinary differential equations.

Definition 13. (Yao and Chen [17]) Let α be a number with $0 < \alpha < 1$. An uncertain differential equation

$$\mathrm{d}X_t = f(t, X_t)\mathrm{d}t + g(t, X_t)\mathrm{d}C_t$$

is said to have an α-path X_t^α if it solves the corresponding ordinary differential equation

$$\mathrm{d}X_t^\alpha = f(t, X_t^\alpha)\mathrm{d}t + |g(t, X_t^\alpha)|\Phi^{-1}(\alpha)\mathrm{d}t$$

where $\Phi^{-1}(\alpha)$ is the inverse uncertainty distribution of a standard normal uncertain variable.

Theorem 4. *(Yao and Chen [17]) Let X_t and X_t^α be the solution and α-path of the uncertain differential equation*

$$\mathrm{d}X_t = f(t, X_t)\mathrm{d}t + g(t, X_t)\mathrm{d}C_t,$$

respectively. Then

$$\mathcal{M}\{X_t \le X_t^\alpha, \forall t\} = \alpha,$$

$$\mathcal{M}\{X_t > X_t^\alpha, \forall t\} = 1 - \alpha.$$

As a corollary, the solution X_t has an inverse uncertainty distribution $\Phi_t^{-1}(\alpha) = X_t^\alpha$.

Extreme values

In this section, we study the extreme values of the solution of an uncertain differential equation, and give their uncertainty distributions. In addition, we design some numerical methods to obtain the uncertainty distributions.

Supremum

Theorem 5. *Let X_t and X_t^α be the solution and α-path of the uncertain differential equation*

$$\mathrm{d}X_t = f(t, X_t)\mathrm{d}t + g(t, X_t)\mathrm{d}C_t,$$

respectively. Then for a strictly increasing function $J(x)$, the supremum

$$\sup_{0 \le t \le s} J(X_t)$$

has an inverse uncertainty distribution

$$\Psi_s^{-1}(\alpha) = \sup_{0 \le t \le s} J\left(X_t^\alpha\right).$$

Proof. Since $J(x)$ is a strictly increasing function, we have

$$\left\{ \sup_{0 \le t \le s} J(X_t) \le \sup_{0 \le t \le s} J\left(X_t^\alpha\right) \right\} \supset \left\{X_t \le X_t^\alpha, \forall t\right\}$$

and

$$\left\{ \sup_{0 \le t \le s} J(X_t) > \sup_{0 \le t \le s} J\left(X_t^\alpha\right) \right\} \supset \left\{X_t > X_t^\alpha, \forall t\right\}.$$

By Theorem 4 and the monotonicity of uncertain measure, we have

$$\mathcal{M}\left\{ \sup_{0 \le t \le s} J(X_t) \le \sup_{0 \le t \le s} J\left(X_t^\alpha\right) \right\} \ge \mathcal{M}\left\{X_t \le X_t^\alpha, \forall t\right\} = \alpha$$

and

$$\mathcal{M}\left\{ \sup_{0 \le t \le s} J(X_t) > \sup_{0 \le t \le s} J\left(X_t^\alpha\right) \right\} \ge \mathcal{M}\left\{X_t > X_t^\alpha, \forall t\right\} = 1 - \alpha.$$

It follows from the duality axiom of uncertain measure that

$$\mathcal{M}\left\{ \sup_{0 \le t \le s} J(X_t) \le \sup_{0 \le t \le s} J\left(X_t^\alpha\right) \right\} = \alpha,$$

i.e., the supremum

$$\sup_{0 \le t \le s} J(X_t)$$

has an inverse uncertainty distribution

$$\Psi_s^{-1}(\alpha) = \sup_{0 \le t \le s} J\left(X_t^\alpha\right).$$

In order to calculate the inverse uncertainty distribution of the supremum, we design a numerical method as below.

Step 1: Fix α in $(0,1)$, and fix h as the step length. Set $i = 0$, $N = s/h$, $X_0^\alpha = X_0$, and $H = J(X_0)$.
Step 2: Employ the recursion formula

$$X_{i+1}^\alpha = X_i^\alpha + f\left(t_i, X_i^\alpha\right) h + g\left(t_i, X_i^\alpha\right) \Phi^{-1}(\alpha) h,$$

and calculate X_{i+1}^α.
Step 3: Set $H \leftarrow \max\left(H, J\left(X_{i+1}^\alpha\right)\right), i \leftarrow i+1$.
Step 4: Repeat Step 2 and Step 3 for N times.
Step 5: The inverse uncertainty distribution of

$$\sup_{0 \le t \le s} J(X_t)$$

is determined by

$$\Psi_s^{-1}(\alpha) = H.$$

□

Theorem 6. *Let X_t be the solution of an uncertain differential equation* $\mathrm{d}X_t = f(t, X_t)\mathrm{d}t + g(t, X_t)\mathrm{d}C_t$. *Assume X_t has an uncertainty distribution $\Phi_t(x)$ at each time t. Then for a strictly increasing function $J(x)$, the supremum*

$$\sup_{0 \le t \le s} J(X_t)$$

has an uncertainty distribution

$$\Psi_s(x) = \inf_{0 \le t \le s} \Phi_t\left(J^{-1}(x)\right).$$

Proof. Since $X_t^\alpha = \Phi_t^{-1}(\alpha)$, we have

$$\mathcal{M}\left\{\sup_{0 \le t \le s} J(X_t) \le \sup_{0 \le t \le s} J\left(\Phi_t^{-1}(\alpha)\right)\right\} = \alpha$$

by Theorem 5. Write

$$x = \sup_{0 \le t \le s} J\left(\Phi_t^{-1}(\alpha)\right),$$

i.e.,

$$\alpha = \inf_{0 \le t \le s} \Phi_t\left(J^{-1}(x)\right).$$

Then we have

$$\mathcal{M}\left\{\sup_{0 \le t \le s} J(X_t) \le x\right\} = \inf_{0 \le t \le s} \Phi_t\left(J^{-1}(x)\right).$$

In other words, the supremum

$$\sup_{0 \le t \le s} J(X_t)$$

has an uncertainty distribution

$$\Psi_s(x) = \inf_{0 \le t \le s} \Phi_t\left(J^{-1}(x)\right).$$

□

Theorem 7. *Let X_t and X_t^α be the solution and α-path of the uncertain differential equation*

$$\mathrm{d}X_t = f(t, X_t)\mathrm{d}t + g(t, X_t)\mathrm{d}C_t,$$

respectively. Then for a strictly decreasing function $J(x)$, the supremum

$$\sup_{0 \le t \le s} J(X_t)$$

has an inverse uncertainty distribution

$$\Psi_s^{-1}(\alpha) = \sup_{0 \le t \le s} J\left(X_t^{1-\alpha}\right).$$

Proof. Since $J(x)$ is a strictly decreasing function, we have

$$\left\{\sup_{0 \le t \le s} J(X_t) \le \sup_{0 \le t \le s} J\left(X_t^{1-\alpha}\right)\right\} \supset \left\{X_t \ge X_t^{1-\alpha}, \forall t\right\}$$

and

$$\left\{ \sup_{0\le t\le s} J(X_t) > \sup_{0\le t\le s} J\left(X_t^{1-\alpha}\right) \right\} \supset \left\{ X_t < X_t^{1-\alpha}, \forall t \right\}.$$

By Theorem 4 and the monotonicity of uncertain measure, we have

$$\mathcal{M}\left\{ \sup_{0\le t\le s} J(X_t) \le \sup_{0\le t\le s} J\left(X_t^{1-\alpha}\right) \right\} \ge \mathcal{M}\left\{ X_t \ge X_t^{1-\alpha}, \forall t \right\} = \alpha$$

and

$$\mathcal{M}\left\{ \sup_{0\le t\le s} J(X_t) > \sup_{0\le t\le s} J\left(X_t^{1-\alpha}\right) \right\} \ge \mathcal{M}\left\{ X_t < X_t^{1-\alpha}, \forall t \right\} = 1-\alpha.$$

It follows from the duality axiom of uncertain measure that

$$\mathcal{M}\left\{ \sup_{0\le t\le s} J(X_t) \le \sup_{0\le t\le s} J\left(X_t^{1-\alpha}\right) \right\} = \alpha,$$

i.e., the supremum

$$\sup_{0\le t\le s} J(X_t)$$

has an inverse uncertainty distribution

$$\Psi_s^{-1}(\alpha) = \sup_{0\le t\le s} J\left(X_t^{1-\alpha}\right).$$

□

Theorem 8. *Let X_t be the solution of an uncertain differential equation $\mathrm{d}X_t = f(t, X_t)\mathrm{d}t + g(t, X_t)\mathrm{d}C_t$. Assume X_t has an uncertainty distribution $\Phi_t(x)$ at each time t. Then for a strictly decreasing function $J(x)$, the supremum*

$$\sup_{0\le t\le s} J(X_t)$$

has an uncertainty distribution

$$\Psi_s(x) = 1 - \sup_{0\le t\le s} \Phi_t\left(J^{-1}(x)\right).$$

Proof. Since $X_t^{1-\alpha} = \Phi_t^{-1}(1-\alpha)$, we have

$$\mathcal{M}\left\{ \sup_{0\le t\le s} J(X_t) \le \sup_{0\le t\le s} J\left(\Phi_t^{-1}(1-\alpha)\right) \right\} = \alpha$$

by Theorem 7. Write

$$x = \sup_{0\le t\le s} J\left(\Phi_t^{-1}(1-\alpha)\right),$$

i.e.,

$$\alpha = 1 - \sup_{0\le t\le s} \Phi_t\left(J^{-1}(x)\right).$$

Then we have

$$\mathcal{M}\left\{ \sup_{0\le t\le s} J(X_t) \le x \right\} = 1 - \sup_{0\le t\le s} \Phi_t\left(J^{-1}(x)\right).$$

In other words, the supremum

$$\sup_{0\le t\le s} J(X_t)$$

has an uncertainty distribution

$$\Psi_s(x) = 1 - \sup_{0\le t\le s} \Phi_t\left(J^{-1}(x)\right).$$

□

Infimum

Theorem 9. *Let X_t and X_t^α be the solution and α-path of the uncertain differential equation*

$$\mathrm{d}X_t = f(t, X_t)\mathrm{d}t + g(t, X_t)\mathrm{d}C_t,$$

respectively. Then for a strictly increasing function $J(x)$, the infimum

$$\inf_{0\le t\le s} J(X_t)$$

has an inverse uncertainty distribution

$$\Upsilon_s^{-1}(\alpha) = \inf_{0\le t\le s} J\left(X_t^\alpha\right).$$

Proof. Since $J(x)$ is a strictly increasing function, we have

$$\left\{\inf_{0\le t\le s} J(X_t) \le \inf_{0\le t\le s} J\left(X_t^\alpha\right)\right\} \supset \left\{X_t \le X_t^\alpha, \forall t\right\}$$

and

$$\left\{\inf_{0\le t\le s} J(X_t) > \inf_{0\le t\le s} J\left(X_t^\alpha\right)\right\} \supset \left\{X_t > X_t^\alpha, \forall t\right\}.$$

By Theorem 4 and the monotonicity of uncertain measure, we have

$$\mathcal{M}\left\{\inf_{0\le t\le s} J(X_t) \le \inf_{0\le t\le s} J\left(X_t^\alpha\right)\right\} \ge \mathcal{M}\left\{X_t \le X_t^\alpha, \forall t\right\} = \alpha$$

and

$$\mathcal{M}\left\{\inf_{0\le t\le s} J(X_t) > \inf_{0\le t\le s} J\left(X_t^\alpha\right)\right\} \ge \mathcal{M}\left\{X_t > X_t^\alpha, \forall t\right\} = 1 - \alpha.$$

It follows from the duality axiom of uncertain measure that

$$\mathcal{M}\left\{\inf_{0\le t\le s} J(X_t) \le \inf_{0\le t\le s} J\left(X_t^\alpha\right)\right\} = \alpha,$$

i.e., the infimum

$$\inf_{0\le t\le s} J(X_t)$$

has an inverse uncertainty distribution

$$\Upsilon_s^{-1}(\alpha) = \inf_{0\le t\le s} J\left(X_t^\alpha\right).$$

In order to calculate the uncertainty distribution of the infimum, we design a numerical method as below.

> Step 1: Fix α in $(0, 1)$, and fix h as the step length. Set $i = 0, N = s/h, X_0^\alpha = X_0$, and $H = J(X_0)$.

Step 2: Employ the recursion formula

$$X_{i+1}^{\alpha} = X_i^{\alpha} + f\left(t_i, X_i^{\alpha}\right) h + g\left(t_i, X_i^{\alpha}\right) \Phi^{-1}(\alpha) h,$$

and calculate X_{i+1}^{α} and $J\left(X_{i+1}^{\alpha}\right)$

Step 3: Set $H \leftarrow \min\left(H, J\left(X_{i+1}^{\alpha}\right)\right), i \leftarrow i+1$.

Step 4: Repeat Step 2 and Step 3 for N times.

Step 5: The inverse uncertainty distribution of $\inf\limits_{0\le t\le s} X_t$ is determined by

$$\Upsilon_s^{-1}(\alpha) = H.$$

□

Theorem 10. *Let X_t be the solution of an uncertain differential equation $\mathrm{d}X_t = f(t,X_t)\mathrm{d}t + g(t,X_t)\mathrm{d}C_t$. Assume X_t has an uncertainty distribution $\Phi_t(x)$ at each time t. Then for a strictly increasing function $J(x)$, the infimum*

$$\inf_{0\le t\le s} X_t$$

has an uncertainty distribution

$$\Upsilon_s(x) = \sup_{0\le t\le s} \Phi_t\left(J^{-1}(x)\right).$$

Proof. Since $X_t^{\alpha} = \Phi_t^{-1}(\alpha)$, we have

$$\mathcal{M}\left\{\inf_{0\le t\le s} J(X_t) \le \inf_{0\le t\le s} J\left(\Phi_t^{-1}(\alpha)\right)\right\} = \alpha$$

by Theorem 9. Write

$$x = \inf_{0\le t\le s} J\left(\Phi_t^{-1}(\alpha)\right),$$

i.e.,

$$\alpha = \sup_{0\le t\le s} \Phi_t\left(J^{-1}(x)\right).$$

Then we have

$$\mathcal{M}\left\{\inf_{0\le t\le s} X_t \le x\right\} = \sup_{0\le t\le s} \Phi_t\left(J^{-1}(x)\right).$$

In other words, the infimum

$$\inf_{0\le t\le s} J(X_t)$$

has an uncertainty distribution

$$\Upsilon_s(x) = \sup_{0\le t\le s} \Phi_t\left(J^{-1}(x)\right).$$

□

Theorem 11. *Let X_t and X_t^{α} be the solution and α-path of the uncertain differential equation*

$$\mathrm{d}X_t = f(t,X_t)\mathrm{d}t + g(t,X_t)\mathrm{d}C_t,$$

respectively. Then for a strictly decreasing function $J(x)$, the infimum

$$\inf_{0\le t\le s} J(X_t)$$

has an inverse uncertainty distribution

$$\Upsilon_s^{-1}(\alpha) = \inf_{0\le t\le s} J\left(X_t^{1-\alpha}\right).$$

Proof. Since $J(x)$ is a strictly decreasing function, we have

$$\left\{\inf_{0\le t\le s} J(X_t) \le \inf_{0\le t\le s} J\left(X_t^{1-\alpha}\right)\right\} \supset \left\{X_t \ge X_t^{1-\alpha}, \forall t\right\}$$

and

$$\left\{\inf_{0\le t\le s} J(X_t) > \inf_{0\le t\le s} J\left(X_t^{1-\alpha}\right)\right\} \supset \left\{X_t < X_t^{1-\alpha}, \forall t\right\}.$$

By Theorem 4 and the monotonicity of uncertain measure, we have

$$\mathcal{M}\left\{\inf_{0\le t\le s} J(X_t) \le \inf_{0\le t\le s} J\left(X_t^{1-\alpha}\right)\right\} \ge \mathcal{M}\left\{X_t \ge X_t^{1-\alpha}, \forall t\right\} = \alpha$$

and

$$\mathcal{M}\left\{\inf_{0\le t\le s} J(X_t) > \inf_{0\le t\le s} J\left(X_t^{1-\alpha}\right)\right\} \ge \mathcal{M}\left\{X_t < X_t^{1-\alpha}, \forall t\right\} = 1-\alpha.$$

It follows from the duality axiom of uncertain measure that

$$\mathcal{M}\left\{\inf_{0\le t\le s} J(X_t) \le \inf_{0\le t\le s} J\left(X_t^{1-\alpha}\right)\right\} = \alpha,$$

i.e., the infimum

$$\inf_{0\le t\le s} J(X_t)$$

has an inverse uncertainty distribution

$$\Upsilon_s^{-1}(\alpha) = \inf_{0\le t\le s} J\left(X_t^{1-\alpha}\right).$$

□

Theorem 12. *Let X_t be the solution of an uncertain differential equation $\mathrm{d}X_t = f(t, X_t)\mathrm{d}t + g(t, X_t)\mathrm{d}C_t$. Assume X_t has an uncertainty distribution $\Phi_t(x)$ at each time t. Then for a strictly decreasing function $J(x)$, the infimum*

$$\inf_{0\le t\le s} J(X_t)$$

has an uncertainty distribution

$$\Upsilon_s(x) = 1 - \inf_{0\le t\le s} \Phi_t\left(J^{-1}(x)\right).$$

Proof. Since $X_t^{1-\alpha} = \Phi_t^{-1}(1-\alpha)$, we have

$$\mathcal{M}\left\{\inf_{0\le t\le s} J(X_t) \le \inf_{0\le t\le s} J\left(\Phi_t^{-1}(1-\alpha)\right)\right\} = \alpha$$

by Theorem 11. Write

$$x = \inf_{0\le t\le s} J\left(\Phi_t^{-1}(1-\alpha)\right),$$

i.e.,

$$\alpha = 1 - \inf_{0 \le t \le s} \Phi_t \left(J^{-1}(x) \right).$$

Then we have

$$\mathcal{M}\left\{ \inf_{0 \le t \le s} X_t \le x \right\} = 1 - \inf_{0 \le t \le s} \Phi_t \left(J^{-1}(x) \right).$$

In other words, the infimum

$$\inf_{0 \le t \le s} J(X_t)$$

has an uncertainty distribution

$$\Upsilon_s(x) = 1 - \inf_{0 \le t \le s} \Phi_t \left(J^{-1}(x) \right).$$

□

First hitting time

In this section, we study the first hitting time of the solution of an uncertain differential equation, and give the uncertainty distributions in different cases.

First hitting time of strictly increasing function of the solution

Theorem 13. *Let X_t and X_t^α be the solution and α-path of the uncertain differential equation*

$$dX_t = f(t, X_t)dt + g(t, X_t)dC_t$$

with an initial value X_0, respectively. Given a strictly increasing function $J(x)$, and a level $z > J(X_0)$, the first hitting time τ_z that $J(X_t)$ reaches z has an uncertainty distribution

$$\Psi(s) = 1 - \inf\left\{ \alpha \in (0,1) \,\middle|\, \sup_{0 \le t \le s} J\left(X_t^\alpha\right) \ge z \right\}.$$

Proof. Write

$$\alpha_0 = \inf\left\{ \alpha \in (0,1) \,\middle|\, \sup_{0 \le t \le s} J\left(X_t^\alpha\right) \ge z \right\}.$$

Since $J(x)$ is a strictly increasing function, we have

$$\{\tau_z \le s\} \supset \{J(X_t) \ge J\left(X_t^{\alpha_0}\right), \forall t\} = \{X_t \ge X_t^{\alpha_0}, \forall t\},$$

$$\{\tau_z > s\} \supset \{J(X_t) < J\left(X_t^{\alpha_0}\right), \forall t\} = \{X_t < X_t^{\alpha_0}, \forall t\}.$$

By Theorem 4 and the monotonicity of uncertain measure, we have

$$\mathcal{M}\{\tau_z \le s\} \ge \mathcal{M}\{X_t \ge X_t^{\alpha_0}, \forall t\} = 1 - \alpha_0,$$

$$\mathcal{M}\{\tau_z > s\} \ge \{X_t < X_t^{\alpha_0}, \forall t\} = \alpha_0.$$

It follows from the duality axiom of uncertain measure that

$$\mathcal{M}\{\tau_z \le s\} = 1 - \alpha_0.$$

This completes the proof. □

For a strictly increasing function $J(x)$, in order to calculate the uncertainty distribution $\Psi(s)$ of the first hitting time τ_z that $J(X_t)$ reaches z when $J(X_0) < z$, we design a numerical method as below.

Step 1: Fix ε as the accuracy, and fix h as the step length. Set $N = s/h$.
Step 2: Employ the recursion formula

$$X_{i+1}^{\varepsilon} = X_i^{\varepsilon} + f\left(t_i, X_i^{\varepsilon}\right) h + g\left(t_i, X_i^{\varepsilon}\right) \Phi^{-1}(\varepsilon) h$$

for N times, and calculate $X_i^{\varepsilon}, i = 1, 2, \cdots, N$. If

$$\max_{1 \le i \le N} J\left(X_i^{\varepsilon}\right) \ge z,$$

then return $1 - \varepsilon$ and stop.
Step 3: Employ the recursion formula

$$X_{i+1}^{1-\varepsilon} = X_i^{1-\varepsilon} + f\left(t_i, X_i^{1-\varepsilon}\right) h + g\left(t_i, X_i^{1-\varepsilon}\right) \Phi^{-1}(1-\varepsilon) h$$

for N times, and calculate $X_i^{1-\varepsilon}, i = 1, 2, \cdots, N$. If

$$\max_{1 \le i \le N} J\left(X_i^{1-\varepsilon}\right) < z,$$

then return ε and stop.
Step 4: Set $\alpha_1 = \varepsilon, \alpha_2 = 1 - \varepsilon$.
Step 5: Set $\alpha = (\alpha_1 + \alpha_2)/2$.
Step 6: Employ the recursion formula

$$X_{i+1}^{\alpha} = X_i^{\alpha} + f\left(t_i, X_i^{\alpha}\right) h + g\left(t_i, X_i^{\alpha}\right) \Phi^{-1}(\alpha) h$$

for N times, and calculate $X_{i+1}^{\alpha}, i = 1, 2, \cdots, N$. If

$$\max_{1 \le i \le N} J\left(X_t^{\alpha}\right) < z,$$

then set $\alpha_1 = \alpha$. Otherwise, set $\alpha_2 = \alpha$.
Step 7: If $|\alpha_2 - \alpha_1| \le \varepsilon$, then return $1 - \alpha$ and stop. Otherwise, go to Step 5.

Theorem 14. *Let X_t be the solution of an uncertain differential equation $\mathrm{d}X_t = f(t, X_t)\mathrm{d}t + g(t, X_t)\mathrm{d}C_t$ with an initial value X_0. Assume X_t has an uncertainty distribution $\Phi_t(x)$ at each time t. Then given a strictly increasing function $J(x)$ and a level $z > J(X_0)$, the first hitting time τ_z that $J(X_t)$ reaches z has an uncertainty distribution*

$$\Psi(s) = 1 - \inf_{0 \le t \le s} \Phi_t\left(J^{-1}(z)\right).$$

Proof. Since the event $\{\tau_z \le s\}$ is equivalent to the event

$$\left\{ \sup_{0 \le t \le s} J(X_t) \ge z \right\}$$

provided $z > J(X_0)$, it follows from Theorem 6 that

$$\begin{aligned} \Psi(s) &= \mathcal{M}\left\{ \sup_{0 \le t \le s} J(X_t) \ge z \right\} \\ &= \mathcal{M}\left\{ \sup_{0 \le t \le s} X_t \ge J^{-1}(z) \right\} \\ &= 1 - \inf_{0 \le t \le s} \Phi_t\left(J^{-1}(z)\right). \end{aligned}$$

This completes the proof. □

Theorem 15. *Let X_t and X_t^α be the solution and α-path of the uncertain differential equation*

$$dX_t = f(t, X_t)dt + g(t, X_t)dC_t$$

with an initial value X_0, respectively. Given a strictly increasing function $J(x)$ and a level $z < J(X_0)$, the first hitting time τ_z that $J(X_t)$ reaches z has an uncertainty distribution

$$\Upsilon(s) = \sup\left\{\alpha \in (0,1) \,\middle|\, \inf_{0\le t\le s} J\left(X_t^\alpha\right) \le z\right\}.$$

Proof. Write

$$\alpha_0 = \sup\left\{\alpha \in (0,1) \,\middle|\, \inf_{0\le t\le s} J\left(X_t^\alpha\right) \le z\right\}.$$

Then

$$\{\tau_z \le s\} \supset \{J(X_t) \le J\left(X_t^{\alpha_0}\right), \forall t\} = \{X_t \le X_t^{\alpha_0}, \forall t\},$$

$$\{\tau_z > s\} \supset \{J(X_t) > J\left(X_t^{\alpha_0}\right), \forall t\} = \{X_t > X_t^{\alpha_0}, \forall t\}.$$

By Theorem 4 and the monotonicity of uncertain measure, we have

$$\mathcal{M}\{\tau_z \le s\} \ge \mathcal{M}\{X_t \le X_t^{\alpha_0}, \forall t\} = \alpha_0,$$

$$\mathcal{M}\{\tau_z > s\} \ge \{X_t > X_t^{\alpha_0}, \forall t\} = 1 - \alpha_0.$$

It follows from the duality axiom of uncertain measure that

$$\mathcal{M}\{\tau_z \le s\} = \alpha_0.$$

This completes the proof. □

For a strictly increasing function $J(x)$, in order to calculate the uncertainty distribution $\Upsilon(s)$ of the first hitting time τ_z that $J(X_t)$ reaches z when $J(X_0) > z$, we design a numerical method as below.

Step 1: Fix ε as the accuracy, and fix h as the step length. Set $N = s/h$.
Step 2: Employ the recursion formula

$$X_{i+1}^\varepsilon = X_i^\varepsilon + f\left(t_i, X_i^\varepsilon\right)h + g\left(t_i, X_i^\varepsilon\right)\Phi^{-1}(\varepsilon)h$$

for N times, and calculate $X_i^\varepsilon, i = 1, 2, \cdots, N$. If

$$\min_{1\le i\le N} J\left(X_i^\varepsilon\right) > z,$$

then return $1-\varepsilon$ and stop.
Step 3: Employ the recursion formula

$$X_{i+1}^{1-\varepsilon} = X_i^{1-\varepsilon} + f\left(t_i, X_i^{1-\varepsilon}\right)h + g\left(t_i, X_i^{1-\varepsilon}\right)\Phi^{-1}(1-\varepsilon)h$$

for N times, and calculate $X_i^{1-\varepsilon}, i = 1, 2, \cdots, N$. If

$$\min_{1\le i\le N} J\left(X_i^{1-\varepsilon}\right) \le z,$$

then return ε and stop.
Step 4: Set $\alpha_1 = \varepsilon, \alpha_2 = 1-\varepsilon$.
Step 5: Set $\alpha = (\alpha_1 + \alpha_2)/2$.

Step 6: Employ the recursion formula

$$X_{i+1}^{\alpha} = X_i^{\alpha} + f\left(t_i, X_i^{\alpha}\right) h + g\left(t_i, X_i^{\alpha}\right) \Phi^{-1}(\alpha) h$$

for N times, and calculate $X_{i+1}^{\alpha}, i = 1, 2, \cdots, N$. If

$$\min_{1 \le i \le N} J\left(X_t^{\alpha}\right) < z,$$

then set $\alpha_1 = \alpha$. Otherwise, set $\alpha_2 = \alpha$.

Step 7: If $|\alpha_2 - \alpha_1| \le \varepsilon$, then return α and stop. Otherwise, go to Step 5.

Theorem 16. *Let X_t be the solution of an uncertain differential equation $\mathrm{d}X_t = f(t, X_t)\mathrm{d}t + g(t, X_t)\mathrm{d}C_t$ with an initial value X_0. Assume X_t has an uncertainty distribution $\Phi_t(x)$ at each time t. Then given a strictly increasing function $J(x)$ and a level $z < J(X_0)$, the first hitting time τ_z that $J(X_t)$ reaches z has an uncertainty distribution*

$$\Psi(s) = \sup_{0 \le t \le s} \Phi_t\left(J^{-1}(z)\right).$$

Proof. Since the event $\{\tau_z \le s\}$ is equivalent to the event

$$\left\{\inf_{0 \le t \le s} J(X_t) \le z\right\}$$

provided $z < J(X_0)$, it follows from Theorem 10 that

$$\begin{aligned} \Psi(s) &= \mathcal{M}\left\{\inf_{0 \le t \le s} J(X_t) \le z\right\} \\ &= \mathcal{M}\left\{\inf_{0 \le t \le s} X_t \le J^{-1}(z)\right\} \\ &= \sup_{0 \le t \le s} \Phi_t\left(J^{-1}(z)\right). \end{aligned}$$

This completes the proof. □

First hitting time of strictly decreasing function of the solution

Theorem 17. *Let X_t and X_t^{α} be the solution and α-path of the uncertain differential equation*

$$\mathrm{d}X_t = f(t, X_t)\mathrm{d}t + g(t, X_t)\mathrm{d}C_t$$

with an initial value X_0, respectively. Given a strictly decreasing function $J(x)$, and a level $z > J(X_0)$, the first hitting time τ_z that $J(X_t)$ reaches z has an uncertainty distribution

$$\Psi(s) = \sup\left\{\alpha \in (0,1) \,\middle|\, \sup_{0 \le t \le s} J\left(X_t^{\alpha}\right) \ge z\right\}.$$

Proof. Write

$$\alpha_0 = \sup\left\{\alpha \in (0,1) \,\middle|\, \sup_{0 \le t \le s} J\left(X_t^{\alpha}\right) \ge z\right\}.$$

Since $J(x)$ is a strictly decreasing function, we have

$$\{\tau_z \le s\} \supset \{J(X_t) \ge J\left(X_t^{\alpha_0}\right), \forall t\} = \{X_t \le X_t^{\alpha_0}, \forall t\},$$

$$\{\tau_z > s\} \supset \{J(X_t) < J\left(X_t^{\alpha_0}\right), \forall t\} = \{X_t > X_t^{\alpha_0}, \forall t\}.$$

By Theorem 4 and the monotonicity of uncertain measure, we have

$$\mathcal{M}\{\tau_z \le s\} \ge \mathcal{M}\{X_t \le X_t^{\alpha_0}, \forall t\} = \alpha_0,$$

$$\mathcal{M}\{\tau_z > s\} \ge \{X_t > X_t^{\alpha_0}, \forall t\} = 1 - \alpha_0.$$

It follows from the duality axiom of uncertain measure that

$$\mathcal{M}\{\tau_z \le s\} = \alpha_0.$$

This completes the proof. □

Theorem 18. *Let X_t be the solution of an uncertain differential equation $\mathrm{d}X_t = f(t, X_t)\mathrm{d}t + g(t, X_t)\mathrm{d}C_t$ with an initial value X_0. Assume X_t has an uncertainty distribution $\Phi_t(x)$ at each time t. Then given a strictly decreasing function $J(x)$ and a level $z > J(X_0)$, the first hitting time τ_z that $J(X_t)$ reaches z has an uncertainty distribution*

$$\Psi(s) = \sup_{0 \le t \le s} \Phi_t\left(J^{-1}(z)\right).$$

Proof. Since the event $\{\tau_z \le s\}$ is equivalent to the event

$$\left\{\sup_{0 \le t \le s} J(X_t) \ge z\right\}$$

provided $z > J(X_0)$, it follows from Theorem 10 that

$$\begin{aligned}\Psi(s) &= \mathcal{M}\left\{\sup_{0 \le t \le s} J(X_t) \ge z\right\}\\ &= \mathcal{M}\left\{\inf_{0 \le t \le s} X_t \le J^{-1}(z)\right\}\\ &= \sup_{0 \le t \le s} \Phi_t\left(J^{-1}(z)\right).\end{aligned}$$

This completes the proof. □

Theorem 19. *Let X_t and X_t^α be the solution and α-path of the uncertain differential equation*

$$\mathrm{d}X_t = f(t, X_t)\mathrm{d}t + g(t, X_t)\mathrm{d}C_t$$

with an initial value X_0, respectively. Given a strictly decreasing function $J(x)$ and a level $z < J(X_0)$, the first hitting time τ_z that $J(X_t)$ reaches z has an uncertainty distribution

$$\Upsilon(s) = 1 - \inf\left\{\alpha \in (0,1) \,\middle|\, \inf_{0 \le t \le s} J\left(X_t^\alpha\right) \le z\right\}.$$

Proof. Write

$$\alpha_0 = \inf\left\{\alpha \in (0,1) \,\middle|\, \inf_{0 \le t \le s} J\left(X_t^\alpha\right) \le z\right\}.$$

Then

$$\{\tau_z \le s\} \supset \{J(X_t) \le J\left(X_t^{\alpha_0}\right), \forall t\} = \{X_t \ge X_t^{\alpha_0}, \forall t\},$$

$$\{\tau_z > s\} \supset \{J(X_t) > J\left(X_t^{\alpha_0}\right), \forall t\} = \{X_t < X_t^{\alpha_0}, \forall t\}.$$

By Theorem 4 and the monotonicity of uncertain measure, we have

$$\mathcal{M}\{\tau_z \leq s\} \geq \mathcal{M}\{X_t \geq X_t^{\alpha_0}, \forall t\} = 1 - \alpha_0,$$

$$\mathcal{M}\{\tau_z > s\} \geq \{X_t < X_t^{\alpha_0}, \forall t\} = \alpha_0.$$

It follows from the duality axiom of uncertain measure that

$$\mathcal{M}\{\tau_z \leq s\} = 1 - \alpha_0.$$

This completes the proof. □

Theorem 20. *Let X_t be the solution of an uncertain differential equation $\mathrm{d}X_t = f(t, X_t)\mathrm{d}t + g(t, X_t)\mathrm{d}C_t$ with an initial value X_0. Assume X_t has an uncertainty distribution $\Phi_t(x)$ at each time t. Then given a strictly decreasing function $J(x)$ and a level $z < J(X_0)$, the first hitting time τ_z that $J(X_t)$ reaches z has an uncertainty distribution*

$$\Psi(s) = 1 - \inf_{0\leq t\leq s} \Phi_t\left(J^{-1}(z)\right).$$

Proof. Since the event $\{\tau_z \leq s\}$ is equivalent to the event

$$\left\{\inf_{0\leq t\leq s} J(X_t) \leq z\right\}$$

provided $z < J(X_0)$, it follows from Theorem 6 that

$$\begin{aligned}\Psi(s) &= \mathcal{M}\left\{\inf_{0\leq t\leq s} J(X_t) \leq z\right\} \\ &= \mathcal{M}\left\{\sup_{0\leq t\leq s} X_t \geq J^{-1}(z)\right\} \\ &= 1 - \inf_{0\leq t\leq s} \Phi_t\left(J^{-1}(z)\right).\end{aligned}$$

This completes the proof. □

Integral

In this section, we study the integral of the solution of an uncertain differential equation, and give its uncertainty distribution. Besides, we design a numerical method to obtain the uncertainty distribution.

Theorem 21. *Let X_t and X_t^α be the solution and α-path of the uncertain differential equation*

$$\mathrm{d}X_t = f(t, X_t)\mathrm{d}t + g(t, X_t)\mathrm{d}C_t,$$

respectively. Assume $J(x)$ is a strictly increasing function. Then the integral

$$\int_0^s J(X_t)\mathrm{d}t$$

has an inverse uncertainty distribution

$$\Psi^{-1}(\alpha) = \int_0^s J\left(X_t^\alpha\right)\mathrm{d}t.$$

Proof. Since $J(x)$ is a strictly increasing function, we have

$$\left\{\int_0^s J(X_t)\mathrm{d}t \leq \int_0^s J\left(X_t^\alpha\right)\mathrm{d}t\right\} \supset \left\{J(X_t) \leq J\left(X_t^\alpha\right), \forall t\right\} = \left\{X_t \leq X_t^\alpha, \forall t\right\}$$

and

$$\left\{\int_0^s J(X_t)\mathrm{d}t > \int_0^s J\left(X_t^\alpha\right)\mathrm{d}t\right\} \supset \left\{J(X_t) > J\left(X_t^\alpha\right), \forall t\right\} = \left\{X_t > X_t^\alpha, \forall t\right\}.$$

By Theorem 4 and the monotonicity of uncertain measure, we have

$$\mathcal{M}\left\{\int_0^s J(X_t)\mathrm{d}t \le \int_0^s J\left(X_t^\alpha\right)\mathrm{d}t\right\} \ge \mathcal{M}\left\{X_t \le X_t^\alpha, \forall t\right\} = \alpha$$

and

$$\mathcal{M}\left\{\int_0^s J(X_t)\mathrm{d}t > \int_0^s J\left(X_t^\alpha\right)\mathrm{d}t\right\} \ge \mathcal{M}\left\{X_t > X_t^\alpha, \forall t\right\} = 1 - \alpha.$$

It follows from the duality axiom of uncertain measure that

$$\mathcal{M}\left\{\int_0^s J(X_t)\mathrm{d}t \le \int_0^s J\left(X_t^\alpha\right)\mathrm{d}t\right\} = \alpha.$$

In other words, the integral

$$\int_0^s J(X_t)\mathrm{d}t$$

has an inverse uncertainty distribution

$$\Psi_s^{-1}(\alpha) = \int_0^s J\left(X_t^\alpha\right)\mathrm{d}t.$$

□

Example 1. Let X_t and X_t^α be the solution and α-path of the uncertain differential equation $\mathrm{d}X_t = f(t, X_t)\mathrm{d}t + g(t, X_t)\mathrm{d}C_t$, respectively. Consider a function $h(t, x) = \exp(-rt)\,x$. Since $h(t, x)$ is strictly increasing with respect to x, the integral

$$\int_0^s h(t, X_t)\mathrm{d}t = \int_0^s \exp(-rt)X_t\mathrm{d}t$$

has an inverse uncertainty distribution

$$\Psi_s^{-1}(\alpha) = \int_0^s h\left(t, X_t^\alpha\right)\mathrm{d}t = \int_0^s \exp(-rt)X_t^\alpha\mathrm{d}t.$$

When $J(x)$ is a strictly increasing function, in order to calculate the uncertainty distribution of the integral of $J(X_t)$, we design a numerical method as below.

Step 1: Fix α in $(0, 1)$, and fix h as the step length. Set $i = 0, N = s/h$, and $X_0^\alpha = X_0$.

Step 2: Employ the recursion formula

$$X_{i+1}^\alpha = X_i^\alpha + f\left(t_i, X_i^\alpha\right)h + g\left(t_i, X_i^\alpha\right)\Phi^{-1}(\alpha)h,$$

and calculate X_{i+1}^α and $J\left(X_{i+1}^\alpha\right)$.

Step 3: Set $i \leftarrow i + 1$.

Step 4: Repeat Step 2 and Step 3 for N times.

Step 5: The inverse uncertainty distribution of

$$\int_0^s J(X_t)\mathrm{d}t$$

is determined by

$$\Psi_s^{-1}(\alpha) = \sum_{i=1}^N J\left(X_i^\alpha\right)h.$$

Theorem 22. *Let X_t and X_t^α be the solution and α-path of the uncertain differential equation*

$$\mathrm{d}X_t = f(t, X_t)\mathrm{d}t + g(t, X_t)\mathrm{d}C_t,$$

respectively. Assume $J(x)$ is a strictly decreasing function. Then the integral

$$\int_0^s J(X_t)\mathrm{d}t$$

has an inverse uncertainty distribution

$$\Upsilon_s^{-1}(\alpha) = \int_0^s J\left(X_t^{1-\alpha}\right)\mathrm{d}t.$$

Proof. Since $J(x)$ is a strictly decreasing function, we have

$$\left\{\int_0^s J(X_t)\mathrm{d}t \le \int_0^s J\left(X_t^{1-\alpha}\right)\mathrm{d}t\right\} \supset \left\{J(X_t) \le J\left(X_t^{1-\alpha}\right), \forall t\right\}$$
$$= \left\{X_t \ge X_t^{1-\alpha}, \forall t\right\}$$

and

$$\left\{\int_0^s J(X_t)\mathrm{d}t > \int_0^s J\left(X_t^{1-\alpha}\right)\mathrm{d}t\right\} \supset \left\{J(X_t) > J\left(X_t^{1-\alpha}\right), \forall t\right\}$$
$$= \left\{X_t < X_t^{1-\alpha}, \forall t\right\}.$$

By Theorem 4 and the monotonicity of uncertain measure, we have

$$\mathcal{M}\left\{\int_0^s J(X_t)\mathrm{d}t \le \int_0^s J\left(X_t^{1-\alpha}\right)\mathrm{d}t\right\} \ge \mathcal{M}\left\{X_t \ge X_t^{1-\alpha}, \forall t\right\} = \alpha$$

and

$$\mathcal{M}\left\{\int_0^s J(X_t)\mathrm{d}t > \int_0^s J\left(X_t^{1-\alpha}\right)\mathrm{d}t\right\} \ge \mathcal{M}\left\{X_t < X_t^{1-\alpha}, \forall t\right\} = 1 - \alpha.$$

It follows from the duality axiom of uncertain measure that

$$\mathcal{M}\left\{\int_0^s J(X_t)\mathrm{d}t \le \int_0^s J\left(X_t^{1-\alpha}\right)\mathrm{d}t\right\} = \alpha.$$

In other words, the integral

$$\int_0^s J(X_t)\mathrm{d}t$$

has an inverse uncertainty distribution

$$\Upsilon_s^{-1}(\alpha) = \int_0^s J\left(X_t^{1-\alpha}\right)\mathrm{d}t.$$

□

Conclusions

This paper considered the solution of an uncertain differential equation, and gave the uncertainty distributions of its extreme values, first hitting time, and integral. In addition, we designed some numerical methods to obtain the uncertainty distributions.

Acknowledgements

This work was supported by National Natural Science Foundation of China Grants No.61273044.

References

1. Kahneman, D, Tversky, A: Prospect theory: An analysis of decision under risk. Econometrica. **47**(2), 263–292 (1979)
2. Liu, B: Why is there a need for uncertainty theory. J. Uncertain Syst. **6**(1), 3–10 (2012)
3. Liu, B: Uncertainty Theory. (2nd ed). Springer, Berlin (2007)
4. Liu, B: Uncertainty Theory: A Branch of Mathematics for Modeling Human Uncertainty. Springer, Berlin (2010)
5. Liu, B: Theory and Practice of Uncertain Programming. (2nd ed). Springer, Berlin (2009)
6. Liu, B: Uncertain risk analysis and uncertain reliability analysis. J. Uncertain Syst. **4**(3), 163–170 (2010)
7. Liu, B: Uncertain set theory and uncertain inference rule with application to uncertain control. J. Uncertain Syst. **4**(2), 83–98 (2010)
8. Liu, B: Uncertain logic for modeling human language. J. Uncertain Syst. **5**(1), 3–20 (2011)
9. Liu, B: Fuzzy process, hybrid process and uncertain process. J. Uncertain Syst. **2**(1), 3–16 (2008)
10. Liu, B: Some research problems in uncertainty theory. J. Uncertain Syst. **3**(1), 3–10 (2009)
11. Liu, B, Yao, K: Uncertain integral with respect to multiple canonical processes. J. Uncertain Syst. **6**(4), 249–254 (2012)
12. Chen, X, Ralescu, DA: Liu process and uncertain calculus. J. Uncertainty Anal. Appl. **1** (2013). Article 3
13. Yao, K: Uncertain calculus with renewal process. Fuzzy Optimization and Decis. Mak. **11**(3), 285–297 (2012)
14. Chen, X, Liu, B: Existence and uniqueness theorem for uncertain differential equations. Fuzzy Optimization and Decis. Mak. **9**(1), 69–81 (2010)
15. Liu, YH: An analytic method for solving uncertain differential equations. J. Uncertain Syst. **6**(4), 243–248 (2012)
16. Yao, K: A type of nonlinear uncertain differential equations with analytic solution (2013). http://orsc.edu.cn/online/110928.pdf
17. Yao, K, Chen, X: A numerical method for solving uncertain differential equations. J. Intell. and Fuzzy Syst. **25**(3), 825–832 (2013). doi:10.3233/IFS-120688
18. Barbacioru, I: Uncertainty functional differential equations for finance. Surv. Math. Appl. **5**, 275–284 (2010)
19. Liu, HJ, Fei, WY: Neutral uncertain delay differential equations. Inf.: An Int. Interdiscip. J. **16**(2), 1225–1232 (2013)
20. Ge, X, Zhu, Y: Existence and uniqueness theorem for uncertain delay differential equations. J. Comput. Inf. Syst. **8**(20), 8341–8347 (2012)
21. Ge, X, Zhu, Y: A necessary condition of optimality for uncertain optimal control problem. Fuzzy Optimization and Decis. Mak. **12**(1), 41–51 (2013)
22. Liu, B: Toward uncertain finance theory. J. Uncertainty Anal. Appl. **1** (2013). Article 1
23. Chen, X: American option pricing formula for uncertain financial market. Int. J. Oper. Res. **8**(2), 32–37 (2011)
24. Peng, J, Yao, K: A new option pricing model for stocks in uncertainty markets. Int. J. Oper. Res. **8**(2), 18–26 (2011)
25. Chen, X, Liu, YH, Ralescu, DA: Uncertain stock model with periodic dividends. Fuzzy Optimization and Decis. Mak. **12**(1), 111–123 (2013b)
26. Chen, X, Gao, J: Uncertain term structure model of interest rate. Soft Computing. **17**(4), 597–604 (2013)
27. Liu, YH, Ralescu, DA, Chen, X: Uncertain currency model and currency option pricing. International Journal of Intelligent Systems, to be published
28. Zhu, Y: Uncertain optimal control with application to a portfolio selection model. Cybern. Syst. **41**(7), 535–547 (2010)
29. Gao, Y: Existence and uniqueness theorem on uncertain differential equations with local Lipschitz condition. J. Uncertain Syst. **6**(3), 223–232 (2012)
30. Yao, K, Gao, J, Gao, Y: Some stability theorems of uncertain differential equation. Fuzzy Optimization and Decis. Mak. **12**(1), 3–13 (2013)
31. Liu, YH, Ha, MH: Expected value of function of uncertain variables. J. Uncertain Syst. **4**(3), 181–186 (2010)
32. Liu, B: Extreme value theorems of uncertain process with application to insurance risk model. Soft Comput. **17**(4), 549–556 (2013)

17

Multifactor uncertain differential equation

Shengguo Li[1], Jin Peng[1*] and Bo Zhang[2]

*Correspondence:
pengjin01@tsinghua.org.cn
[1] Institute of Uncertain Systems, Huanggang Normal University, Hubei 438000, China
Full list of author information is available at the end of the article

Abstract

This paper proposes a type of multifactor uncertain differential equation within the framework of uncertainty theory. The analytic solutions of four special types of multifactor uncertain differential equations are first discussed. Then, a numerical method for solving general multifactor uncertain differential equation is presented. Finally, under the Lipschitz condition and linear growth condition, it is proved that the multifactor uncertain differential equation has a unique solution.

Keywords: Uncertainty theory; Canonical process; Uncertainty differential equation; Numerical method

Introduction

Uncertainty theory is a tool to study the indeterminacy phenomena in human systems, which was founded by Liu [1] in 2007. It was refined by Liu [2] and has become an axiomatic system via normality, duality, subadditivity, and product axioms of uncertain measure. Up to now, many branches of mathematics emerged based on uncertainty theory, such as mathematical programming [3], uncertain set and uncertain inference [4], uncertain logic [5], uncertain risk [6,7], and uncertain insurance [8].

Uncertain process is essentially a sequence of uncertain variables indexed by time which was first introduced by Liu [9]. After that, a significant uncertain process called canonical process was designed by [10]. The canonical process is a stationary independent increment process with Lipschitz continuous sample paths. Meanwhile, uncertain calculus with respect to canonical process called Liu calculus was developed by Liu [10]. In order to describe the evolution of uncertain phenomenon with some jumps, Liu [9] proposed the uncertain renewal process. Afterward, Yao [11] presented the uncertain calculus with respect to renewal process called the Yao calculus. Recently, Yao [12] proposed multidimensional uncertain calculus with Liu process, Chen [13] studied the uncertain calculus with finite variation processes. More research about uncertain process can be found in references [14-16].

Uncertain differential equation was proposed by Liu [9], which is an important tool to deal with uncertain dynamic systems. Different from stochastic differential equation driven by a Wiener process [17], uncertain differential equation is a type of differential equation driven by uncertain process. In order to know well uncertain differential equation, many researchers did a lot of work. Chen and Liu [18] proved an existence and uniqueness theorem of solution under global Lipschitz condition and proposed an analytic solution for linear uncertain differential equation. Gao [19] gave an existence and uniqueness theorem with local Lipschitz condition. In 2009, Liu [10] gave a concept of stability of uncertain differential equation. After that, Yao et al. [20] proved some stability theorems of uncertain differential equation. In addition, Sheng and Wang [21] investigated the stability in pth moment for uncertain differential equation, Liu et al. [22] studied the almost sure stability, and Yao et al. [23] showed the stability in mean. In order to obtain the solution of uncertain differential equation, Liu [24] and Yao [25] provided the analytic solutions for some special nonlinear uncertain differential equations, respectively. Yao and Chen [26] presented a numerical method for solving uncertain differential equation when it is difficult to obtain analytic solution. Yao [27] also discussed the extreme values and integral of solution of uncertain differential equation.

Uncertain differential equation was first applied in finance by Liu [10] in 2009. Meanwhile, Liu [10] presented an uncertain stock model in uncertain financial market and proved the European option pricing formulas. After that, Chen [28] gave the America option pricing formulas. Besides, Peng and Yao [29] presented another uncertain stock model and corresponding option pricing formulas. Liu [30] discussed some possible applications of uncertain differential equations to financial markets. Li and Peng [31] proposed a stock model with uncertain stock diffusion. Liu et al. [32] built an uncertain currency model and proved the currency option pricing. Jiao and Yao [33] considered an interest rate model in uncertain environment. Yao [34] proved a no-arbitrage theorem for uncertain stock model. In addition, uncertain differential equation was also applied in uncertain optimal control [35] and uncertain differential game [36].

The extensions of uncertain differential equation also attracted the attention of scholars. Several recent contributions in the extension literature have studied this question in many directions. Yao [11] suggested the uncertain differential equation with jumps. Ge and Zhu [37] discussed the backward uncertain differential equation. Barbacioru [38], Ge and Zhu [39], and Liu and Fei [40] focused on the uncertain delay differential equation. Yao [12] proposed the multidimentional uncertain differential equation via multidimensional uncertain calculus. Ji and Zhou [41] proved an existence and uniqueness theorem of solution for multidimensional uncertain differential equation. Yao [42] studied the higher order uncertain differential equation.

Usually, the uncertain factor influencing dynamic systems is not alone. In 2012, Liu and Yao [43] extended uncertain integral from single canonical process to multiple ones. This provides a motivation to consider the concept of uncertain differential equation driven by multiple uncertain processes. In this paper, we present a type of uncertain differential equation driven by multiple canonical processes which can be regarded as a generalization of the uncertain differential equation proposed by Liu [9].

The rest of the paper is organized as follows. Some preliminary concepts of uncertainty theory and uncertain calculus are recalled in the 'Preliminary' section. After that, the multifactor uncertain differential equation is presented. Following that, a numerical method is introduced. In addition, an existence and uniqueness theorem is proved. Finally, a brief summary is given.

Preliminary

In this section, uncertainty theory and uncertain calculus are introduced and some basic concepts are given.

Uncertainty theory

Let Γ be a nonempty set and $\mathcal{L}$ a σ-algebra over Γ. Each element Λ in $\mathcal{L}$ is called an event. A set function $\mathcal{M}$ from $\mathcal{L}$ to $[0, 1]$ ia called uncertain measure if it satisfies the following axioms:

(1) (Normality axiom) $\mathcal{M}\{\Gamma\} = 1$ for the universal set Γ;
(2) (Duality axiom) $\mathcal{M}\{\Lambda\} + \mathcal{M}\{\Lambda^c\} = 1$ for any $\Lambda \in \mathcal{L}$;
(3) (Subadditivity axiom) for every countable sequence of events $\Lambda_1, \Lambda_2, \cdots$, we have:

$$\mathcal{M}\left\{\bigcup_{i=1}^{\infty} \Lambda_i\right\} \leq \sum_{i=1}^{\infty} \mathcal{M}\{\Lambda_i\}.$$

The triplet $(\Gamma, \mathcal{L}, \mathcal{M})$ is called an uncertain space. In order to obtain an uncertain measure of compound event, Liu [10] defined a product uncertain measure which produces the fourth axiom of uncertainty theory:

(4) (Product axiom) Let $(\Gamma_k, \mathcal{L}_k, \mathcal{M}_k)$ be uncertain spaces for $k = 1, 2, \cdots$ The product uncertain measure $\mathcal{M}$ is an uncertain measure on the product σ-algebra $\mathcal{L}_1 \times \mathcal{L}_2 \times \cdots$ satisfying:

$$\mathcal{M}\left\{\prod_{k=1}^{\infty} \Lambda_k\right\} = \min_{1 \leq k \leq \infty} \mathcal{M}_k\{\Lambda_k\}$$

where Λ_k are arbitrarily chosen events from $\mathcal{L}_k$ for $k = 1, 2, \cdots$, respectively.

An uncertain variable is defined as a measurable function from an uncertain space $(\Gamma, \mathcal{L}, \mathcal{M})$ to the set of real numbers, i.e., for any Borel set B of real numbers, the set:

$$\{\xi \in B\} = \{\gamma \in \Gamma | \xi(\gamma) \in B\}$$

is an event.

The uncertainty distribution $\Phi : \Re \to [0, 1]$ of an uncertain variable ξ is defined by Liu [1] as:

$$\Phi(x) = \mathcal{M}\left\{\gamma \in \Gamma \mid \xi(\gamma) \leq x\right\},$$

and the inverse function Φ^{-1} is called the inverse uncertainty distribution of ξ.

An uncertain variable ξ is called normal if it has a normal uncertainty distribution:

$$\Phi(x) = \left(1 + \exp\left(\frac{\pi(e - x)}{\sqrt{3}\sigma}\right)\right)^{-1}, \quad x \in \Re$$

denoted by $\mathcal{N}(e, \sigma)$ where e and σ are real numbers with $\sigma > 0$.

The expected value of uncertain variable ξ is defined by Liu [1] as:

$$E[\xi] = \int_0^{+\infty} \mathcal{M}\{\xi \geq r\} \mathrm{d}r - \int_{-\infty}^{0} \mathcal{M}\{\xi \leq r\} \mathrm{d}r$$

provided that at least one of the two integrals is finite. The variance of ξ is defined as $V[\xi] = E[(\xi - E[\xi])^2]$.

Let $\xi_1, \xi_2, \cdots, \xi_n$ be independent uncertain variables with uncertainty distributions $\Phi_1, \Phi_2, \cdots, \Phi_n$, respectively. Liu [2] proved that if $f(x_1, x_2, \cdots, x_n)$ is a strictly increasing function with respect to $x_1, x_2, \cdots, x_m$ and strictly decreasing with respect to $x_{m+1}, x_{m+2}, \cdots, x_n$, then $\xi = f(\xi_1, \xi_2, \cdots, \xi_n)$ is an uncertain variable with inverse uncertainty distribution:

$$\Psi^{-1}(\alpha) = f(\Phi_1^{-1}(\alpha), \cdots, \Phi_m^{-1}(\alpha), \Phi_{m+1}^{-1}(1-\alpha), \cdots, \Phi_n^{-1}(1-\alpha)).$$

Furthermore, the expected value of uncertain variable $\xi = f(\xi_1, \xi_2, \cdots, \xi_n)$ was obtained by Liu and Ha [44] as follows:

$$E[\xi] = \int_0^1 f(\Phi_1^{-1}(\alpha), \cdots, \Phi_m^{-1}(\alpha), \Phi_{m+1}^{-1}(1-\alpha), \cdots, \Phi_n^{-1}(1-\alpha))d\alpha.$$

Uncertain calculus

Definition 1. *(Liu [9]) Let T be an index set and let $(\Gamma, \mathcal{L}, \mathcal{M})$ be an uncertain space. An uncertain process is a measurable function from $T \times (\Gamma, \mathcal{L}, \mathcal{M})$ to the set of real numbers, i.e., for each $t \in T$ and any Borel set B of real numbers, the set:*

$$\{X_t \in B\} = \{\gamma \in \Gamma \mid X_t(\gamma) \in B\}$$

is an event.

Definition 2. *(Liu [10]) An uncertain process C_t is said to be a canonical process if*
(i) $C_0 = 0$ and almost all sample paths are Lipschitz continuous;
(ii) C_t has stationary and independent increments;
(iii) every increment $C_{s+t} - C_t$ is a normal uncertain variable with expected value 0 and variance t^2.

Definition 3. *(Liu [10]) Let X_t be an uncertain process and C_t be a canonical process. For any partition of closed integral $[a, b]$ with $a = t_1 < t_2 < \cdots < t_{k+1} = b$, the mesh is written as:*

$$\Delta = \max_{1 \le i \le k} \mid t_{i+1} - t_i \mid .$$

Then, Liu integral of X_t with respect to C_t is:

$$\int_a^b X_t dC_t = \lim_{\Delta \to 0} \sum_{i=1}^{k} X_{t_i}(C_{t_{i+1}} - C_{t_i})$$

provided that the limit exists almost surely and is finite. In this case, the uncertain process X_t is said to be integrable.

Example 1. *Let $f(t)$ be a continuous function with respect to t. Then, the uncertain integral:*

$$\int_0^s f(t)dC_t$$

is a normal uncertain variable at each time s, and:

$$\int_0^s f(t)dC_t \sim \mathcal{N}\left(0, \int_0^s \mid f(t) \mid dt\right).$$

Definition 4. *(Liu [9]) Suppose C_t is a canonical process, and f, g are some given functions. Then,*

$$dX_t = f(t, X_t)dt + g(t, X_t)dC_t \tag{1}$$

is called an uncertain differential equation.

The uncertain differential with respect to canonical processes $C_{1t}, C_{2t}, \cdots, C_{nt}$ is defined by Liu and Yao [43] as follows.

Definition 5. *(Liu and Yao [43]) Let $C_{1t}, C_{2t}, \cdots, C_{nt}$ be canonical processes and let Z_t be an uncertain process. If there exist uncertain processes μ_t and $\sigma_{1t}, \sigma_{2t}, \cdots, \sigma_{nt}$ such that:*

$$Z_t = Z_0 + \int_0^t \mu_s ds + \sum_{i=1}^{n} \int_0^t \sigma_{is} dC_{is} \tag{2}$$

for any $t \geq 0$, then, we say Z_t has an uncertain differential:

$$dZ_t = \mu_t dt + \sum_{i=1}^{n} \sigma_{it} dC_{it}. \tag{3}$$

In this case, Z_t is called a differentiable uncertain process with drift μ_t and diffusions $\sigma_{1t}, \sigma_{2t}, \cdots, \sigma_{nt}$.

Theorem 1. *(Liu and Yao [43]) (Fundamental Theorem of Uncertain Calculus) Let $C_{1t}, C_{2t}, \cdots, C_{nt}$ be canonical processes. If $h(t, c_1, c_2, \cdots, c_n)$ is a continuously differentiable function, then the uncertain process $Z_t = h(t, C_{1t}, C_{2t}, \cdots, C_{nt})$ is differentiable and has an uncertain differential:*

$$dZ_t = \frac{\partial h}{\partial t}(t, C_{1t}, C_{2t}, \cdots, C_{nt})dt + \sum_{i=1}^{n} \frac{\partial h}{\partial c_i}(t, C_{1t}, C_{2t}, \cdots, C_{nt})dC_{it}.$$

Multifactor uncertain differential equation

Usually, the uncertain factor influencing dynamic systems is not alone. In order to model the dynamic systems with multiple factors, this section will extend the uncertain differential equation driven by single canonical process to one driven by multiple independent canonical processes.

Definition 6. *(Liu [45]) Uncertain processes $X_{1t}, X_{2t}, \cdots, X_{nt}$ are said to be independent if for any positive integer k and any times $t_1, t_1, \cdots, t_k$, the uncertain vectors:*

$$\boldsymbol{\xi}_i = \left(X_{it_1}, X_{it_2}, \cdots, X_{it_k}\right), i = 1, 2, \cdots, n$$

are independent, i.e., for any Borel sets $B_1, B_2, \cdots, B_n$ of k-dimensional real vectors, we have:

$$\mathcal{M}\left\{\bigcap_{i=1}^{n}(\boldsymbol{\xi}_i \in B_i)\right\} = \bigwedge_{i=1}^{n} \mathcal{M}\{\boldsymbol{\xi}_i \in B_i\}.$$

Theorem 2. *(Liu [45]) Let $X_{1t}, X_{2t}, \cdots, X_{nt}$ be independent uncertain processes with regular uncertainty distributions $\Phi_{1t}, \Phi_{2t}, \cdots, \Phi_{nt}$, respectively. If the function $f(x_1, x_2, \cdots, x_n)$ is strictly increasing with respect to $x_1, x_2, \cdots, x_m$ and strictly decreasing with respect to $x_{m+1}, x_{m+2}, \cdots, x_n$, then:*

$$X_t = f(X_{1t}, X_{2t}, \cdots, X_{nt})$$

is an uncertain variable with inverse uncertainty distribution:

$$\Phi_t^{-1}(\alpha) = f(\Phi_{1t}^{-1}(\alpha), \cdots, \Phi_{mt}^{-1}(\alpha), \Phi_{m+1,t}^{-1}(1-\alpha), \cdots, \Phi_{nt}^{-1}(1-\alpha)).$$

Definition 7. *Suppose $C_{1t}, C_{2t}, \cdots, C_{nt}$ are independent canonical processes, and f, $g_1, g_2, \cdots, g_n$ are some given functions. Then:*

$$dX_t = f(t, X_t)dt + \sum_{i=1}^{n} g_i(t, X_t)dC_{it} \tag{4}$$

is called an uncertain differential equation with respect to $C_{1t}, C_{2t}, \cdots, C_{nt}$. A solution is an uncertain process X_t that satisfies Equation 4 identically in t.

The uncertain differential Equation 4 is equivalent to the uncertain integral equation:

$$X_s = X_0 + \int_0^s f(t, X_t)dt + \sum_{i=1}^{n} \int_0^s g_i(t, X_t)dC_{it}. \tag{5}$$

Example 2. *Let a, b and c be real numbers, and let C_{1t}, C_{2t} be independent canonical processes. The uncertain differential equation:*

$$dX_t = adt + bdC_{1t} + cdC_{2t} \tag{6}$$

has a solution:

$$X_t = X_0 + at + bC_{1t} + cC_{2t}. \tag{7}$$

Theorem 3. *Let μ_t, $\nu_{1t}, \nu_{2t}, \cdots, \nu_{nt}$ be integrable uncertain processes and let $C_{1t}, C_{2t}, \cdots, C_{nt}$ be independent canonical processes. Then, the uncertain differential equation:*

$$dX_t = \mu_t X_t dt + \sum_{i=1}^{n} \nu_{it} X_t dC_{it} \tag{8}$$

has a solution:

$$X_t = X_0 \exp\left(\int_0^t \mu_s ds + \sum_{i=1}^{n} \int_0^t \nu_{is} dC_{is}\right). \tag{9}$$

Proof. At first, the original uncertain differential equation is equivalent to:

$$\frac{dX_t}{X_t} = \mu_t dt + \sum_{i=1}^{n} \nu_{it} dC_{it}.$$

It follows from the fundamental theorem of uncertain calculus that:

$$d\ln X_t = \frac{dX_t}{X_t} = \mu_t dt + \sum_{i=1}^{n} \nu_{it} dC_{it}$$

and then:

$$\ln X_t = \ln X_0 + \int_0^t \mu_s ds + \sum_{i=1}^{n} \int_0^t \nu_{is} dC_{is}.$$

Therefore, the uncertain differential Equation 8 has a solution (9).

Example 3. *Let a, b, and c be real numbers, and let C_{1t} and C_{2t} be independent canonical processes. The uncertain differential equation:*

$$dX_t = aX_t dt + bX_t dC_{1t} + cX_t dC_{2t} \tag{10}$$

has a solution:

$$X_t = X_0 \exp\left(at + bC_{1t} + cC_{2t}\right). \tag{11}$$

Theorem 4. *Let μ_{1t}, μ_{2t}, $\nu_{1t}, \nu_{2t}, \cdots, \nu_{nt}$ and $\omega_{1t}, \omega_{2t}, \cdots, \omega_{nt}$ be integrable uncertain processes. Assume $C_{1t}, C_{2t}, \cdots, C_{nt}$ are independent canonical processes, then the uncertain differential equation:*

$$dX_t = (\mu_{1t}X_t + \mu_{2t})dt + \sum_{i=1}^{n}(\nu_{it}X_t + \omega_{it})dC_{it} \tag{12}$$

has a solution:

$$X_t = U_t\left(X_0 + \int_0^t \frac{\mu_{2s}}{U_s}ds + \sum_{i=1}^{n}\int_0^t \frac{\omega_{is}}{U_s}dC_{is}\right) \tag{13}$$

where:

$$U_t = \exp\left(\int_0^t \mu_{1s}ds + \sum_{i=1}^{n}\int_0^t \nu_{is}dC_{is}\right). \tag{14}$$

Proof. Define two uncertain processes U_t and V_t via uncertain differential equations,

$$dU_t = \mu_t U_t dt + \sum_{i=1}^{n} \nu_{it}U_t dC_{it},$$

$$dV_t = \frac{\mu_{2t}}{U_t}dt + \sum_{i=1}^{n}\frac{\omega_{it}}{U_t}dC_{it}.$$

It follows from the integration by parts that:

$$d(U_tV_t) = V_t dU_t + U_t dV_t = (\mu_{1t}U_tV_t + \mu_{2t}dt) + \sum_{i=1}^{n}(\nu_{it}U_tV_t + \omega_{it})C_{it}.$$

That is, the uncertain process $X_t = U_tV_t$ is a solution of the uncertain differential Equation (12). Note that:

$$U_t = U_0 \exp\left(\int_0^t \mu_{1s}ds + \sum_{i=1}^{n}\int_0^t \nu_{is}dC_{is}\right),$$

$$V_t = V_0 + \int_0^t \frac{\mu_{2s}}{U_s}dt + \sum_{i=1}^{n}\int_0^t \frac{\omega_{is}}{U_s}dC_{is}.$$

Taking $U_0 = 1$ and $V_0 = X_0$, we get the solutions (13) and (14). The theorem is proved.

Note that $n = 1$, the uncertain differential Equation 12 degenerates to the linear uncertain differential equation in Chen and Liu [18].

Example 4. *Let m, a, σ, and ω be real numbers and let C_{1t} and C_{2t} be independent canonical processes. The uncertain differential equation:*

$$dX_t = (m - aX_t)dt + \sigma dC_{1t} + \omega dC_{2t} \tag{15}$$

has the solution:

$$X_t = \exp(-at)\left(X_0 + \frac{m}{a}(\exp(at) - 1) + \int_0^t \sigma \exp(as)dC_{1s} + \int_0^t \omega \exp(as)dC_{2s}\right) \tag{16}$$

provided that $a \neq 0$.

Example 5. *Let m, σ, and ω be real numbers and let C_{1t} and C_{2t} be independent canonical processes. The uncertain differential equation:*

$$dX_t = mdt + \sigma X_t dC_{1t} + \omega X_t dC_{2t} \tag{17}$$

has the solution:

$$X_t = \exp(\sigma C_{1t} + \omega C_{2t})\left(X_0 + \int_0^t m \exp(\sigma C_{1s} + \omega C_{2s})ds\right). \tag{18}$$

Theorem 5. *Let f be a function of two variables and let $\sigma_{1t}, \sigma_{2t}, \cdots, \sigma_{nt}$ be integrable uncertain processes. Assume $C_{1t}, C_{2t}, \cdots, C_{nt}$ are independent canonical processes, then the uncertain differential equation:*

$$dX_t = f(t, X_t)dt + \sum_{i=1}^{n} \sigma_{it} X_t dC_{it} \tag{19}$$

has a solution:

$$X_t = Y_t^{-1} Z_t \tag{20}$$

where:

$$Y_t = \exp\left(-\sum_{i=1}^{n} \int_0^t \sigma_{is} dC_{is}\right) \tag{21}$$

and Z_t is the solution of uncertain differential equation:

$$dZ_t = Y_t f\left(t, Y_t^{-1} Z_t\right) dt \tag{22}$$

with initial value $Z_0 = X_0$.

Proof. By the fundamental theorem of uncertain calculus, the uncertain process Y_t has an uncertain differential:

$$dY_t = -\exp\left(-\sum_{i=1}^{n} \int_0^t \sigma_{is} dC_{is}\right) \sum_{i=1}^{n} \sigma_{it} dC_{it} = -Y_t \sum_{i=1}^{n} \sigma_{it} dC_{it}.$$

It follows from the integration by parts that:

$$d(X_t Y_t) = X_t dY_t + Y_t dX_t = -X_t Y_t \sum_{i=1}^{n} \sigma_{it} dC_{it} + Y_t f(t, X_t) + X_t Y_t \sum_{i=1}^{n} \sigma_{it} dC_{it}.$$

That is,

$$d(X_t Y_t) = Y_t f(t, X_t).$$

Defining $Z_t = X_t Y_t$, we obtain $X_t = Y_t^{-1} Z_t$ and $dZ_t = Y_t f\left(t, Y_t^{-1} Z_t\right)$. Furthermore, since $Y_0 = 1$, the initial value Z_0 is just X_0. The theorem is proved.

Note that $n = 1$, the uncertain differential Equation 19 degenerates to the nonlinear uncertain differential equation in Liu [24].

Example 6. *Let $\sigma_1, \sigma_2, \cdots, \sigma_n$ be real numbers and let $C_{1t}, C_{2t}, \cdots, C_{nt}$ be independent canonical processes. Consider the uncertain differential equation:*

$$dX_t = f(t, X_t)dt + \sum_{i=1}^{n} \sigma_i X_t dC_{it}. \tag{23}$$

Theorem 5 shows that:

$$Y_t = \exp\left(-\sum_{i=1}^{n} \sigma_i C_{is}\right)$$

and:

$$X_t = \exp\left(\sum_{i=1}^{n} \sigma_i C_{is}\right) Z_t$$

where Z_t is the solution of uncertain differential equation:

$$dZ_t = \exp\left(-\sum_{i=1}^{n} \sigma_i C_{is}\right) f\left(t, \exp\left(\sum_{i=1}^{n} \sigma_i C_{is}\right) Z_t\right) dt$$

with initial value $Z_0 = X_0$. Taking $f(t, X_t) = X_t^{\alpha}$, $\alpha \neq 1$, we can obtain:

$$dZ_t^{1-\alpha} = (1-\alpha) \exp\left((1-\alpha)\sum_{i=1}^{n} \sigma_i C_{it}\right) dt$$

and:

$$X_t = \exp\left(\sum_{i=1}^{n} \sigma_i C_{it}\right) \left(X_0^{1-\alpha} + (1-\alpha)\int_0^t \exp\left((1-\alpha)\sum_{i=1}^{n} \sigma_i C_{is}\right) ds\right)^{\frac{1}{1-\alpha}}.$$

Theorem 6. *Let $g_1, g_2, \cdots, g_n$ be functions of two variables and let α_t be an integrable uncertain process. Assume $C_{1t}, C_{2t}, \cdots, C_{nt}$ are independent canonical processes, then the uncertain differential equation:*

$$dX_t = \alpha_t X_t dt + \sum_{i=1}^{n} g_i(t, X_t) dC_{it} \tag{24}$$

has a solution:

$$X_t = Y_t^{-1} Z_t \tag{25}$$

where:

$$Y_t = \exp\left(-\int_0^t \alpha_s ds\right) \tag{26}$$

and Z_t is the solution of uncertain differential equation:

$$dZ_t = Y_t \sum_{i=1}^{n} g_i\left(t, Y_t^{-1} Z_t\right) dt \tag{27}$$

with initial value $Z_0 = X_0$.

Proof. It follows from the fundamental theorem of uncertain calculus that:

$$dY_t = -\exp\left(-\int_0^t \alpha_s ds\right) \alpha_t dt = -Y_t \alpha_t dt.$$

Using the integration by parts, we have the following:

$$d(X_tY_t) = X_t dY_t + Y_t dX_t = -X_tY_t\alpha_t dt + Y_t\alpha_t X_t dt + Y_t\sum_{i=1}^{n} g_i(t, X_t)dC_{it}.$$

That is,

$$d(X_tY_t) = Y_t\sum_{i=1}^{n} g_i(t, X_t)dC_{it}.$$

Define $Z_t = X_tY_t$, then $X_t = Y_t^{-1}Z_t$ and $dZ_t = Y_t\sum_{i=1}^{n} g_i\left(t, Y_t^{-1}Z_t\right)dC_{it}$. In addition, since $Y_0 = 1$, the initial value Z_0 is just X_0. The theorem is proved.

Note that $n = 1$, the uncertain differential Equation 24 degenerates to the nonlinear uncertain differential equation in Liu [24].

Example 7. *Let α , b, c, and β be real numbers with $\beta \neq 1$, and let $C_{1t}, C_{2t}, \cdots, C_{nt}$ be independent canonical processes. Consider the uncertain differential equation:*

$$dX_t = \alpha X_t dt + bX_t^{\beta}dC_{1t} + cX_t^{\beta}dC_{2t}. \tag{28}$$

At first,

$$Y_t = \exp(-\alpha t)$$

and Z_t satisfies uncertain differential equation:

$$dZ_t = b\exp((\beta - 1)\alpha t)Z_t^{\beta}dC_{1t} + c\exp((\beta - 1)\alpha t)Z_t^{\beta}dC_{2t}.$$

Since $\beta \neq 1$, we have:

$$dZ_t^{1-\alpha} = (1 - \beta)(b\exp((\beta - 1)\alpha t)dC_{1t} + c\exp((\beta - 1)\alpha t)dC_{2t}).$$

It follows from the fundamental theorem of uncertain calculus that:

$$Z_t^{1-\alpha} = Z_0^{1-\alpha} + (1 - \beta)\left(b\int_0^t \exp((\beta - 1)\alpha s)dC_{1s} + c\int_0^t \exp((\beta - 1)\alpha s)dC_{2s}\right).$$

Theorem 6 says the uncertain differential equation has a solution:

$$X_t = \exp(\alpha t)\left(X_0^{1-\alpha} + (1-\beta)(b\int_0^t \exp((\beta - 1)\alpha s)dC_{1s} + c\int_0^t \exp((\beta - 1)\alpha s)dC_{2s}\right)^{\frac{1}{1-\beta}}.$$

Numerical method

However, in many cases, it is difficult to find analytic solutions of uncertain differential equations. Yao and Chen [26] presented a numerical method called Yao-Chen method to obtain the inverse uncertainty distribution of solution.

Yao-Chen formula

Definition 8. *(Yao and Chen [26]) Let α be a number with $0 < \alpha < 1$. An uncertain differential equation*

$$dX_t = f(t, X_t)dt + g(t, X_t)dC_t \tag{29}$$

is said to have an α-path X_t^{α} if it solves the corresponding ordinary differential equation:

$$dX_t^{\alpha} = f\left(t, X_t^{\alpha}\right)dt + \mid g\left(t, X_t^{\alpha}\right)\mid \Phi^{-1}(\alpha)dt \tag{30}$$

where $\Phi^{-1}(\alpha)$ *is the inverse standard normal uncertainty distribution, i.e.,*

$$\Phi^{-1}(\alpha) = \frac{\sqrt{3}}{\pi} \ln \frac{\alpha}{1-\alpha}.$$

Theorem 7. *(Yao-Chen Formula [26]) Assume that* $f, g_1, g_2, \cdots, g_n$ *are continuous functions of two variables. Let* X_t *and* X_t^α *be the solution and* α*-path of the uncertain differential equation:*

$$dX_t = f(t, X_t)dt + g(t, X_t)dC_t,$$

respectively. Then:

$$\mathcal{M}\left\{X_t \leq X_t^\alpha, \forall t\right\} = \alpha,$$

$$\mathcal{M}\left\{X_t > X_t^\alpha, \forall t\right\} = 1 - \alpha.$$

Theorem 8. *(Yao and Chen [26]) Assume that* $f, g_1, g_2, \cdots, g_n$ *are continuous functions of two variables. Let* X_t *and* X_t^α *be the solution and* α*-path of the uncertain differential equation:*

$$dX_t = f(t, X_t)dt + g(t, X_t)dC_t,$$

respectively. Then, the solution X_t *has an inverse uncertainty distribution:*

$$\Psi_t^{-1}(\alpha) = X_t^\alpha.$$

Generalization

In this subsection, we generalize the Yao-Chen formula to the multifactor uncertain differential equation.

Definition 9. *Let* α *be a number with* $0 < \alpha < 1$, *and let* $C_{1t}, C_{2t}, \cdots, C_{nt}$ *be independent canonical processes. An uncertain differential equation:*

$$dX_t = f(t, X_t)dt + \sum_{i=1}^{n} g_i(t, X_t)dC_{it} \tag{31}$$

is said to have an α*-path* X_t^α *if it solves the corresponding ordinary differential equation:*

$$dX_t^\alpha = f\left(t, X_t^\alpha\right)dt + \sum_{i=1}^{n} \mid g_i\left(t, X_t^\alpha\right) \mid \Phi^{-1}(\alpha)dt \tag{32}$$

where $\Phi^{-1}(\alpha)$ *is the inverse uncertainty distribution of standard normal uncertain variable* $\mathcal{N}(0, 1)$, *i.e.,*

$$\Phi^{-1}(\alpha) = \frac{\sqrt{3}}{\pi} \ln \frac{\alpha}{1-\alpha}, \quad 0 < \alpha < 1.$$

Example 8. *Let a, b, and c be real numbers. The uncertain differential equation:*

$$dX_t = adt + bdC_{1t} + cdC_{2t}, \ X_0 = 0$$

has an α*-path:*

$$X_t^\alpha = at + (\mid b \mid + \mid c \mid)\Phi^{-1}(\alpha).$$

Lemma 9. *Assume that $f(t,x)$ and $g(t,x)$ are continuous functions. Let $\phi(t)$ be a solution of the ordinary differential equation:*

$$\frac{dx}{dt} = f(t,x)dt + K \mid g(t,x) \mid, \; x(0) = x_0$$

where K is a real number. Let $\psi(t)$ be a solution of the ordinary differential equation:

$$\frac{dx}{dt} = f(t,x)dt + k(t)g(t,x), \; x(0) = x_0$$

where $k(t)$ is a real function.

(i) *If $k(t)g(t,x) \leq K \mid g(t,x) \mid$ for $t \in [0, T]$, then $\psi(T) \leq \phi(T)$,*
(ii) *If $k(t)g(t,x) > K \mid g(t,x) \mid$ for $t \in [0, T]$, then $\psi(T) > \phi(T)$.*

Theorem 10. *Assume that $f, g_1, g_2, \cdots, g_n$ are continuous functions of two variables and $C_{1t}, C_{2t}, \cdots, C_{nt}$ are independent canonical processes. Let X_t and X_t^α be the solution and α-path of the uncertain differential equation:*

$$dX_t = f(t,X_t)dt + \sum_{i=1}^{n} g_i(t,X_t)dC_{it},$$

respectively. Then:

$$\mathcal{M}\left\{X_t \leq X_t^\alpha, \forall t\right\} = \alpha,$$

$$\mathcal{M}\left\{X_t > X_t^\alpha, \forall t\right\} = 1 - \alpha.$$

Proof. For each α-path X_t^α, we construct sets as follows,

$$T_i^+ = \left\{t \mid g_i\left(t, X_t^\alpha\right) \geq 0\right\},$$

$$T_i^- = \left\{t \mid g_i\left(t, X_t^\alpha\right) < 0\right\},$$

$i = 1, 2, \cdots, n$. It is obvious that $T_i^+ \cap T_i^- = \emptyset$ and $T_i^+ \cup T_i^- = [0, +\infty)$ for each $1 \leq i \leq n$.
Write:

$$\Lambda_{i1}^+ = \left\{\gamma \mid \frac{dC_{it}(\gamma)}{dt} \leq \Phi^{-1}(\alpha) \text{ for } t \in T_i^+\right\},$$

$$\Lambda_{i1}^- = \left\{\gamma \mid \frac{dC_{it}(\gamma)}{dt} \geq \Phi^{-1}(1-\alpha) \text{ for } t \in T_i^-\right\},$$

$i = 1, 2, \cdots, n$, where Φ^{-1} is the inverse uncertainty distribution of $\mathcal{N}(0,1)$. Since T_i^+ and T_i^- are disjoint sets and C_{it} have independent increments, we get:

$$\mathcal{M}\left\{\Lambda_{i1}^+\right\} = \alpha, \quad \mathcal{M}\left\{\Lambda_{i1}^-\right\} = \alpha \quad \mathcal{M}\left\{\Lambda_{i1}^+ \cap \Lambda_{i1}^-\right\} = \alpha.$$

For any $\gamma \in \Lambda_{i1}^+ \cap \Lambda_{i1}^-$, we always have:

$$g_i(t, X_t(\gamma))\frac{dC_{it}(\gamma)}{dt} \leq \mid g_i\left(t, X_t^\alpha\right) \mid \Phi^{-1}(\alpha), \forall t, \; i = 1, 2, \cdots, n.$$

Let $\Lambda_1^+ \cap \Lambda_1^- = \bigcap_{i=1}^{n}(\Lambda_{i1}^+ \cap \Lambda_{i1}^-)$. Because $C_{1t}, C_{2t}, \cdots, C_{nt}$ are independent and $\mathcal{M}\{\Lambda_{i1}^+ \cap \Lambda_{i1}^-\} = \alpha$, $i = 1, 2, \cdots, n$, we have:

$$\mathcal{M}\{\Lambda_1^+ \cap \Lambda_1^-\} = \mathcal{M}\left\{\bigcap_{i=1}^{n}(\Lambda_{i1}^+ \cap \Lambda_{i1}^-)\right\} = \min_{1\le i\le n} \mathcal{M}\{\Lambda_{i1}^+ \cap \Lambda_{i1}^-\} = \alpha.$$

Then, for any $\gamma \in \Lambda_1^+ \cap \Lambda_1^-$, we have:

$$\sum_{i=1}^{n} g_i(t, X_t(\gamma))\frac{dC_{it}(\gamma)}{dt} \le \sum_{i=1}^{n} | g_i(t, X_t^\alpha) | \Phi^{-1}(\alpha), \forall t.$$

The Lemma 9 shows that $X_t \le X_t^\alpha$ for all t, so $\Lambda_1^+ \cap \Lambda_1^- \subset \{X_t \le X_t^\alpha, \forall t\}$. Hence:

$$\mathcal{M}\{X_t \le X_t^\alpha, \forall t\} \ge \mathcal{M}\{\Lambda_1^+ \cap \Lambda_1^-\} = \alpha. \tag{33}$$

On the other hand, write:

$$\Lambda_{i2}^+ = \left\{\gamma \mid \frac{dC_{it}(\gamma)}{dt} > \Phi^{-1}(\alpha) \text{ for } t \in T_i^+\right\},$$

$$\Lambda_{i2}^- = \left\{\gamma \mid \frac{dC_{it}(\gamma)}{dt} < \Phi^{-1}(1-\alpha) \text{ for } t \in T_i^-\right\},$$

$i = 1, 2, \cdots, n$. Since T_i^+ and T_i^- are disjoint sets and C_{it} has independent increments, we get:

$$\mathcal{M}\{\Lambda_{i2}^+\} = 1-\alpha, \quad \mathcal{M}\{\Lambda_{i2}^-\} = 1-\alpha, \quad \mathcal{M}\{\Lambda_{i2}^+ \cap \Lambda_{i2}^-\} = 1-\alpha.$$

For any $\gamma \in \Lambda_{i2}^+ \cap \Lambda_{i2}^-$, we always have:

$$g_i(t, X_t(\gamma))\frac{dC_{it}(\gamma)}{dt} > | g_i(t, X_t^\alpha) | \Phi^{-1}(\alpha), \forall t, \; i = 1, 2, \cdots, n.$$

Let $\Lambda_2^+ \cap \Lambda_2^- = \bigcap_{i=1}^{n}(\Lambda_{i2}^+ \cap \Lambda_{i2}^-)$. Because $C_{1t}, C_{2t}, \cdots, C_{nt}$ are independent and $\mathcal{M}\{\Lambda_{i2}^+ \cap \Lambda_{i2}^-\} = 1-\alpha$, $i = 1, 2, \cdots, n$, we have:

$$\mathcal{M}\{\Lambda_2^+ \cap \Lambda_2^-\} = \mathcal{M}\left\{\bigcap_{i=1}^{n}(\Lambda_{i2}^+ \cap \Lambda_{i2}^-)\right\} = \min_{1\le i\le n} \mathcal{M}\{\Lambda_{i2}^+ \cap \Lambda_{i2}^-\} = 1-\alpha.$$

Then, for any $\gamma \in \Lambda_2^+ \cap \Lambda_2^-$, we have:

$$\sum_{i=1}^{n} g_i(t, X_t(\gamma))\frac{dC_{it}(\gamma)}{dt} > \sum_{i=1}^{n} | g_i(t, X_t^\alpha) | \Phi^{-1}(\alpha), \forall t.$$

The Lemma 9 shows that $X_t > X_t^\alpha$ for any t, so $\Lambda_2^+ \cap \Lambda_2^- \subset \{X_t > X_t^\alpha, \forall t\}$. Hence:

$$\mathcal{M}\{X_t > X_t^\alpha, \forall t\} \ge \mathcal{M}\{\Lambda_{i2}^+ \cap \Lambda_{i2}^-\} = 1-\alpha. \tag{34}$$

Since $\{X_t \le X_t^\alpha, \forall t\}$ and $\{X_t \not\le X_t^\alpha, \forall t\}$ are opposite events with each other. It follows from the duality axiom that:

$$\mathcal{M}\{X_t \le X_t^\alpha, \forall t\} + \mathcal{M}\{X_t \not\le X_t^\alpha, \forall t\} = 1.$$

In addition, $\{X_t > X_t^\alpha, \forall t\} \subset \{X_t \not\le X_t^\alpha, \forall t\}$ means that:

$$\mathcal{M}\{X_t \le X_t^\alpha, \forall t\} + \mathcal{M}\{X_t > X_t^\alpha, \forall t\} \le 1. \tag{35}$$

Thus, the results follow from (33), (34), and (35).

Theorem 11. *Assume that $f, g_1, g_2, \cdots, g_n$ are continuous functions of two variables and $C_{1t}, C_{2t}, \cdots, C_{nt}$ are independent canonical processes. Let X_t and X_t^α be the solution and α-path of the uncertain differential equation:*

$$dX_t = f(t, X_t)dt + \sum_{i=1}^{n} g_i(t, X_t)dC_{it},$$

respectively. Then, the solution X_t has an inverse uncertainty distribution:

$$\Psi_t^{-1}(\alpha) = X_t^\alpha.$$

Proof. Obviously, $\{X_t \le X_t^\alpha\} \supset \{X_s \le X_s^\alpha, \forall s\}$. It follows from the monotonicity theorem and Theorem 10 that:

$$\mathcal{M}\{X_t \le X_t^\alpha\} \ge \mathcal{M}\{X_s \le X_s^\alpha, \forall s\} = \alpha. \tag{36}$$

Similarly, we also obtain:

$$\mathcal{M}\{X_t > X_t^\alpha\} \ge \mathcal{M}\{X_s > X_s^\alpha, \forall s\} = 1 - \alpha. \tag{37}$$

Besides, by using the duality axiom, we have:

$$\mathcal{M}\{X_t \le X_t^\alpha\} + \mathcal{M}\{X_t > X_t^\alpha\} = 1. \tag{38}$$

It follows from (36), (37), and (38) that:

$$\Psi_t^{-1}(\alpha) = X_t^\alpha.$$

Example 9. *Let a, b, and c be real numbers and let C_{1t} and C_{2t} be independent canonical processes. The uncertain differential equation:*

$$dX_t = aX_t dt + bX_t dC_{1t} + cX_t dC_{2t}, \; X_0 = 1 \tag{39}$$

has a solution:

$$X_t = \exp(at + bC_{1t} + cC_{2t})$$

with an inverse uncertainty distribution:

$$\Psi_t^{-1}(\alpha) = \exp\left(at + (\mid b \mid + \mid c \mid)\Phi^{-1}(\alpha)\right).$$

Based on the previous theorem, the Yao-Chen method can be generalized to the multifactor uncertain differential equation as follows.

Step 1: Fix α on $(0, 1)$.

Step 2: Solve the corresponding ordinary differential equation:

$$dX_t^\alpha = f\left(t, X_t^\alpha\right)dt + \sum_{i=1}^{n} \mid g_i\left(t, X_t^\alpha\right) \mid \Phi^{-1}(\alpha)dt$$

and obtain X_t^α, for example, we can choose the recursion formula:

$$X_{i+1}^\alpha = X_i^\alpha + f\left(t_i, X_i^\alpha\right)h + \sum_{j=1}^{n} \mid g_j\left(t_i, X_i^\alpha\right) \mid \Phi^{-1}(\alpha)h$$

where $\Phi^{-1}(\alpha)$ is the inverse standard normal uncertainty distribution and h is the step length.

Step 3: The inverse uncertainty distribution of X_t is obtained.

Example 10. *In order to illustrate the numerical method, let us consider an uncertain differential equation:*

$$dX_t = X_t dt + X_t dC_{1t} + X_t dC_{2t},\ X_0 = 1 \tag{40}$$

whose solution is $X_t = \exp(t + C_{1t} + C_{2t})$. *The Matlab Uncertainty Toolbox (http://orsc.edu.cn/liu/resources.htm) may solve this equation successfully and obtain an inverse uncertainty distribution of* X_t *at* $t = 1/2$ *shown in Figure 1.*

Existence and uniqueness theorem

This section will give an existence and uniqueness theorem of solution for the multifactor uncertain differential equation under Lipschitz condition and linear growth condition.

Lemma 12. *(Chen and Liu [18]) Let* C_t *be a canonical process, and* X_t *an integrable uncertain process on* $[a, b]$ *with respect to t. Then, the inequality:*

$$\left| \int_a^b X_t(\gamma) dC_t(\gamma) \right| \leq K(\gamma) \int_a^b | X_t(\gamma) | \, dt$$

holds, where $K(\gamma)$ *is the Lipschitz constant of the sample path* $X_t(\gamma)$.

Theorem 13. *Let* $f, g_1, g_2, \cdots, g_n$ *be functions of two variables and let* $C_{1t}, C_{2t}, \cdots, C_{nt}$ *be independent canonical processes. Then, the uncertain differential equation:*

$$dX_t = f(t, X_t)dt + \sum_{i=1}^{n} g_i(t, X_t)dC_{it}$$

has a unique solution if the coefficients $f, g_1, g_2, \cdots, g_n$ *satisfy the Lipschitz condition:*

$$| f(t,x) - f(t,y) | + \sum_{i=1}^{n} | g_i(t,x) - g_i(t,y) | \leq L | x - y |, \text{ for all } x, y \in \Re,\ t \geq 0 \tag{41}$$

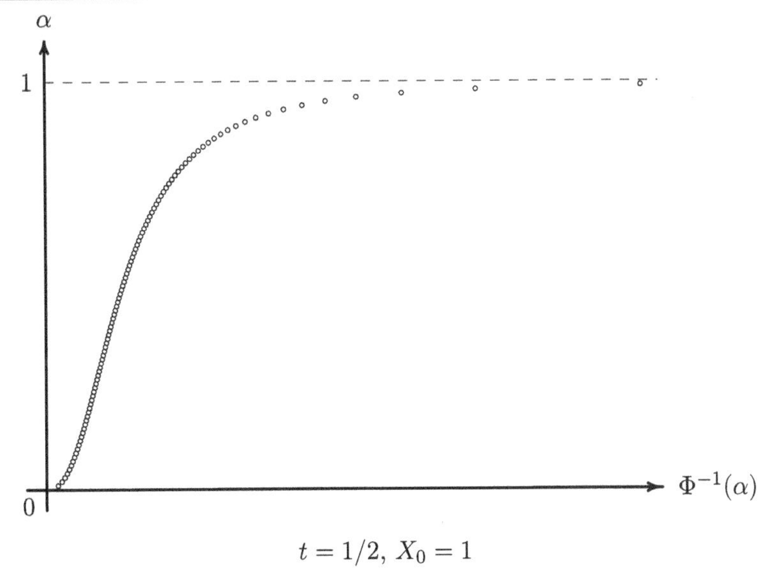

Figure 1 $t = 1/2, X_0 = 1$.

and linear growth condition:

$$| f(t,x) | + \sum_{i=1}^{n} | g_i(t,x) | \leq L(1+ | x |), \text{ for all } x \in \Re,\ t \geq 0 \tag{42}$$

for some constant L. Moreover, the solution is sample-continuous.

Proof. We first prove the existence of solution by a successive approximation method. Define $X_t^{(0)} = X_0$, and:

$$X_t^{(n)} = X_0 + \int_0^t f\left(s, X_s^{(n-1)}\right) ds + \sum_{i=1}^{n} \int_0^t g_i\left(s, X_s^{(n-1)}\right) dC_{is}$$

for $n = 1, 2, \cdots, n$ and write:

$$D_t^n(\gamma) = \max_{0 \leq s \leq t} | X_s^{(n+1)}(\gamma) - X_s^{(n)}(\gamma) |$$

for each $\gamma \in \Gamma$.

We claim that:

$$D_t^n(\gamma) \leq (1+ | X_0 |) \frac{L^{n+1}(1+K_\gamma)^{n+1}}{(n+1)!} t^{n+1}$$

where $K_\gamma = \sum_{i=1}^{n} K_{i\gamma}$, and $K_{i\gamma}$ is the Lipschitz constant to the sample path $C_{it}(\gamma)$, $i = 1, 2, \cdots, n$.

For $n = 0$, we have:

$$\begin{aligned} D_t^{(0)}(\gamma) &= \max_{0 \leq s \leq t} | \int_0^s f(v, X_0) dv + \sum_{i=1}^{n} \int_0^s g_i(v, X_0) dC_{iv}(\gamma) | \\ &\leq \int_0^t | f(v, X_0) | dv + \sum_{i=1}^{n} K_{i\gamma} \int_0^t | g_i(v, X_0) | dv \\ &\leq (1+ | X_0 |) L(1+K_\gamma) t \end{aligned}$$

where the first inequality comes from Lemma 12, the second comes from the linear growth condition.

This confirms the claim for $n = 0$. Next, we assume the claim is true for some $n-1$. Then:

$$\begin{aligned}
D_t^{(n)}(\gamma) &= \max_{0\le s\le t} \mid \int_0^s (f(v, X_v^{(n)}(\gamma)) - f\left(v, X_v^{(n-1)}(\gamma)\right) dv \\
&+ \sum_{i=1}^n \int_0^s \left(g_i(v, X_v^{(n)}(\gamma)\right) - g_i\left(v, X_v^{(n-1)}(\gamma)\right) dC_{iv}(\gamma) \mid \\
&\le \int_0^t \mid \left(f(v, X_v^{(n)}(\gamma)\right) - f\left(v, X_v^{(n-1)}(\gamma)\right) \mid dv \\
&+ \sum_{i=1}^n \int_0^t \mid \left(g_i(v, X_v^{(n)}(\gamma)\right) - g_i\left(v, X_v^{(n-1)}(\gamma)\right) \mid dC_{iv}(\gamma) \\
&\le L \int_0^t \mid X_v^{(n)}(\gamma) - X_v^{(n-1)}(\gamma) \mid dv \\
&+ \sum_{i=1}^n K_{i\gamma} \int_0^t \mid \left(g_i(v, X_v^{(n)}(\gamma)\right) - g_i\left(v, X_v^{(n-1)}(\gamma)\right) \mid dv \\
&\le L \int_0^t \mid X_v^{(n)}(\gamma) - X_v^{(n-1)}(\gamma) \mid dv \\
&+ L \sum_{i=1}^n K_{i\gamma} \int_0^t \mid X_v^{(n)}(\gamma) - X_v^{(n-1)}(\gamma) \mid dv \\
&\le L(1+K_\gamma) \int_0^t \mid X_v^{(n)}(\gamma) - X_v^{(n-1)}(\gamma) \mid dv \\
&\le L(1+K_\gamma) \int_0^t (1+ \mid X_0 \mid) \frac{L^n(1+K_\gamma)^n}{n!} v^n dv \\
&= (1+ \mid X_0 \mid) \frac{L^{n+1}(1+K_\gamma)^{n+1}}{(n+1)!} t^{n+1}.
\end{aligned}$$

It follows from Weierstrassąŕ criterion that, for each sample γ, the paths $X_t^{(k)}(\gamma)$ converges uniformly on any given interval $[0, T]$. Write the limit by $X_t(\gamma)$ that is just a solution:

$$X_t = X_0 + \int_0^t f(s, X_s) ds + \sum_{i=1}^n \int_0^t g_i(s, X_s) dC_{is}.$$

Next, we prove that the solution is unique. Assume that X_t and $X_t^\star$ are solutions. The Lipschitz condition and linear growth condition show:

$$\mid X_t(\gamma) - X_t^\star(\gamma) \mid \le L(1+K_\gamma) \int_0^t \mid X_v(\gamma) - X_v^\star(\gamma) \mid dv.$$

It follows from Gronwall inequality that:

$$\mid X_t(\gamma) - X_t^\star(\gamma) \mid \le 0 \cdot \exp(L(1+K_\gamma)).$$

Hence, $X_t = X_t^\star$. The uniqueness is proved.

At last, we will prove the sample-continuity of X_t. For each $\gamma \in \Gamma$, by the above proof, we get:

$$X_t(\gamma) \le \sum_{n=0}^{+\infty} (1+ \mid X_0 \mid) \frac{L^{n+1}(1+K_\gamma)^{n+1}}{(n+1)!} t^{n+1} = (1+ \mid X_0 \mid) \exp(L(1+K_\gamma)t).$$

Suppose $0 < s < t$, we have:

$$
\begin{aligned}
| X_t(\gamma) - X_s(\gamma) | &= | \int_s^t f(v, X_v)dv + \sum_{i=1}^{n} \int_s^t g_i(v, X_v)dC_{iv} | \\
&\leq \int_s^t | f(v, X_v) | dv + \sum_{i=1}^{n} \int_s^t | g_i(v, X_v) | dC_{iv} \\
&\leq \int_s^t | f(v, X_v) | dv + \sum_{i=1}^{n} K_{i\gamma} \int_s^t | g_i(v, X_v) | dv \\
&\leq (1 + K_\gamma)L(1 + | X_v(\gamma) |)(t - s) \\
&\leq (1 + K_\gamma)L(1 + (1 + | X_0 |) \exp(L(1 + K_\gamma)t))(t - s).
\end{aligned}
$$

Thus $| X_t(\gamma) - X_s(\gamma) | \to 0$ as $s \to t$. Hence, X_t is sample-continuous. The theorem is proved.

Note that $n = 1$, the existence and uniqueness theorem degenerates to the one in Chen and Liu [18].

Conclusions

Uncertain differential equation is an important tool to deal with dynamic systems in uncertain environments. In this paper, the multifactor uncertain differential equation was proposed. Four special types of multifactor uncertain differential equations were studied and the corresponding analytic solutions were given. For general multifactor uncertain differential equation, a numerical method was provided for obtaining the solution. Also, an existence and uniqueness theorem that the multifactor uncertain differential equation has a unique solution was proved. The proposed multifactor uncertain differential equation can be used to describe the multifactor stock model in uncertain market.

Acknowledgments

This work is supported by the Projects of the Humanity and Social Science Foundation of Ministry of Education of China (No.13YJA630065), the Key Project of Hubei Provincial Natural Science Foundation (No.2012FFA065).

Author details

[1]Institute of Uncertain Systems, Huanggang Normal University, Hubei 438000, China. [2]School of Statistics and Mathematics, Zhongnan University of Economics and Law, Hubei 430073, China.

References

1. Liu, B: Uncertainty Theory. 2nd ed. Springer-Verlag, Berlin (2007)
2. Liu, B: Uncertainty Theory: A Branch of Mathematics for Modeling Human Uncertainty. Springer-Verlag, Berlin (2010)
3. Liu, B: Theory and Practice of Uncertain Programming. 2nd ed. Springer-Verlag, Berlin (2009)
4. Liu, B: Uncertain set theory and uncertain inference rule with application to uncertain control. J Uncertain Syst. **4**(2), 83–98 (2010)
5. Liu, B: Uncertain logic for modeling human language. J Uncertain Syst. **5**(1), 3–20 (2011)
6. Liu, B: Uncertain risk analysis and uncertain reliability analysis. J. Uncertain Syst. **4**(3), 163–170 (2010)
7. Peng, J: Risk metrics of loss function for uncertain system. Fuzzy Optimization Decis. Making. **12**, 53–64 (2013)
8. Li, S, Peng, J, Zhang, B: The uncertain premium principle based on the distortion function. Insurance:. Math. Econ. **53**, 317–324 (2013)
9. Liu, B: Fuzzy process, hybrid process and uncertain process. J. Uncertain Syst. **2**, 3–16 (2008)
10. Liu, B: Some research problems in uncertainty theory. J. Uncertain Syst. **3**, 3–10 (2009)
11. Yao, K: Uncertain calculus with renewal process. Fuzzy Optimization Decis. Making. **11**, 285–297 (2012)
12. Yao, K: Multi-dimensional uncertain calculus with Liu process. J. Uncertain Syst. **8**(4), 244–254 (2014)
13. Chen, X: Uncertain calculus with finite variation processes (2015)
14. Gao, Y: Variation analysis of semi-canonical process. Math. Comput. Model. **53**, 1983–1989 (2011)

15. Liu, B: Extreme value theorems of uncertain process with application to insurance risk model. Soft Comput. **17**, 549–556 (2013)
16. Zhang, T, Chen, B: Multi-dimensional canonical process. Information: Int. Interdiscip. J. **16**(2A), 1025–1030 (2013)
17. Itô, K: On stochastic differential equations. Mem. Am. Math. Soc. **4**, 1–51 (1951)
18. Chen, X, Liu, B: Existence and uniqueness theorem for uncertain differential equations. Fuzzy Optimization Decis. Making. **9**, 69–81 (2010)
19. Gao, Y: Existence and uniqueness theorem on uncertain differential equations with local Lipschitz condition. J. Uncertain Syst. **6**(3), 223–232 (2012)
20. Yao, K, Gao, J, Gao, Y: Some stability theorems of uncertain differential equation. Fuzzy Optimization Decis. Making. **12**(1), 3–13 (2013)
21. Sheng, Y, Wang, C: Stability in p-th moment for uncertain differential equation. J. Intell. Fuzzy Syst. **26**(3), 1263–1271 (2014)
22. Liu, H, Ke, H, Fei, Y: Almost sure stability for uncertain differential equation. Fuzzy Optimization Decis. Making. **13**(4), 463–473 (2014)
23. Yao, K, Ke, H, Sheng, Y: Stability in mean for uncertain differential equation (2015)
24. Liu, Y: An analytic method for solving uncertain differential equations. J. Uncertain Syst. **6**, 244–249 (2012)
25. Yao, K: A type of uncertain differential equations with analytic solution. J. Uncertainty Anal. Appl. 1, Article, 8 (2013)
26. Yao, K, Chen, X: A numerical method for solving uncertain differential equation. J. Intell. Fuzzy Syst. **25**, 825–832 (2013)
27. Yao, K: Extreme values and integral of solution of uncertain differential equation. J. Uncertainty Anal. Appl. 1, Article, 2 (2013)
28. Chen, X: American option pricing formula for uncertain financial market. Int. J. Oper. Res. **8**, 32–37 (2011)
29. Peng, J, Yao, K: A new option pricing model for stocks in uncertainty markets. Int. J. Oper. Res. **8**, 18–26 (2011)
30. Liu, B: Toward uncertain finance theory. J. Uncertainty Anal. Appl. 1, Article, 1 (2013)
31. Li, S, Peng, J: A new stock model for option pricing in uncertain environment. Iran. J. Fuzzy Syst. **11**, 27–42 (2014)
32. Liu, H, Chen, X, Ralescu, D: Uncertain currency model and currency option pricing. Int. J. Intell. Syst. **30**(1), 40–51 (2015)
33. Jiao, D, Yao, K: An interest rate model in uncertain environment (2015)
34. Yao, K: A no-arbitrage theorem for uncertain stock model. Fuzzy Optimization Decis. Making (2015). doi:10.1007/s10700-014-9198-9
35. Zhu, Y: Uncertain optimal control with application to a portfolio selection model. Cybernet. Syst. **41**(7), 535–547 (2010)
36. Yang, X, Gao, J: Uncertain differential games with application to capitalism. J. Uncertainty Anal. Appl. 1, Article, 17 (2013)
37. Ge, X, Zhu, Y: A necessary condition of optimality for uncertain optimal control problem. Fuzzy Optimization Decis. Making. **12**(1), 41–51 (2013)
38. Barbacioru, I: Uncertainty functional differential equations for finance. Surv. Math. Appl. **5**, 275–284 (2010)
39. Ge, X, Zhu, Y: Existence and uniqueness theorem for uncertain delay differential equations. J. Comput. Inform. Syst. **8**(20), 8341–8347 (2012)
40. Liu, H, Fei, W: Neutral uncertain delay differential equations. Information: Int. Interdiscip. J. **16**(2), 1225–1232 (2013)
41. Ji, X, Zhou, J: Multi-dimensional uncertain differential equation: existence and uniqueness of solution. Fuzzy Optimization Decis. Making (2015). doi:10.1007/s10700-015-9210-z
42. Yao, K: Higher order uncertain differential equation. http://orsc.edu.cn/online/141222.pdf (2014)
43. Liu, B: Yao, K: Uncertain integral with respect to multiple canonical processes. J. Uncertain Syst. **6**, 250–255 (2012)
44. Liu, Y, Ha, M: Expected value of function of uncertain variables. J. Uncertain Syst. **4**, 181–186 (2010)
45. Liu, B: Uncertainty distribution and independence of uncertain processes. Fuzzy Optimization Decis. Making. **13**(3), 259–271 (2014)

18

Existence and uniqueness of the solution to uncertain fractional differential equation

Yuanguo Zhu

Correspondence:
ygzhu@njust.edu.cn
School of Science, Nanjing University of Science and Technology, 210094 Nanjing, China

Abstract

The paper proves an existence and uniqueness theorem of the solution to an uncertain fractional differential equation by Banach fixed point theorem under Lipschitz and linear growth conditions. Then, the paper presents an existence theorem for the solution of an uncertain fractional differential equation by Schauder fixed point theorem under continuity condition.

Keywords: Uncertainty theory; Fractional derivative; Uncertain fractional differential equation; Existence; Uniqueness

Introduction

Fractional differential equation has been a classical research field of differential equations. The study of fractional differential equations attracted many mathematicians, physicists, and engineers. Fractional differential equations have important applications such as in rheology, viscoelasticity, electrochemistry, and electromagnetism. A fractional differential equation is a differential equation including fractional derivatives. There are several kinds of fractional derivatives such as the Riemann-Liouville type, Caputo type, Grünwald-Letnikov type, and Riesz type. Some references about fractional differential equations may be seen in [1-8].

Stochastic fractional differential equations were used to model dynamical systems affected by random noises [9-15]. Generally, there are two types of stochastic fractional differential equations. One is of the form

$$D^{\alpha}x(t) = f(t, x(t)) + g(t, x(t))\frac{\mathrm{d}W(t)}{\mathrm{d}t}$$

where $D^{\alpha}x(t)$ denotes the fractional derivative of the function $x(t)$, and $W(t)$ is the Wiener process. The other is of the form

$$\mathrm{d}x(t) = f(t, x(t))\mathrm{d}t + g(t, x(t))\mathrm{d}B_t^H$$

where B_t^H is the fractional Brownian motion.

Recently in [16], the concept of uncertain fractional differential equations was introduced based on the uncertainty theory. The Riemann-Liouville type of uncertain fractional differential equation

$$D^{p}X_t = f(t, X_t) + g(t, X_t)\frac{\mathrm{d}C_t}{\mathrm{d}t}$$

and the Caputo type of uncertain fractional differential equation

$$^{c}D^{p}X_t = f(t, X_t) + g(t, X_t)\frac{dC_t}{dt}$$

were dealt with where C_t is the canonical Liu process. The solutions were provided by the Mittag-Leffler function for linear uncertain fractional differential equations. An application to the price of a zero-coupon bond with a maturity date was presented.

Analytical solutions were presented only for some special uncertain fractional differential equations in [16]. In order to understand what kinds of uncertain fractional differential equations have solutions, we in this paper will give some sufficient conditions to guarantee the existence of solutions of uncertain fractional differential equations. That is, we will show the existence and uniqueness of solutions for uncertain fractional differential equations. The structure of the paper is as follows: Firstly, some concepts and results in uncertainty theory will be reviewed. Then, the fractional derivatives and uncertain fractional differential equations will be recalled. Finally, an existence and uniqueness theorem and an existence theorem will be proved.

Preliminary

Uncertainty theory was founded by Liu in 2007 [17] and refined in 2010 [18]. Basic concepts in uncertainty theory include uncertain measure, uncertainty space, product uncertain measure, and uncertain variable. Let Γ be a nonempty set and $\mathcal{L}$ be a σ-algebra over Γ. Each element $\Lambda \in \mathcal{L}$ is called an event. Set function $\mathcal{M}$ defined on $\mathcal{L}$ is called an uncertain measure if it satisfies three axioms: (normality) $\mathcal{M}\{\Gamma\} = 1$, (duality) $\mathcal{M}\{\Lambda\} + \mathcal{M}\{\Lambda^c\} = 1$ for any event Λ, and (countable subadditivity) $\mathcal{M}\left\{\bigcup_{i=1}^{\infty} \Lambda_i\right\} \leq \sum_{i=1}^{\infty} \mathcal{M}\{\Lambda_i\}$ for every countable sequence of events $\Lambda_1, \Lambda_2, \cdots$. The triplet $(\Gamma, \mathcal{L}, \mathcal{M})$ is called an uncertainty space. The product uncertain measure $\mathcal{M}$ is an uncertain measure satisfying $\mathcal{M}\left\{\prod_{i=1}^{\infty} \Lambda_k\right\} = \bigwedge_{i=1}^{\infty} \mathcal{M}_k\{\Lambda_k\}$, where $(\Gamma_k, \mathcal{L}_k, \mathcal{M}_k)$ are uncertainty spaces and Λ_k are arbitrarily chosen events from $\mathcal{L}_k$ for $k = 1, 2, \cdots$, respectively. An uncertain variable is defined as a function ξ from an uncertainty space $(\Gamma, \mathcal{L}, \mathcal{M})$ to the set R of real numbers such that $\{\xi \in B\}$ is an event for any Borel set B.

The uncertainty distribution $\Phi : R \to [0, 1]$ of an uncertain variable ξ is defined by $\Phi(x) = \mathcal{M}\{\xi \leq x\}$ for $x \in R$. The expected value of an uncertain variable ξ is defined by

$$E[\xi] = \int_0^{+\infty} \mathcal{M}\{\xi \geq r\}dr - \int_{-\infty}^{0} \mathcal{M}\{\xi \leq r\}dr$$

provided that at least one of the two integrals is finite. The variance of ξ is defined by $V[\xi] = E\left[(\xi - E[\xi])^2\right]$.

A normal uncertain variable with expected value e and variance σ^2 has the uncertainty distribution

$$\Phi(x) = \left(1 + \exp\left(\frac{\pi(e-x)}{\sqrt{3}\sigma}\right)\right)^{-1}, \ x \in R$$

which is denoted by $\xi \sim \mathcal{N}(e, \sigma)$.

The uncertain variables $\xi_1, \xi_2, \cdots \xi_m$ are said to be independent [19] if

$$\mathcal{M}\left\{\bigcap_{i=1}^{m}(\xi_i \in B_i)\right\} = \min_{1 \leq i \leq m} \mathcal{M}\{\xi_i \in B_i\}$$

for any Borel sets $B_1, B_2, \cdots B_m$ of real numbers. For numbers a and b, $E[a\xi + b\eta] = aE[\xi] + bE[\eta]$ if ξ and η are independent uncertain variables.

Liu [20] defined uncertain process as a function X_t from $S \times (\Gamma, \mathcal{L}, \mathcal{M})$ to the set of real numbers where S is a totally ordered set such that $\{X_t \in B\}$ is an event for any Borel set B at each time $t \in S$.

Definition 1. *[19] An uncertain process C_t is called a canonical Liu process if it satisfies the following: (i) $C_0 = 0$ and almost all sample paths are Lipschitz continuous, (ii) C_t has stationary and independent increments, and (iii) every increment $C_{s+t} - C_s$ is a normal uncertain variable with expected value 0 and variance t^2, denoted by $C_{s+t} - C_s \sim \mathcal{N}(0, t)$.*

For any partition of closed interval $[a, b]$ with $a = t_1 < t_2 < \cdots < t_{k+1} = b$, the mesh is written as $\Delta = \max_{1 \le i \le k} |t_{i+1} - t_i|$. Then, the uncertain integral of X_t with respect to C_t is defined by Liu [19] as

$$\int_a^b X_t \mathrm{d}C_t = \lim_{\Delta \to 0} \sum_{i=1}^{k} X_{t_i} \cdot \left(C_{t_{i+1}} - C_{t_i}\right)$$

provided that the limit exists almost surely and is finite. If there exist two uncertain processes μ_t and σ_t such that $Z_t = Z_0 + \int_0^t \mu_s \mathrm{d}s + \int_0^t \sigma_s \mathrm{d}C_s$ for any $t \ge 0$, then we say Z_t has an uncertain differential $\mathrm{d}Z_t = \mu_t \mathrm{d}t + \sigma_t \mathrm{d}C_t$. An uncertain differential equation driven by a canonical Liu process C_t is defined as

$$\mathrm{d}X_t = f(t, X_t)\mathrm{d}t + g(t, X_t)\mathrm{d}C_t \tag{1}$$

where f and g are two given functions. A solution X_t of the uncertain differential equation is equivalent to a solution of the uncertain integral equation

$$X_t = X_0 + \int_0^t f(s, X_s)\mathrm{d}s + \int_0^t g(s, X_s)\mathrm{d}C_s.$$

For an uncertain differential equation (1) in a multidimensional case, X_t is a multidimensional state, C_t is a multidimensional canonical Liu process, $f(t, X_t)$ is a vector-valued function, and $g(t, X_t)$ is a matrix-valued function.

An existence and uniqueness theorem of solution for the uncertain differential equation (1) was proved by Chen and Liu [21]. Meanwhile, Chen and Liu [21] obtained an analytic solution to linear uncertain differential equations. Liu [22] and Yao [23] presented some methods for solving nonlinear uncertain differential equations. Yao and Chen [24] introduced a numerical method for solving the uncertain differential equation. Some extensions of the uncertain differential equation were studied such as the uncertain delay differential equation by Barbacioru [25], Ge and Zhu [26], and Liu and Fei [27] and the backward uncertain differential equation by Ge and Zhu [28]. The uncertain differential equation has been applied in some fields such as uncertain finance [29], uncertain optimal control [30], and uncertain differential game [31].

Lemma 1. *[28] Suppose that C_t is an l-dimensional canonical Liu process, and Y_t is an integrable $n \times l$-dimensional uncertain process on $[a,b]$ with respect to t. Then, the inequality*

$$\left\| \int_a^b Y_t(\gamma)dC_t(\gamma) \right\|_\infty \leq K_\gamma \int_a^b \|Y_t(\gamma)\|_\infty dt$$

holds, where K_γ is the Lipschitz constant of the sample path $C_t(\gamma)$ with the norm $\|\cdot\|_\infty$.

Uncertain fractional differential equations

In the sequel, we will always assume $p \in (0,1]$. The Riemann-Liouville type of uncertain fractional differential equation and the Caputo type of uncertain fractional differential equation in the one-dimensional case were introduced in [16]. Now we state those concepts in a multidimensional case. Let $C_t = (C_{1t}, C_{2t}, \cdots, C_{lt})^\tau$ be an l-dimensional canonical Liu process.

Definition 2. *Suppose that $f : [0,\infty) \times R^n \to R^n$ and $g : [0,\infty) \times R^n \to R^{n\times l}$ are two functions. Then,*

$$D^p X_t = f(t, X_t) + g(t, X_t)\frac{dC_t}{dt} \tag{2}$$

is called an uncertain fractional differential equation of the Riemann-Liouville type. A solution of (2) with the initial condition

$$\lim_{t\to 0+} t^{1-p}X_t = x_0$$

is an uncertain process X_t such that

$$X_t = t^{p-1}x_0 + \frac{1}{\Gamma(p)}\int_0^t (t-s)^{p-1}f(s,X_s)ds + \frac{1}{\Gamma(p)}\int_0^t (t-s)^{p-1}g(s,X_s)dC_s \tag{3}$$

holds almost surely.

Definition 3. *Suppose that $f : [0,\infty) \times R^n \to R^n$ and $g : [0,\infty) \times R^n \to R^{n\times l}$ are two functions. Then,*

$${}^cD^p X_t = f(t, X_t) + g(t, X_t)\frac{dC_t}{dt} \tag{4}$$

is called an uncertain fractional differential equation of the Caputo type. A solution of (4) is an uncertain process X_t such that

$$X_t = X_0 + \frac{1}{\Gamma(p)}\int_0^t (t-s)^{p-1}f(s,X_s)ds + \frac{1}{\Gamma(p)}\int_0^t (t-s)^{p-1}g(s,X_s)dC_s \tag{5}$$

holds almost surely.

Remark 1. (i) The pth Riemann-Liouville fractional order derivative of the function $u : [0,T] \to R^n$ is defined by

$$D^p u(t) = \frac{1}{\Gamma(1-p)}\frac{d}{dt}\int_0^t (t-s)^{-p}u(s)ds, \quad t > 0.$$

(ii) The pth Caputo fractional order derivative of the function $u:[0,T]\to R^n$ is defined by

$$^{c}D^{p}u(t)=\frac{1}{\Gamma(1-p)}\int_0^t (t-s)^{-p}u'(s)\mathrm{d}s,\quad t>0$$

where $u'(s)$ is the first-order derivative of $u(s)$.
(iii) The relation between the Riemann-Liouville and Caputo fractional order derivatives is

$$D^{p}u(t)={}^{c}D^{p}u(t)+\frac{t^{-p}}{\Gamma(1-p)}u(0).$$

(iv) The gamma function

$$\Gamma(\alpha)=\int_0^{\infty} t^{\alpha-1}e^{-t}\mathrm{d}t,\quad \alpha>0$$

has the properties

$$\Gamma(\alpha+1)=\alpha\Gamma(\alpha),\ \alpha>0;\ \Gamma(1)=1;\ \Gamma\left(\frac{1}{2}\right)=\sqrt{\pi}.$$

Existence and uniqueness

For simplicity, we use $|\cdot|$ to denote a norm in R^n or $R^{n\times l}$. Let $D_{[a,b]}$ denote the space of continuous R^n-valued functions on $[a,b]$, which is a Banach space with the norm

$$\|x_t\|=\max_{t\in[a,b]}|x_t|,\quad \text{for } x_t\in D_{[a,b]}.$$

Give two functions $f(t,x):[0,T]\times R^n\to R^n$, and $g(t,x):[0,T]\times R^n\to R^{n\times l}$. Now we introduce the following mapping Φ on $D_{[0,T]}$: for $X_t\in D_{[0,T]}$,

$$\Phi(X_t)=t^{p-1}x_0+\frac{1}{\Gamma(p)}\int_0^t (t-s)^{p-1}f(s,X_s)\mathrm{d}s+\frac{1}{\Gamma(p)}\int_0^t (t-s)^{p-1}g(s,X_s)\mathrm{d}C_s \quad (6)$$

where x_0 is a given initial state.

Lemma 2. *For uncertain process $X_t\in D_{[0,T]}$, the mapping Ψ defined by*

$$\begin{aligned}\Psi(X_t)&=\left(t^{p-1}-\tilde{a}\right)x_0+X_a+\frac{1}{\Gamma(p)}\int_a^t (t-s)^{p-1}f(s,X_s)\mathrm{d}s\\&+\frac{1}{\Gamma(p)}\int_a^t (t-s)^{p-1}g(s,X_s)\mathrm{d}C_s,\quad t>a\geq 0\end{aligned}\quad (7)$$

is sample-continuous where $\tilde{a}=a^{p-1}$ if $a>0$ or 1 if $a=0$, and f and g satisfy the linear growth condition

$$|f(t,x)|+|g(t,x)|\leq L(1+|x|),\quad \forall x\in R^n,\quad t\in[0,+\infty)$$

where L is a positive constant.

Proof. In fact, for $\gamma \in \Gamma$ and $t > r > a$, we have

$$
\begin{aligned}
&|\Psi(X_t(\gamma)) - \Psi(X_r(\gamma))| \\
&= \left| \left(t^{p-1} - r^{p-1}\right) x_0 + \frac{1}{\Gamma(p)} \int_r^t (t-s)^{p-1} f(s, X_s(\gamma)) \mathrm{d}s \right. \\
&\quad + \frac{1}{\Gamma(p)} \int_r^t (t-s)^{p-1} g(s, X_s(\gamma)) \mathrm{d}C_s(\gamma) \\
&\quad + \frac{1}{\Gamma(p)} \int_a^r \left[(t-s)^{p-1} - (r-s)^{p-1}\right] f(s, X_s(\gamma)) \mathrm{d}s \\
&\quad \left. + \frac{1}{\Gamma(p)} \int_a^r \left[(t-s)^{p-1} - (r-s)^{p-1}\right] g(s, X_s(\gamma)) \mathrm{d}C_s(\gamma) \right| \\
&\le \left(r^{p-1} - t^{p-1}\right) |x_0| + \frac{1}{\Gamma(p)} \int_r^t (t-s)^{p-1} |f(s, X_s(\gamma))| \mathrm{d}s \\
&\quad + \frac{1}{\Gamma(p)} \left| \int_r^t (t-s)^{p-1} g(s, X_s(\gamma)) \mathrm{d}C_s(\gamma) \right| \\
&\quad + \frac{1}{\Gamma(p)} \int_a^r \left[(t-s)^{p-1} - (r-s)^{p-1}\right] |f(s, X_s(\gamma))| \mathrm{d}s \\
&\quad + \frac{1}{\Gamma(p)} \left| \int_a^r \left[(t-s)^{p-1} - (r-s)^{p-1}\right] g(s, X_s(\gamma)) \mathrm{d}C_s(\gamma) \right| \\
&\le \left(r^{p-1} - t^{p-1}\right) |x_0| + \frac{1}{\Gamma(p)} \int_r^t (t-s)^{p-1} |f(s, X_s(\gamma))| \mathrm{d}s \\
&\quad + \frac{K_\gamma}{\Gamma(p)} \int_r^t (t-s)^{p-1} |g(s, X_s(\gamma))| \mathrm{d}s \\
&\quad + \frac{1}{\Gamma(p)} \int_a^r \left[(t-s)^{p-1} - (r-s)^{p-1}\right] |f(s, X_s(\gamma))| \mathrm{d}s \\
&\quad + \frac{K_\gamma}{\Gamma(p)} \int_a^r \left[(t-s)^{p-1} - (r-s)^{p-1}\right] \left|g(s, X_s(\gamma))\right| \mathrm{d}s \quad \text{(by Lemma 1)} \\
&\le \left(r^{p-1} - t^{p-1}\right) |x_0| + \frac{L}{\Gamma(p+1)} (1 + \|X_t(\gamma)\|)(1 + K_\gamma) \left[(t-a)^p - (r-a)^p\right]
\end{aligned}
$$

by the linear growth condition. Thus, $|\Psi(X_t(\gamma)) - \Psi(X_r(\gamma))| \to 0$ as $|t-r| \to 0$. That is, $\Psi(X_t)$ is sample-continuous. □

Theorem 1. *(Existence and uniqueness) The uncertain fractional differential equation (2) (or (4)) has a unique solution X_t in $[0, +\infty)$ if the coefficients $f(t,x)$ and $g(t,x)$ satisfy the Lipschitz condition*

$$|f(t,x) - f(t,y)| + |g(t,x) - g(t,y)| \le L|x-y|, \quad \forall x, y \in R^n, \quad t \in [0, +\infty)$$

and the linear growth condition

$$|f(t,x)| + |g(t,x)| \le L(1 + |x|), \quad \forall x \in R^n, \quad t \in [0, +\infty)$$

where L is a positive constant. Furthermore, X_t is sample-continuous.

Proof. We only prove the theorem for the uncertain fractional differential equation (2). A similar process of the proof is suitable to the uncertain fractional differential equation (4). Let $T > 0$ be an arbitrarily given number, and let Φ be a mapping defined by (6) on $D_{[0,T]}$.

Give $\gamma \in \Gamma$. For $\lambda \in [0, T)$, assume $c > 0$ such that $\lambda + c \leq T$. Define a mapping ψ on $D_{[\lambda,\lambda+c]}$: for $X_t \in D_{[\lambda,\lambda+c]}$, $t \in [\lambda, \lambda + c]$,

$$\psi(X_t) = \left(t^{p-1} - \tilde{\lambda}\right) x_0 + X_\lambda + \frac{1}{\Gamma(p)} \int_\lambda^t (t-s)^{p-1} f(s, X_s) \mathrm{d}s + \frac{1}{\Gamma(p)} \int_\lambda^t (t-s)^{p-1} g(s, X_s) \mathrm{d}C_s$$

where $\tilde{\lambda} = \lambda^{p-1}$ if $\lambda > 0$ or 1 if $\lambda = 0$. For $X_t(\gamma) \in D_{[\lambda,\lambda+c]}$, we know that $\psi(X_t(\gamma)) \in D_{[\lambda,\lambda+c]}$ by Lemma 2.

Let $X_t(\gamma), Y_t(\gamma) \in D_{[\lambda,\lambda+c]}$. For any $t \in [\lambda, \lambda + c]$, we have

$$\begin{aligned}
\|\psi(X_t(\gamma)) - \psi(Y_t(\gamma))\| &= \max_{t \in [\lambda,\lambda+c]} |\psi(X_t(\gamma)) - \psi(Y_t(\gamma))| \\
&\leq \max_{t \in [\lambda,\lambda+c]} \left| \frac{1}{\Gamma(p)} \int_\lambda^t (t-s)^{p-1} \left[f(s, X_s(\gamma)) - f(s, Y_s(\gamma)) \right] \mathrm{d}s \right. \\
&\quad \left. + \frac{1}{\Gamma(p)} \int_\lambda^t (t-s)^{p-1} \left[g(s, X_s(\gamma)) - g(s, Y_s(\gamma)) \right] \mathrm{d}C_s(\gamma) \right| \\
&\leq \max_{t \in [\lambda,\lambda+c]} \left\{ \frac{1}{\Gamma(p)} \int_\lambda^t (t-s)^{p-1} |f(s, X_s(\gamma)) - f(s, Y_s(\gamma))| \mathrm{d}s \right. \\
&\quad \left. + \frac{K_\gamma}{\Gamma(p)} \int_\lambda^t (t-s)^{p-1} |g(s, X_s(\gamma)) - g(s, Y_s(\gamma))| \mathrm{d}s \right\} \\
&\leq \frac{(1+K_\gamma)L}{\Gamma(p)} \max_{t \in [\lambda,\lambda+c]} \int_\lambda^t (t-s)^{p-1} |X_s(\gamma) - Y_s(\gamma)| \mathrm{d}s \quad \text{(by Lipschitz condition)} \\
&\leq \frac{(1+K_\gamma)Lc^p}{\Gamma(p+1)} \|X_t(\gamma) - Y_t(\gamma)\|.
\end{aligned}$$

Let $\rho(\gamma) = \left(1 + K_\gamma\right) Lc^p / \Gamma(p+1)$. By taking a suitable $c = c(\gamma) > 0$ such that $\rho(\gamma) \in (0, 1)$. That is, ψ is a contraction mapping on $D_{[\lambda,\lambda+c]}$. Thus, by the well-known Banach fixed point theorem, we have a unique fixed point $X_t(\gamma)$ of ψ in $D_{[\lambda,\lambda+c]}$. Furthermore, $X_t(\gamma) = \lim_{k\to\infty} \psi(X_{t,k}(\gamma))$ where

$$X_{t,k}(\gamma) = \psi(X_{t,k-1}(\gamma)), \; k = 1, 2, \cdots$$

for any given $X_{t,0}(\gamma) = x_t \in D_{[\lambda,\lambda+c]}$.

Assume that $[0, c], [c, 2c], \cdots, [kc, T]$ are the subsets of $[0, T]$ with $kc < T \leq (k+1)c$. The above process implies that the mapping ψ has a unique fixed point $X_t^{(i+1)}(\gamma)$ with $X_{ic}^{(i+1)}(\gamma) = X_{ic}^{(i)}(\gamma)$ on the interval $[ic, (i+1)c]$ for $i = 0, 1, 2, \cdots, k$, where we set $(k+1)c = T$. Define $X_t(\gamma)$ on the interval $[0, T]$ by setting

$$X_t(\gamma) = \begin{cases} X_t^{(1)}(\gamma), & t \in [0, c], \\ X_t^{(2)}(\gamma), & t \in [c, 2c], \\ \cdots & \\ X_t^{(k+1)}(\gamma), & t \in [kc, T]. \end{cases}$$

It is easy to see that $X_t(\gamma)$ is the unique fixed point of Φ defined by (6) in $D_{[0,T]}$. In addition, $X_t(\gamma) = \lim_{k\to\infty} \Phi(X_{t,k}(\gamma))$ where

$$X_{t,k}(\gamma) = \Phi(X_{t,k-1}(\gamma)),\ k = 1, 2, \cdots$$

for any given $X_{t,0}(\gamma) = x_t \in D_{[0,T]}$. Since $X_{t,k}$ are uncertain vectors for $k = 1, 2, \cdots$, we know that X_t is an uncertain vector by Theorem 3 in the Appendix. It follows from the arbitrariness of $T > 0$ that X_t is the unique solution of uncertain fractional differential equation (2). Furthermore, since $X_t(\gamma)$ is in $D_{[0,T]}$, X_t is sample-continuous. The theorem is proved. □

If the functions f and g do not satisfy the Lipschitz condition and linear growth condition, we present the following existence theorem just for continuous f and g.

Theorem 2. *(Existence) Let $f(t,x)$ and $g(t,x)$ be continuous in*

$$G = [0,T] \times \left\{x \in R^n : |x - x_0| \le b\right\}.$$

Then, uncertain fractional differential equation of the Caputo type (4) has a solution X_t in $t \in [0,T]$ with the crisp initial condition $X_0 = x_0 \in R^n$.

Proof. For any $\gamma \in \Gamma$, let $c > 0$ be a positive number such that

$$\frac{M\left(1+K_\gamma\right)}{\Gamma(p+1)} c^p = b$$

where K_γ is the Lipschitz constant of the canonical Liu process C_t, and $M = \max_{(t,x)\in G} |f(t,x)| \vee |g(t,x)|$. Denote

$$H = \left\{X_t(\gamma) \in D_{[0,h]} : X_t \text{ is an uncertain vector and } \|X_t(\gamma) - x_0\| \le \frac{M(1+K_\gamma)}{\Gamma(p+1)} h^p\right\}$$

where $h = \min\{T, c\}$.

It is easy to see that H is a closed convex set. Define a mapping Φ on H by

$$\begin{aligned}\Phi(X_t(\gamma)) = x_0 &+ \frac{1}{\Gamma(p)}\int_0^t (t-s)^{p-1} f(s, X_s(\gamma))\mathrm{d}s \\ &+ \frac{1}{\Gamma(p)}\int_0^t (t-s)^{p-1} g(s, X_s(\gamma))\mathrm{d}C_s(\gamma), \quad 0 \le t \le h. \end{aligned} \tag{8}$$

For $X_t(\gamma) \in H$, we have

$$\begin{aligned}\|\Phi(X_t(\gamma)) - x_0\| &= \max_{0\le t\le h} |\Phi(X_t(\gamma)) - x_0| \\ &\le \max_{0\le t\le h}\left\{\frac{1}{\Gamma(p)}\int_0^t (t-s)^{p-1}|f(s, X_s(\gamma))|\mathrm{d}s \right.\\ &\quad \left. + \frac{K_\gamma}{\Gamma(p)}\int_0^t (t-s)^{p-1}|g(s, X_s(\gamma))|\mathrm{d}s\right\} \\ &\le \max_{0\le t\le h} \frac{M(1+K_\gamma)}{\Gamma(p)}\int_0^t (t-s)^{p-1}\mathrm{d}s \\ &\le \frac{M(1+K_\gamma)}{\Gamma(p+1)} h^p. \end{aligned} \tag{9}$$

That is $\Phi(X_t(\gamma)) \in H$, and the mapping Φ is bounded uniformly in $X_t(\gamma) \in H$. In addition, for $0 \leq t_1 < t_2 \leq h$, it is easy to verify

$$
\begin{aligned}
|\Phi(X_{t_1}(\gamma)) - \Phi(X_{t_2}(\gamma))| = & \left| \frac{1}{\Gamma(p)} \int_{t_1}^{t_2} (t_2 - s)^{p-1} f(s, X_s(\gamma)) \mathrm{d}s \right. \\
& + \frac{1}{\Gamma(p)} \int_{t_1}^{t_2} (t_2 - s)^{p-1} g(s, X_s(\gamma)) \mathrm{d}C_s(\gamma) \\
& + \frac{1}{\Gamma(p)} \int_0^{t_1} \left[(t_2 - s)^{p-1} - (t_1 - s)^{p-1} \right] f(s, X_s(\gamma)) \mathrm{d}s \\
& \left. + \frac{1}{\Gamma(p)} \int_0^{t_1} \left[(t_2 - s)^{p-1} - (t_1 - s)^{p-1} \right] g(s, X_s(\gamma)) \mathrm{d}C_s(\gamma) \right| \\
\leq & \frac{M\left(1 + K_\gamma\right)}{\Gamma(p+1)} \left(t_2^p - t_1^p\right)
\end{aligned}
$$

which comes to a conclusion that Φ is equicontinuous for $X_t(\gamma) \in H$ in $[0, h]$. It follows from the Ascoli-Arzela theorem that Φ is a compact mapping on H.

Let $X_{t,i}(\gamma)$ converge to $X_t(\gamma)$ in H as $i \to \infty$. That is, $X_{t,i}(\gamma)$ converges to $X_t(\gamma)$ uniformly in $t \in [0, h]$. Thus,

$$
\begin{aligned}
\Phi(X_{t,i}(\gamma)) & = \frac{1}{\Gamma(p)} \int_0^t (t - s)^{p-1} f(s, X_{s,i}(\gamma)) \mathrm{d}s \\
& \quad + \frac{1}{\Gamma(p)} \int_0^t (t - s)^{p-1} g(s, X_{s,i}(\gamma)) \mathrm{d}C_s(\gamma) \\
& \to \frac{1}{\Gamma(p)} \int_0^t (t - s)^{p-1} f(s, X_s(\gamma)) \mathrm{d}s \\
& \quad + \frac{1}{\Gamma(p)} \int_0^t (t - s)^{p-1} g(s, X_s(\gamma)) \mathrm{d}C_s(\gamma) \\
& = \Phi(X_t(\gamma))
\end{aligned}
$$

uniformly in $t \in [0, h]$. This shows that Φ is continuous on H.

It follows from the Schauder fixed point theorem that Φ has a fixed point $X_t(\gamma)$ on H. Hence,

$$
\begin{aligned}
X_t(\gamma) = x_0 & + \frac{1}{\Gamma(p)} \int_0^t (t - s)^{p-1} f(s, X_s(\gamma)) \mathrm{d}s \\
& + \frac{1}{\Gamma(p)} \int_0^t (t - s)^{p-1} g(s, X_s(\gamma)) \mathrm{d}C_s(\gamma)
\end{aligned} \tag{10}
$$

for $t \in [0, h]$. By the extension method, there exists $X_t(\gamma)$ satisfying (10) in $t \in [0, T]$. That is, X_t is a solution of (4). Therefore, the conclusion of the theorem is proved. □

Conclusions

For uncertain fractional differential equations of the Riemann-Liouville type and Caputo type, an existence and uniqueness theorem of solution was presented by employing the Banach fixed point theorem under sufficient conditions that the drift and diffusion functions are linear growing and Lipschitzian. If the drift and diffusion functions are just continuous, an existence theorem of solution for the uncertain fractional differential equation of the Caputo type was proved by employing the Schauder fixed point theorem. The existence and uniqueness of solution to the uncertain fractional differential equation will give a theoretical foundation for studying the stability of uncertain fractional differential equations and uncertain finance.

Appendix

Theorem 3. *Let* $\xi_1, \xi_2, \cdots$ *be uncertain variables and* $\lim_{i\to\infty} \xi_i = \xi$ *almost surely. Then,* ξ *is an uncertain variable.*

Proof. Let

$$\mathcal{B} = \{B \subset R : \{\xi \in B\} \text{is an event}\}.$$

Then, $\mathcal{B}$ is a σ-algebra. For $a, b \in R$, since

$$\xi = \sup_{n\geq 1} \inf_{i\geq n} \xi_i = \inf_{n\geq 1} \sup_{i\geq n} \xi_i,$$

we have

$$\{\xi < b\} = \left\{\sup_{n\geq 1} \inf_{i\geq n} \xi_i < b\right\} = \bigcup_{k\geq 1} \bigcap_{n\geq 1} \bigcup_{i\geq n} \{\xi_i \leq b - \epsilon_k\},$$

$$\{\xi > a\} = \left\{\inf_{n\geq 1} \sup_{i\geq n} \xi_i > a\right\} = \bigcup_{k\geq 1} \bigcap_{n\geq 1} \bigcup_{i\geq n} \{\xi_i \geq a + \epsilon_k\}$$

where $\{\epsilon_k\}$ is a sequence of positive numbers converging decreasingly to zero. Since ξ_i are uncertain variables for all i, we know that $\{\xi_i \leq b-\epsilon_k\}$ and $\{\xi_i \geq a+\epsilon_k\}$ are events. Hence, $\{\xi < b\}$ and $\{\xi > a\}$ are events and then $\{a < \xi < b\}$ is an event. That is, $(a, b) \in \mathcal{B}$. Since the smallest σ-algebra containing all open intervals of R is just Borel algebra over R, the class $\mathcal{B}$ contains all Borel sets. That is, for any Borel set B, $\{\xi \in B\}$ is an event. Therefore, ξ is an uncertain variable. □

Acknowledgements
This work is supported by the National Natural Science Foundation of China (No. 61273009).

References

1. Ahmad, B, Nieto, JJ: Existence results for a coupled system of nonlinear fractional differential equations with three-point boundary conditions. Comput. Math. Appl. **58**, 1838–1843 (2009)
2. Kilbas, AA, Srivastava, HM, Trujillo, JJ: Theory and Applications of Fractional Differential Equations. Elsevier Science BV (2006)
3. Lakshmikantham, V, Vatsala, AS: Basic theory of fractional differential equations. Nonlinear Anal. TMA. **69**, 2677–2682 (2008)
4. Lakshmikantham, V, Leela, S, Vasundhara Devi, J: Theory of Fractional Dynamic Systems. Cambridge Academic Publishers, Cambridge (2009)
5. Miller, KS, Ross, B: An Introduction to the Fractional Calculus and Differential Equations. John Wiley, New York (1993)
6. Oldham, KB, Spanier, J: The Fractional Calculus. Academic Press, New York (1974)
7. Podlubny, I: Fractional Differential Equation. Academic Press, San Diego (1999)
8. Samko, SG, Kilbas, AA, Marichev, OI: Fractional Integrals and Derivatives, Theory and Applications. Gordon and Breach Science Publishers, New York (1993)
9. Ciotir, I, Răscanu, A: Viability for differential equations driven by fractional Brownian motion. J. Differential Equations. **247**, 1505–1528 (2009)
10. Cui, J, Yan, L: Existence result for fractional neutral stochastic integro-differential equations with infinite delay. J Phys. A: Math. Theor. **44**(33), 335201 (2011)
11. El-Borai, MM, El-Said El-Nadi, K, Fouad, HA: On some fractional stochastic delay differential equations. Comput. Math. Appl. **59**, 1165–1170 (2010)
12. Mandelbrot, BB, Van Ness, JW: Fractional Brownian motions, fractional noises and applications. SIAM Rev. **10**, 422–437 (1968)
13. Nualart, D, Răscanu, A: Differential equations driven by fractional Brownian motion. Collectanea Mathematica. **53**, 55–81 (2002)
14. Ren, Y, Bi, Q, Sakthivel, R: Stochastic functional differential equations with infinite delay driven by G-Brownian motion. Math. Methods Appl. Sci. **36**(13), 1746–1759 (2013)
15. Sakthivel, R, Revathi, P, Ren, Y: Existence of solutions for nonlinear fractional stochastic differential equations. Nonlinear Anal. TMA. **81**, 70–86 (2013)

16. Zhu, Y: Uncertain fractional differential equations and an interest rate model. Math. Methods Appl. Sci (2014). doi:10.1002/mma.3335
17. Liu, B: Uncertainty Theory. 2nd ed. Springer-Verlag, Berlin (2007)
18. Liu, B: Uncertainty Theory: A Branch of Mathematics for Modeling Human Uncertainty. Springer-Verlag, Berlin (2010)
19. Liu, B: Some research problems in uncertainy theory. J Uncertain Syst. **3**(1), 3–10 (2009)
20. Liu, B: Fuzzy process, hybrid process and uncertain process. J Uncertain Syst. **2**(1), 3–16 (2008)
21. Chen, X, Liu, B: Existence and uniqueness theorem for uncertain differential equations. Fuzzy Optimization Decis. Making. **9**(1), 69–81 (2010)
22. Liu, Y: An analytic method for solving uncertain differential equations. J. Uncertain Syst. **6**(4), 244–249 (2012)
23. Yao, K: A type of nonlinear uncertain differential equations with analytic solution. J. Uncertainty Anal. Appl. **1**, Article 8 (2013)
24. Yao, K, Chen, X: A numerical method for solving uncertain differential equations. J. Intell. Fuzzy Syst. **25**(3), 825–832 (2013)
25. Bărbăcioru, IC: Uncertainty functional differential equations for finance. Surveys Mathematics Appl. **5**, 275–284 (2010)
26. Ge, X, Zhu, Y: Existence and uniqueness theorem for uncertain delay differential equations. J. Comput. Inf. Syst. **8**(20), 8341–8347 (2012)
27. Liu, H, Fei, W: Neutral uncertain delay differential equations. Inf. Int. Interdiscip. J. **16**(2), 1225–1232 (2013)
28. Ge, X, Zhu, Y: A necessary condition of optimality for uncertain optimal control problem. Fuzzy Optimization Decis. Making. **12**(1), 41–51 (2013)
29. Liu, B: Toward uncertain finance theory. J. Uncertainty Anal. Appl. **1**, Article 1 (2013)
30. Zhu, Y: Uncertain optimal control with application to a portfolio selection model. Cybern. Syst. Int. J. **41**(7), 535–547 (2010)
31. Yang, X, Gao, J: Uncertain differential games with application to capitalism. J Uncertainty Anal. App. **1**, Article 17 (2013)

Permissions

All chapters in this book were first published in JUAA, by Springer; hereby published with permission under the Creative Commons Attribution License or equivalent. Every chapter published in this book has been scrutinized by our experts. Their significance has been extensively debated. The topics covered herein carry significant findings which will fuel the growth of the discipline. They may even be implemented as practical applications or may be referred to as a beginning point for another development.

The contributors of this book come from diverse backgrounds, making this book a truly international effort. This book will bring forth new frontiers with its revolutionizing research information and detailed analysis of the nascent developments around the world.

We would like to thank all the contributing authors for lending their expertise to make the book truly unique. They have played a crucial role in the development of this book. Without their invaluable contributions this book wouldn't have been possible. They have made vital efforts to compile up to date information on the varied aspects of this subject to make this book a valuable addition to the collection of many professionals and students.

This book was conceptualized with the vision of imparting up-to-date information and advanced data in this field. To ensure the same, a matchless editorial board was set up. Every individual on the board went through rigorous rounds of assessment to prove their worth. After which they invested a large part of their time researching and compiling the most relevant data for our readers.

The editorial board has been involved in producing this book since its inception. They have spent rigorous hours researching and exploring the diverse topics which have resulted in the successful publishing of this book. They have passed on their knowledge of decades through this book. To expedite this challenging task, the publisher supported the team at every step. A small team of assistant editors was also appointed to further simplify the editing procedure and attain best results for the readers.

Apart from the editorial board, the designing team has also invested a significant amount of their time in understanding the subject and creating the most relevant covers. They scrutinized every image to scout for the most suitable representation of the subject and create an appropriate cover for the book.

The publishing team has been an ardent support to the editorial, designing and production team. Their endless efforts to recruit the best for this project, has resulted in the accomplishment of this book. They are a veteran in the field of academics and their pool of knowledge is as vast as their experience in printing. Their expertise and guidance has proved useful at every step. Their uncompromising quality standards have made this book an exceptional effort. Their encouragement from time to time has been an inspiration for everyone.

The publisher and the editorial board hope that this book will prove to be a valuable piece of knowledge for researchers, students, practitioners and scholars across the globe.

List of Contributors

Baoding Liu
Uncertainty Theory Laboratory, Department of Mathematical Sciences, Tsinghua University, Beijing 100084, China

Sundaram Sampath
Department of Statistics, University of Madras, Chennai 600005, India

Balu Ramya
Department of Statistics, University of Madras, Chennai 600005, India

Markus P Rumpfkeil
Department of Mechanical and Aerospace Engineering, University of Dayton, Dayton, Ohio, USA

Xiaowei Chen
Department of Risk Management and Insurance, Nankai University, Tianjin 300071, China

Dan A Ralescu
Department of Mathematical Sciences, University of Cincinnati, Cincinnati, OH 45221-0025, USA

Asrat Atsedeweyn Andargie
Department of Statistics, Andhra University, Visakhapatnam 530003, India

K Srinivasa Rao
Department of Statistics, Andhra University, Visakhapatnam 530003, India

Sankar Kumar Roy
Department of Applied Mathematics with Oceanology and Computer Programming, Vidyasagar University, Midnapore, West Bengal 721102, India

Prasanta Mula
ISRO Satellite Centre, Old Airport Road, Vimanapura Post, Bangalore 560017, India

Kai Yao
School of Management, University of Chinese Academy of Sciences, Beijing 100190, China

Debasis Bhattacharya
Visva-Bharati University, Santiniketan 731235, India

Soma Roychowdhury
Indian Institute of Social Welfare and Business Management, Calcutta 700073, India

Kenjiro Yanagi
Graduate School of Science and Engineering, Yamaguchi University, Une, Yamaguchi 755-8611, Japan

Shigeru Furuichi
College of Humanities and Science, Nihon University, Tokyo 156-8550, Japan

Ken Kuriyama
Faculty of Education, Bukkyo University, Kyoto 603-8301, Japan

Ismat Beg
Centre for Mathematics and Statistical Sciences, Lahore School of Economics, Lahore 53200,Pakistan

Tabasam Rashid
Department of Science and Humanities, National University of Computer and Emerging Sciences, Lahore 54000, Pakistan
University of Management and Technology, Lahore 54000, Pakistan

Supak Phiangsungnoen
Department of Mathematics, Faculty of Science, King Mongkut's University of Technology Thonburi (KMUTT), 126 Pracha Uthit Rd., Bang Mod, Thung Khru, Bangkok 10140, Thailand

Wutiphol Sintunavarat
Department of Mathematics and Statistics, Faculty of Science and Technology, Thammasat University Rangsit Center, Pathumthani 12121, Thailand

Poom Kumam
Department of Mathematics, Faculty of Science, King Mongkut's University of Technology Thonburi (KMUTT), 126 Pracha Uthit Rd., Bang Mod, Thung Khru, Bangkok 10140, Thailand

Rupak Bhattacharyya
Department of Applied Science & Humanities, Global Institute of Management & Technology, Krishnagar, West Bengal 741102, India

Amitava Chatterjee
Department of Mathematics, National Institute of Technology, Durgapur, West Bengal 713209, India

Samarjit Kar
Department of Mathematics, National Institute of Technology, Durgapur, West Bengal 713209, India

Xianqiao Tong
Department of Mechanical Engineering, Virginia Tech, 800 Drillfield Dr, Blacksburg, VA 24061, USA

Tomonari Furukawa
Department of Mechanical Engineering, Virginia Tech, 800 Drillfield Dr, Blacksburg, VA 24061, USA

Hugh Durrant-Whyte
Australian Centre for Field Robotics (ACFR), Rose St, Sydney 2006, Australia

Abhijit Baidya
Department of Mathematics, National Institute of Technology, Agartala, Jirania 799055, West Tripura, India

Uttam Kumar Bera
Department of Mathematics, National Institute of Technology, Agartala, Jirania 799055, West Tripura, India

Manoranjan Maiti
Department of Applied Mathematics, Vidyasagar University, Midnapore 721102, West Bengal, India

Kai Yao
Department of Mathematical Sciences, Tsinghua University, Beijing 100084, China

Shengguo Li
Institute of Uncertain Systems, Huanggang Normal University, Hubei 438000, China

Jin Peng
Institute of Uncertain Systems, Huanggang Normal University, Hubei 438000, China

Bo Zhang
School of Statistics and Mathematics, Zhongnan University of Economics and Law, Hubei 430073, China

Yuanguo Zhu
School of Science, Nanjing University of Science and Technology, 210094 Nanjing, China